Destination Marketing and Management: Theories and Applications

Destination Marketing and Management: Theories and Applications

Editor
Kornelia Marshall

Destination Marketing and Management: Theories and Applications

Edited by **Kornelia Marshall**

ISBN: 978-1-68117-250-7
Library of Congress Control Number: 2016934793

© 2017 by
SCITUS Academics LLC,
www.scitusacademics.com
Box No. 4766, 616 Corporate Way,
Suite 2, Valley Cottage,
NY 10989

This book contains information obtained from highly regarded resources. Copyright for individual articles remains with the authors as indicated. All chapters are distributed under the terms of the Creative Commons Attribution License, which permits unrestricted use, distribution, and reproduction in any medium, provided the original author and source are credited.

Notice

Reasonable efforts have been made to publish reliable data and views articulated in the chapters are those of the individual contributors, and not necessarily those of the editors or publishers. Editors or publishers are not responsible for the accuracy of the information in the published chapters or consequences of their use. The publisher believes no responsibility for any damage or grievance to the persons or property arising out of the use of any materials, instructions, methods or thoughts in the book. The editors and the publisher have attempted to trace the copyright holders of all material reproduced in this publication and apologize to copyright holders if permission has not been obtained. If any copyright holder has not been acknowledged, please write to us so we may rectify.

Preface

The term "destination" refers broadly to an area where tourism is a relatively important activity and where the economy may be significantly influenced by tourism revenues.Most tourism activities take place at a destination, and destination serves as a fundamental unit of analysis in any modelling of the tourism system. Destination marketing is the process of communicating with potential visitors to influence their destination preference, intention to travel and ultimately their final destination and product choices. Destination marketing is a major part of the 'Implementation' process; it is the articulation and communication of the values, vision and competitive attributes of the destination. The actions implemented in the destination marketing phase should be underpinned by the findings of the 'Destination Planning', process and the subsequent 'Destination Development' activities.Managing tourism destinations is an important part of controlling tourism's environmental impacts. Destination management can include land use planning, business permits and zoning controls, environmental and other regulations, business association initiatives, and a host of other techniques to shape the development and daily operation of tourism-related activities.However, destination marketing and management is a complex subject that requires a comprehensive, holistic and systematic approach. From the demand side, travellers have a choice of available destinations; from the supply side, destination marketing organizations are competing for attention from a highly competitive marketplace. This book, Destination Marketing and Management: Theories and Applications, provides a comprehensive understanding of the concept and scope of the tourism industry in general and of destination marketing and management in particular, as they are situated in their particular policy, planning, economic, geographical and historical contexts.

Table of Contents

Chapter 1	An Entropy-Perspective Study on the Sustainable Development Potential of Tourism Destination Ecosystem in Dunhuang, China	1
Chapter 2	Comparative Study of Functions Affecting the Behavioral Patterns of Tourists in Iran and America's Tourism Marketing Plan Utilizing SWOT Model	45
Chapter 3	Determinants of Destination Management System (DMS) and Tourism Industry Assessment of Madagascar	69
Chapter 4	Destination management in New Zealand: Structures and functions	87
Chapter 5	Progress and prospects for event tourism research	127
Chapter 6	Destination Marketing Organizations and Climate Change—The Need for Leadership and Education	275
Chapter 7	Destination Image Perception of Tourists to Guangzhou—Based on Content Analysis of Online Travels	299

Chapter 8	An Evaluation of Destination Management Systems in Madagascar with Aspect of Tourism Sector	319
Chapter 9	Biosphere Reserve as a Learning Tourism Destination: Approaches from Tasik Chini	335
Chapter 10	Tourism Waste Management in the European Union: Lessons Learned from Four Popular EU Tourist Destinations	363
	Index	393

CHAPTER 1

An Entropy-Perspective Study on the Sustainable Development Potential of Tourism Destination Ecosystem in Dunhuang, China

Huihui Feng [1,2], Xingpeng Chen [1,*], Peter Heck [2] and Hong Miao [3]

[1] College of Earth and Environmental Sciences, Lanzhou University, Lanzhou 730000, China
[2] Institute for Applied Material Flow Management (IfaS), Trier University of Applied Science, Birkenfeld 55761, Germany
[3] School of Resources and Environment, Ningxia University, Yinchuan 750000, China

ABSTRACT

This paper analyzed the characteristic of the tourism destination ecosystem from perspective of entropy in Dunhuang City. Given these circumstances, an evaluation index system that considers the potential of sustainable development was formed based on dissipative structure and entropy change for the tourism destination ecosystem. The sustainable development potential evaluation model for tourism destination ecosystem was built up based on information entropy. Then, we analyzed each indicator impact for the sustainable development potential and proposed some measures for the tourism destination ecosystem. The conclusions include: (a) the requirements of Dunhuang tourism destination ecosystem on the natural ecosystem continuously grew between 2000 and 2012; (b) The sustainable development potential of the Dunhuang tourism destination ecosystem was on an oscillation upward trend during the study period, which is dependent on government attention, and pollution problems were improved.

KEYWORDS

sustainable tourism; tourism destination ecosystem; entropy change; entropy; China

1. INTRODUCTION

Sustainable tourism (ST) is an important part of global sustainable development (SD) [1]. The rapid development of tourism has brought great benefits to tourism destinations, while a variety of other problems are emerging, such as, resources and environmental issues and poor management of the tourism industry. Generally, as a reception carrier, the tourism destination of tours concentrates all elements of tourism on an effective framework, which is the most vital part for examining the impact of tourism. Hence, research for sustainable development of tourism destinations could improve the overall efficiency of tourism and optimize the ecological services related to the tourism. Until now, a number of organizations and academics have paid attention to this topic as well as achieved great progress in research methods and practices.

Particularly, their research focus on the following contents [2,3,4,5,6,7,8,9,10,11,12]: (a) The concept of ST [13,14,15,16, 17,18, 19, 20,21,22], for example, Hunter suggested that the concept of ST be redefined in terms of an over-arching paradigm which incorporates a range of approaches to the tourism/environment system within destination areas [13]. Swarbrooke [14] and Aall [17] noted that ST is not just about protecting the environment; it is also concerned with economic viability and social justice, and a suitable balance must be established between these three

1. Introduction

dimensions to guarantee its long-term sustainability. Sharpley point that there has been significant differences between the concepts of ST and SD, the principles and objectives of SD cannot be transposed onto the specific context of tourism [15]. Hardy argued that ST has traditionally given more focus to aspects related to the environment and economic development, which should be more focused on community involvement [4]. Gianna noted the need to very clearly distinguish between the concept of ST and the idea of tourism as one possible tool to support sustainability at multiple levels [16]. Saarinen concluded that perspectives of the resource-based tradition and the community-based tradition have their advantages in different use contexts and they can complement each other, but in respect to the idea of sustainability and the future challenges of humanity, they all share the same major limitation, which is the strong focus on the local scale [18]. Moyle and McLennan noted that the frequency of occurrence of sustainability as a concept has slightly increased in strategies over the past decade. At the same time, there has been a shift in the conceptualisation of sustainability, with thinking evolving from nature-based, social and triple bottom line concepts toward a focus on climate change, responsibility, adaption and transformation [20]; (b) The indicators of ST [23,24,25,26,27,28,29,30,31,32,33,34,35,36,37,38,39,40,41], for example, McElroy constructed a "Composite Tourism Penetration Index" from per capita visitor spending, daily visitor densities per 1000 population and hotel rooms per square kilometer. They tested it on 20 small Caribbean islands and yielded three levels of increasing penetration [26]. McCool and Moisey provided a tourism industry perspective on what items could be sustained and what indicators should be used to monitor for sustainability policies [27]. Wang analyzed the principle of indicators of ST, constructed the indicators of ST, the indicator weight and selected the comprehensive evaluation method [40]. Ward and Butler investigated how to monitor sustainable tourism development (STD) in Samoa. It described some of the methodological considerations and processes involved in the development of STD indicators and particularly highlighted the importance of formulating clear objectives before trying to identify indicators, the value of establishing a multi-disciplinary advisory panel, and the necessity of designing an effective and flexible implementation framework for converting indicator results into management action [31]. Ko proposed that the "Barometer of tourism sustainability" (BTS) model represents the comprehensive level of tourism sustainability in a given destination, combining human and natural indicators into an index of sustainable tourism development, without trading one off against the other. The "AMOEBA of tourism sustainability indicators" (ATSI) model was introduced to complement the BTS analysis and to illustrate individual levels of sustainability of tourism indicators [24]. Chris and Sirakaya employed a modified Delphi technique to constructed indicators from political, social, ecological, economic, technological and cultural dimensions for community tourism development (CTD) [25]. Schianetz and Kavanagh proposed the methodological framework for the selection and evaluation of sustainability indicators for tourism destinations, the systemic indicator system (SIS) this framework takes the interrelatedness of sociocultural, economic and environmental issues into account [33]. Reddy engaged in the identification, selection and evaluation of sustainability indicators for rapid assessment of tourism development in Andaman and Nicobar Islands of India (ANI). These indicators are developed and assessed mainly for developed countries and evaluated a feasible bottom-up approach based on local knowledge [38]. Blancas and Gonzalez introduced an indicator system to evaluate sustainability in established coastal tourism destinations, as well as developing a new synthetic indicator to simplify the measurement of sustainability and facilitate the comparative analysis of destination ranking [30]. Buckley suggested that the indicators of ST should include: population, peace, prosperity, pollution, protection [6]. Oyola and Blancas presented an indicator system to evaluate ST at cultural destinations. Also, they suggested a method based on goal programming to construct composite indicators. Then, they proposed three basic practical uses for these indicators: the formulation of general action plans at a regional level, the definition of short-term strategies for destinations and the establishment of destination benchmarking practices [36]. Delgado and Saarinenc examined the significant role of indicators based on literature review in tourism planning and management. The indicator type (set or index) needs to be carefully selected depending on the situation under analysis and the purpose underpinning the study. However, indicator effectiveness to achieve the ideals of sustainable tourism development is affected by the ambiguity in the definition of the concept of ST and problems associated with data availability and baseline knowledge. The main challenge is to overcome strategic guidelines and political and theoretical proposals of indicators and achieve practical applications for the sustainable development of tourism [23]; (c) Ecological security and environment carrying capacity for tourism [42,43,44,45,46,47,48,49,50,51,52,53,54], for instance, Ahn used the limits of acceptable change (LAC) framework as a guide to examine and inform the process of ST on a regional scale. Also, he examined resident attitudes toward tourism development in general, toward desirable types of tourism services, toward local conditions and finally, toward perceptions about if and how conditions might change due to tourism [42]. Gössling provided a methodological framework for the calculation of ecological footprints (EF) related to leisure tourism. Based on the example of the Seychelles, it reveals the statistical obstacles that have to be overcome in the calculation process and discusses the strengths and weaknesses of such an approach [44]. Hunter attempted to connect, conceptually, the realms of ST and EF thinking. It is argued that primary research should focus on calculating the TEF associated with individual tourism products, throughout the product's life-cycle. As well as bringing another dimension to our understanding of tourism's actual ecological demand, it is also argued that the concept of the TEF may be used to clarify theoretical aspects of the sustainable tourism debate, helping to rejuvenate this debate in the process [47]. Cui put out the tourism bearing capacity index and its arithmetic model of operation. He defined the tourism environmental bearing capacity as the bearing intensity of tourism destinations during a period which does not do harm to the present and future people in its current state and which can be accepted by the residents. The bearing intensity of tourism destinations mainly includes the tourist density, the tourism land use intensity and the tourism income value [50]. Xiao constructed the models of general ecological security coefficient (GESC) of island tourism destination and special ecological security coefficient (SESC) of island tourism destinations, and then the assessment framework and judgment criterion were proposed on the ecological security of island tourist destination (ESITD) and island tourist sustainable development (ITSD). Furthermore, the models of island tourist ecological footprints were established based on the idea of EF and an empirical analysis of Zhoushan Islands, China was conducted [51]. Salerno and Viviano describes how the concept of Tourism Carrying Capacity (TCC) has shifted from a uni-dimensional approach to incorporating environmental, social and political aspects. Then, an empirical analysis of internationally popular protected area used by trekkers, the Mt. Everest Region, was conducted [52]. Zhong examined the applicability of the model to China's Zhangjiajie National Forest Park. At the same time, both external and internal factors affecting the park's tourism development as well as the environmental, social, and economic changes of the area are also discussed [53]; (d) The development pattern of ST [55,56,57,58,59,60] was examined by Rodríguez along with an analysis of the life cycle of Tenerife (Canary Islands, Spain), and two types of strategic decisions are considered: the political–legal decisions of the regional government to regulate tourism activity and the decisions to regrade supply, developed by the administrative institutions related to tourism activity in this destination [55]. Keitumetse devised a Community-Based Cultural Heritage Resources Management (COBACHREM) model that merges the technical and academic approaches to illustrate a symbiosis between cultural and natural resources for sustainable resources conservation at community levels [57]. Rizio explored a forest ecosystem and identified its potential flows of utility, addressing those which best satisfy tourism activities and recreational purposes; to identify the most appropriate tools to manage the flows of utility based on sustainable principles which integrate tourism activities [58]; (e) With regards to perception of residents and visitors [61,62,63,64,65,66, 67,68, 69,70,71,72,73,74,75], for example, Xuan studied residents' perceptions of the economic, socio-culture, environment impacts of tourism and residents' attitudes to tourism development in Hainan and Sanya, China. It was concluded that the residents are more aware of positive tourism impacts than negative impacts, and they support tourism development with some reasonable attitudes [61]. Choi and Sirakaya too developed and validated a scale assessing residents' attitudes toward sustainable tourism (SUS-TAS). Then, they administered a 51-item scale of resident attitudes toward sustainable community tourism and 800 households in a small tourism community in Texas. Psychometric properties of SUSTAS along with its practical and theoretical implications are discussed within the framework of sustainable tourism development [65]. Nicholas and Thapa examined visitors' perspectives and support for sustainable tourism development in the Pitons Management Area (PMA) in St. Lucia. Specifically, the focus was on visitors' environmental, economic, and social attitudes based on a sustainable tourism development framework and the effect and best predictive validity of attitudes on support for sustainable tourism development were explored [68]. Bimonte and Punzo analysed how distinct groups of residents, characterised by different levels of involvement in tourism-related activities, perceived the tourism phenomenon, and to check whether there exists a latent or potential ground for conflicts between groups of residents [75]. Cottrell and Vaske examined the relative influence of four sustainability dimensions (environmental, economic, socio-cultural, and institutional) in predicting resident satisfaction with sustainable tourism development in Frankenwald Nature Park, Germany. Structural equation modeling supported the hypothesis that all four dimensions were significant predictors of satisfaction. The economic dimension was the strongest predictor, followed by the institutional, social, and environmental dimensions. Findings indicate that all four dimensions should be included for a holistic approach to planning and monitoring sustainable tourism development [69]. Sörensson and Friedrichs used importance-performance analysis (IPA) to examine the performance of one particular tourist destination with regard to social and environmental sustainability, and to establish whether international tourists and national tourists differ in the sustainability factors they consider important [70]. Dorcheh and Mohamed reviewed existing literature on local perceptions of tourism development and its process. It also discusses influential theories and explains the social exchange theory as an effective framework for sustainable cultural tourism [71]. Miller and Merrilees examined tourists' pro-environmental behaviours in four major categories: recycling; green transport use; sustainable energy/material use (lighting/water usage), and green food consumption. They explored five major antecedents to those categories: habitual behaviour, environmental attitudes, facilities available, a need to take a break from environmental duties, and sense of tourist social responsibility. Also, the poorly understood belief that pro-environmental behaviour weakens when residents become tourists was examined [73]; (f) With regards to research for stakeholders [76,77,78,79,80,81,82,83,84], Hardy and Beeton argued that the nexus involves an understanding of stakeholder perceptions, and applies this to the Daintree region of Far North Queensland, Australia, to determine whether tourism in the region is operating in a sustainable or maintainable manner. In order to do this, an iterative approach was taken and local people, operators, regulators and tourists were interviewed, and content analysis applied to management and strategic documents for the region. The results illustrate the importance of understanding stakeholder perceptions in facilitating sustainable tourism [83]. Timur and Getz examined the concept of sustainable tourism development in urban destinations. Both qualitative and quantitative data were employed, from interviews and questionnaires undertaken in Victoria and Calgary, Canada, and San Francisco, USA. Respondents representing the three clusters of the tourism industry, local government and the host environment were examined on their interpretation of "sustainable tourism", sustainability goals and barriers to achieving sustainable tourism in urban destinations. Results revealed important similarities and differences among key stakeholders, and particularly a lack of appreciation for a triple bottom line approach among the tourism industry respondents [81]. Vellecco and Mancino demonstrate that in lacking shared responsibility, conflicts and tensions inside the local community paralyse innovative environmental behaviors when they ought really to be turned into opportunities for debate so that shared strategies and solutions may be identified in three Italian areas [78]. Holden found that although stakeholders shared positive perceptions of the economic benefits of tourism, its continued use for sustainable development is uncertain. Key challenges include a lack of confidence in the economic certainty of tourism and its use for out-migration, a maturing tourism market, and challenges to the local control of natural resources with external hegemonic forces [80]. Dabphet and Scott explored the diffusion of the sustainable tourism development concept among stakeholders in the tourism destination of Kret Island, Thailand. It is argued that both interpersonal and media communication and the identification of key actors in the community are needed to effectively diffuse sustainable tourism ideas among destination stakeholders. The results validate the use of diffusion theory as a means to understand the transfer of the sustainable tourism development concept among stakeholders, and they also provide information useful for the design of information dissemination programs [82]; (g) In regard to relevant policy for ST [85,86,87,88,89,90,91,92,93,94], for instance, Farsari and Butler explored policies for sustainable tourism development and potential interrelationships between policy considerations. Such policies have been characterized as ad hoc and incremental, lacking a clear orientation towards sustainable development, and the complex relationships underpinning them have rarely been considered in decision-making for sustainable tourism [87]. Dodds and Butler found that although respondents were aware of sustainable tourism, the individual advantage from exploiting shared pooled or shared resources is often perceived as being greater than the potential long-term shared losses that result from the deterioration of such resources, which means that there is little motivation for individual actors (whether governments, elected officials, or individual operators), to invest or engage in protection or conservation for more sustainable tourism [86]. Yasarataa and Altinay noted key political actors' interests and how to mitigate personal interests to facilitate and maintain sustainable tourism development in small states North Cyprus, Turkish [92]. Whitford and Ruhanen recommends that there cannot be a "one size fits all" framework for indigenous tourism development to suit all circumstances. Policies need to draw upon indigenous diversity and, in a consistent, collaborative, coordinated and integrated manner, provide the mechanisms and capacity-building to facilitate long-term sustainable indigenous tourism [93]. Solstrand suggest that the environmental and socio-cultural sustainability of marine angling tourism (MAT) in Norway and Iceland requires a complex socio-ecological systems perspective, with interactive governance strategies leading management policies. Sustainability requires that a management strategy not only focuses on the economic aspects; priority must also be given to minimizing multi-stakeholder conflicts and providing sufficient resource data to protect vulnerable fish stocks [89]. Xu and Sofield examined the situation that little guidance is provided to promote sustainability principles in tourism development strategies in China. In the future, a pro-active sustainability approach should be integrated with environmental concerns to allow tourism to participate constructively in the national transformation to a sustainable society [90].

Due to tourism research involved in geography, ecology, environmental science, sociology and so on, also combined with different scales covering the micro to the macro [2], the research focus on the multi-index comprehensive evaluation method (MICEM), tourism carrying capacity (TCC), tourism environment impacts assess (TEIA), ecological footprint analysis of tourism (EFT), life cycle assessment (LCA), limits of acceptable change (LAC) and Geographic Information Systems (GIS) [95,96,97]. The MICEM could quantify the level of tourism sustainable development, which employed AHP and Delphi method by the level of sustainable development and other potential targets. However, selection indicator and its weights usually by personal decision-making [33,39,98,99,100,101]. TCC could comprehensively measure the carrying capacity of tourism destinations such as ecology, resources, psychological and space, which employed remote sensing (RS), field measurements, questionnaire and Delphi method, etc. However, it has a characteristic of randomness and subjectivity when assessing environmental carrying capacity [102,103,104]. TEIA is an effectively method to analysis the effect of tourism on ecological environments by mathematical statistical analysis methods form microscopic view, which construct assessment index system and select assessment model based on environmental background to monitoring the feedback mechanism for impact of tourism environmental. However, it's usually ignoring the effectiveness of monitoring and feedback mechanism [105,106]. EFT constructs the tourism ecological footprint according to various data of per capita consumption by bottom-up questionnaire and investigates statistics. The consequent was directly comparable based on productive land area; it is a global standards value [56]. It is suitable to research on a small-scale since the ecological burden is likely to be passed on by interregional trade in tourism destination [43,47,107]. LCA could identify the stage of development of the tourism destination and solve some problems [49]. It is difficult to quantify the environmental problems in sustainable development [108,109]. The theoretical framework for LAC include identifying the issues and concerns, and defining and describing the types of tourism opportunities, etc. in the planning area. It could solve the contradiction between development of tourism and conservation of resources, which is mostly influenced by the decisions of programmer makers and managers [42,110]. GIS are now recognized widely as a valuable tool for which applications for regional tourism planning have not mushroomed as in other fields. This is also reflected in the field of sustainable tourism. Nonetheless, sustainable tourism decision-making and carrying capacity estimation has a lot to benefit from using such technologies. GIS can be used for managing the various information needs, estimating indicators, and generally assisting decision making in the planning phase, as well as, in the monitoring and evaluation phases [111]. Therefore, researchers have used diverse methodologies with more quantitative analysis, as for each method there are certain advantages and disadvantages [112]. Current studies are using more comprehensive approaches. Previous literatures show that a tourist destination is a relatively complete artificial ecosystem; the ecological is a basis for sustainable development [113]. However, relevant research has failed to exhaustively analyze the structures, considering the functions and evolutionary mechanisms of the compound tourism destination ecosystem. This is indeed a shortcoming of those research studies. Entropy, as a measure of system dissipation or disorganization, has been used to analyze social systems in various contexts [114,115,116,117,118]. In relation to tourism [119,120,121,122,123], for example, Bailey noted that Social Entropy Theory (SET) was a very general macro sociological systems theory [114]. Kenneth and Bailey point out that the most recent applications of entropy are in social entropy theory and macro accounting theory [115]. Stepanic jr and Stefancic hold that the established level of analogy between certain characteristics of social systems and part of thermodynamic formalism in the simplified model encourage one to assume that even deeper analogies might be drawn to construct more complete and detailed models of social systems [116]. Wilson describes entropy in urban and regional modelling introducing a new framework for constructing spatial interaction and associated location models [117]. Cabral and Augusto summarized entropy multifaceted character with regard to its implications for urban sprawl, and propose a framework to apply the concept of entropy to urban sprawl for monitoring and management. Hao point out that the phenomenon of the increase of entropy also exists in the tourism destination's ecological system [121]. Zhao proposed the conception and mainly indicates that research can broaden insights on tourism systems' carrying capacities through entropy change analysis from the view of the tourism system's entropy principle under the tourism dissipative structure mechanism [122]. Qian noted that the tourism environment system was an open system. It is unceasingly exchanging material and energy with the outside. It is impossible to achieve the absolute balance through introduction of the negative entropy flow [123]. Relevant research has indicated that tourism destination ecosystems are a typical dissipative structure. Therefore, there is certain feasibility in analyzing its evolution and sustainable development potential from the perspective of entropy. Given the analysis above and based on the relevant former research, this study based on the structure, function and characteristic of tourism destination ecosystems, applied information entropy theory for Dunhuang city which combines analysis entropy with information entropy to establish a tourist destination quantitative ecosystem model for evaluating the potential of sustainable development of Dunhuang tourism. It could offset the disadvantage of indistinct strategies and lack of specificity in some degrees in the sustainable development of tourism destinations.

2. STUDY AREA

Dunhuang City is located in the border area of Gansu, Qinghai and Xinjiang Provinces, as the western end of the Hexi Corridor in Gansu Province,

China. It belongs to a typical arid oasis region with unique geographical, historical and cultural status. In history, Dunhuang was an important hub on the ancient Silk Road and the point of integration of Eastern with Western civilization. To some extent, it is known as "human Dunhuang" owing to the intersection and coordination of the world's four major cultural systems [124]. It is rich in tourism resources, with Mogao Grottoes called the "Pearl of Oriental Art", Mingsha Mountain known as "desert spectacle", Crescent Lake and other tourism resources. All these places promote the development of tourism resources. In 2012, the number of visitor arrivals was 0.312 million people-times and the total income was 2.687 billion Yuan [125]. However, this explosive growth has brought a series of severe problems for cultural heritage, such as, tourists' periodic overload. Because of those intensive human activities, the weak regional environment, and the global climate, there has been a shortage of water resources, as the core of the regional ecological problems, which continue to worsen [126]. Therefore, it is very necessary to study the potential of sustainable development of the Dunhuang tourism destination ecosystem.

3. METHODOLOGY

Entropy, firstly proposed by German physicist R Clausius [127,128], is the unique macroscopic quantity in thermodynamics and statistical physics. In 1948, Shannon introduced this concept into information theory and named it "information entropy" [129]. Generally speaking, information entropy theory is based on probability and statistics to reflect the degree of disorder and quantify the evolution direction of the system. When analyzing the complexity and uncertainty of problems, it can be used as a multi-dimensional method to quantify and determine the comprehensive benefits [130,131]. According to theory of dissipative structures by I. Prigogine [132,133,134], an open system, which is far from equilibrium during the process of exchanging matter and energy with outside environments, has the tendency of entropy growth; hence, this system, only by constantly introducing negative entropy from the outside flow in order to offset internal positive entropy, will finally have a new and ordered direction for further development. That means that the large entropy of the system corresponds to the low degree of order and vice versa.

3.1. Entropy Change and Dissipative Structure of the Tourism Destination Ecosystem

The tourism destination ecosystem is a special ecosystem of areas with rich tourism resources and occurrences, that is established based on the original nature or artificial ecosystem during tourism development [121]. As the spatial carrier between tourist activities and the ecological environment, tourism definition of ecosystems involves the continuous exchange of materials, energy and information with external environments. It also makes irreversible the non-equilibrium processes inside the system, which is always producing positive entropy, inflowing negative entropy with the characteristics of openness, which is far from equilibrium, nonlinearity and ordered fluctuation [122,123]. Tourism destination ecosystem is a typical dissipative self-organizing system that possesses dissipative characteristics and analyzes the evolutionary process of entropy changes.

The tourism destination ecosystem development and evolution is led mainly by the evolution of its socioeconomic ecosystem under normal conditions. Given this progression, this study analyzes the interactions between the tourism destination socioeconomic ecosystem and its natural ecosystem and other regions by analyzing the entropy change process of the tourism destination socioeconomic ecosystem. This involves analyzing the evolutionary process and developmental trend of the tourism destination ecosystem, as well as evaluating the sustainable development potential of the tourism destination ecosystem.

According to the dissipative structure theory [135], there are two parts of entropy changes in tourism destination ecosystems. The first one is the entropy flow caused by the tourism destination socioeconomic ecosystem's exchange of materials, energy and information with external environments etc.; it reflects the carrying capacity of the tourism destination nature ecosystem for its socioeconomic ecosystem. Another part is the entropy

3. Methodology

production caused by irreversible non-equilibrium processes inside the system, which reflects its regeneration potential and could indicate the vitality of the tourism destination ecosystem. The total entropy change of the system is the summation of entropy production and entropy flow, reflecting the overall level of development in the tourism destination ecosystem [136]. The environment is affected by the development of tourism and associated activities as part of the evolution in becoming a tourism destination ecosystem. This is a result of the increase in disorder of entropy production, which is caused by de-vegetation, water pollution, soil fertility, air quality degradation, biodiversity decline and the assimilation features of traditional culture, etc. inside of the tourism destination ecosystem. The total entropy changes of the system and disordered parameters will increase if the tourism destination ecosystem does not exchange moderate amounts of material, energy and information with external environments, so the entropy flow does not offset entropy production. This will result in some negative effects for the tourism destination ecosystem, such as an increase in disorder within the system, lack of power, regulatory failure and weaker functioning.

The analysis of entropy changes in tourism destination ecosystems describes the state of tourism system and changes during the exchange of recourses with an external system. The amount or size of entropy not only expresses the level of internal system resources' effective utilization, but also reflects the elastic changes of system affordability. The change in size of the system entropy refers to higher or lower effective utilization rates in the evolutionary process of the tourism destination ecosystem's exchange of materials and energy with external environments [122].

3.2. Indicator System Establishment

The indicator system is an effective tool for measuring and evaluating the tourism sustainable development level. There are lots of indicators that play a more important role for tourism sustainable development. Depending on the principles of scientific city, comprehensiveness, dynamics, hierarchy, maneuverability and perceptiveness [137,138], using the references from Indicators of Sustainable Development for Tourism Destinations: A Guidebook (WTO, 2004) [139], European Tourism Indicator System For

Sustainable Destinations (EU, 2013) [140], ecological civilization city construction indicator system of Dunhuang city and related research results [23,24,25,26,27,28,29,30,31,32,33,34,35,36,37,38,39,40,41,141], the article establishes tourism destination ecosystem sustainable development analysis and indicators system evaluation according to three aspects. They are the structure, function and characteristic of tourism destination ecosystems; the entropy production and entropy flow in the process of system operation; and the ecological environment pollution and destruction during the tourism industry development within the system. The article selected two parts of the entropy production and entropy flow; four aspects that are the supportive entropy input index, the stressful entropy output index, the consumption metabolism index of entropy and the regenerate metabolism index of entropy; 29 representative index (Table 1).

Supportive entropy input index: Mainly embodies attractiveness and bearing capacity of the tourism destination ecosystem. The tourism destination satisfied the tourists' demands by supporting resources, infrastructures and services, support foundation of tourism development, and promotes communication with the outside and for internal operations. Therefore, here we select the related index that can reflect tourism destination development status quo and potential of its development (basic infrastructure, available water resources, etc.).

Stressful entropy output index: Expresses tourism activities putting pressure on the tourism destination ecosystem. During the evolution process of tourism destination ecosystem and tourist industry development, tourists and their consumption (transport, hotels, sightseeing, catering) and tourist industry energy consumption have direct or indirect influence on the tourism local ecological environment, and puts some pressure on system development, as well as slowing down the positive evolution speed of the system.

Consumption metabolism index of entropy: During the evolution process of the tourism destination ecosystem, the discharge of wastes, pollutants produced and a series of ecological problems from tourism activities to some extent weaken the sustainable development potential of the system.

3. Methodology

Table 1. Index system hierarchy of sustainable development potential evaluation.

Criterion	Sub-Criterion	Indicators	Measurements
Entropy flow	Supportive entropy input index (A)	Number of travel agencies (A1)	unit
		Number of direct engaged persons in tourism industry (A2)	person
		Number of star-rated hotels (A3)	unit
		Number of beds in star-rated hotels (A4)	bed
		Passenger-kilometers by highways (A5)	10^4 passenger-km
		Passenger-kilometers by railways (A6)	10^4 passenger-km
		Passenger-kilometers by civil aviation (A7)	10^4 passenger-km
		Annual water supply (A8)	10^4 t
	Stressful entropy output index (B)	Number of visitor arrivals (B1)	10^4 person-times
		Transport expenditure as percentage of tourism expenditure (B2)	%
		sightseeing expenditure as percentage of tourism expenditure (B3)	%
		Hotels expenditure as percentage of tourism expenditure (B4)	%
		Catering expenditure as Percentage of tourism expenditure (B5)	%
		Water used by tourists (B6)	10^4 t
Entropy production	Consumption metabolism index of entropy (C)	Total wastewater discharged (C1)	10^4 t
		Industrial wastewater discharged (C2)	10^4 t
		Emission of disulfide (C3)	t
		solid wastes discharged (C4)	10^4 t
		waste discharge by tourists (C5)	t
		Carbon emission by tourism (C6)	t
	Regenerate metabolism index of entropy (D)	Number of training institutions for tourism (D1)	unit
		Direct engaged persons in tourism industry as percentage of employees (D2)	%
		Tourism GDP (D3)	10^4 Yuan
		Tourism GDP as percentage of GDP (D4)	%
		Investment in anti-pollution Projects as percentage of GDP (D5)	%
		Proportion of industrial solid waste treated and utilized (D6)	%
		Rate of harmless garbage disposal (D7)	%
		Green coverage rate in developed areas (D8)	%
		Gardens per capita (D9)	m^2

Regenerate metabolism index of entropy: Mainly shows human being's governance capacity of the tourism destination ecosystem. The waste discharge by tourists is above the system bearing capacity, so that some pollution cannot be purified by the system itself; therefore, the system must rely on artificial management policies and scientific technologies. That is why human beings invest into environmental pollution management as a recovery function of tourism destination ecosystem's sustainable development.

According to the established evolution indexes of the tourism destination ecosystem, there is the calculation formula for entropy flow, entropy production and total entropy changes (Table 2).

Table 2. Symbols and formulae of entropy flow, entropy production and total entropy change.

Objective	Symbols and Formula	Means
Supportive entropy input	$\Delta_e S_1$	Disorder of system
Stressful entropy output	$\Delta_e S_2$	Disorder of system
Consumption metabolism of entropy	$\Delta_i S_2$	Disorder of system
Regenerate metabolism of entropy	$\Delta_i S_1$	Disorder of system
Entropy flow	$\Delta_e S_2 - \Delta_e S_1$	Coordination of system
Entropy production	$\Delta_i S_2 - \Delta_i S_1$	Vitality of system
Total entropy change	$(\Delta_e S_2 - \Delta_e S_1) + (\Delta_i S_2 - \Delta_i S_1)$	Order and health of system

3.3. Depending on Entropy Information Evaluation Model Establishment

Based on information entropy's benefits, information entropy's evaluation model is widely used in many scientific fields [136,142]. The tourism research focuses on the following: measuring the weight of an indicator based on the entropy method [143]; analyzing the characteristics and development measures based on information entropy, theory of entropy and dissipative structure [121,122,123,126,144]. More qualitative analysis was be used and a few calculation methods for entropy of tourism systems. According to the information entropy of Shannon, if we used random variables X represents uncertainty in the system, the discrete random variable could be supposed as x and its value is $X = \{x_1, x_2, ..., x_n\}$ ($n \geq 2$), the probability for each value is $P = \{p_1, p_2, ..., p_n\}$ ($0 \leq p_i \leq 1$, $i = 1, 2, ..., n$). $\Sigma P_i = 1$. The information entropy can be described as follows [131,145,146]:

3. Methodology

$$S = -\sum_{i=1}^{n} P_i \ln(P_i) \quad (1)$$

where S the is information entropy of an uncertain system, P_i is the probability of the random state variable X in the uncertain system.

3.3.1. Measurement for Entropy Flow and Entropy Production of Tourism Destination Ecosystem

According to the measurement models for information entropy theory, we compute a formula for the entropy flow and entropy production based on information entropy theory and models for each year, then analyzed the complexity, coordination order and health for tourism destination ecosystems. Measurement of n indictors in m years, ΔS represents the four types of entropy based on information entropy [130,131,145], i.e., the input supportive type of entropy ($\Delta_e S_1$), the output pressure type of entropy ($\Delta_e S_2$), the consumption metabolic type of entropy ($\Delta_i S_2$) and the regeneration metabolic type of entropy ($\Delta_i S_1$).

$$\Delta S = -\frac{1}{\ln m} \sum_{i=1}^{n} \frac{q_{ij}}{q_j} \ln \frac{q_{ij}}{q_j} \quad (2)$$

where ΔS represents the four types of entropy, q_{ij} is the standardized value of calculated from the raw data, q_j is sum for standardized value of index in j year, m is sum for the number of appraisal events and n is the number of indicators, i is each index, $q_i = \sum_{i=1}^{n} q_{ij}$ ($i=1,2,\ldots,n; j=1,2,\ldots,m$)

If the number of index is n, and the number of appraisal events is m, then E_i denotes the indicator-based information of indicator i and can be derived thus:

$$E_i = -\frac{1}{\ln m} \sum_{j=1}^{m} \frac{q_{ij}}{q_i} \ln \frac{q_{ij}}{q_i}$$

(3)

where E_i is the information entropy of indicator, q_{ij} is the standardized value calculated from the raw data and q_i is sum for standardized value all appraisal events in i index, $q_i = \sum_{i=1}^{n} q_{ij}$ ($i=1,2,\ldots,n; j=1,2,\ldots,m$)

According to the entropy weighting method [142], the entropy weight of i indicator is defined as:

$$Q_i = (1 - E_i) \bigg/ \left(n - \sum_{i=1}^{n} E_i \right)$$

(4)

where Q_i is the entropy weight of i, E_i is the indicator-based information entropy of indicator i, n is the number of indicator and $\sum_{i=1}^{n} Q_i = 1$, $0 \leq Q_i \leq 1$, $n \geq 2$.

The entropy weight of an indicator is not the most important coefficient of the indicator in regard to decision-making issues. It is instead the relative degree of competition with other indicators when a set of evaluation objects is given and the evaluation indicators are determined, the entropy weighting value is closely related to the evaluation objects. From the information perspective, the entropy weight of an evaluation indicator represents how much useful information an indicator can provide [131,145]. When the entropy weighting of an indicator is larger than other indicators in the evaluation index system for the sustainable development of tourism destination ecosystems, the useful information provided by the indicator could have a greater impact on the system than the other indicators [131].

3.3.2. Sustainable Development Evaluation Model of the Tourism Destination Ecosystem

The index weight was calculated by information entropy, and then integrated to the value of normalization calculated a comprehensive score of values [130,131,146]:

$$G = \sum Q_i X_i \tag{5}$$

where: G is an appraisal score, Q_i is the weighting factor derived from information entropy (described below), X_i is the standardized value between 0 and 1 generated from raw data for each indicator. A larger value of G indicates a better state of the tourism destination ecosystem and a better potential of the tourism destination ecosystem for sustainable development.

4. DATA SOURCES AND PROCESSING METHOD

4.1. Data Sources

Related data applied in this study were extracted from the 10th Five-Year Statistical Yearbook of Dunhuang City [147], 11th Five-Year Statistical Yearbook of Dunhuang City [148], Statistical Yearbook of Dunhuang City between 2011 and 2013 [149], Environmental Quality Bulletin of Dunhuang City between 2006 and 2012 [150]. Some data are obtained from the interview and questionnaire.

4.2. Data Processing Methods

This study adopted the standardize deviation to processing data and the score between [0, 1] when analyzing the evolution and development of the tourism destination socioeconomic ecosystem. The following aspects should be noted in processing data [131,146]: (a) As the entropy change model has used the four types of entropy for vector quantization, there is no need to distinguish between positive and negative indicators to standardize the data processing; (b) The assessment model for the sustainable development potential of the tourism destination ecosystem, which is based on information entropy, has not used vector quantization on different types of indicators, the data processing must distinguish between positive and negative indicators.

For the four indicators, the input supportive type of entropy ($\Delta_e S_1$) and regeneration metabolic type of entropy ($\Delta_i S_1$) are positive indicators, the bigger value means more coordination of the system. The output pressure type of entropy ($\Delta_e S_2$) and the consumption metabolic type of entropy ($\Delta_i S_2$) are negative indicators, the bigger value means less coordination of the system. The normalization methods for the positive and negative indicators are listed below:

Normalization method for positive indicators:

$$X'_{ij} = X_{ij} / Max(X_i) \tag{6}$$

Normalization method for negative indicators:

$$X'_{ij} = Min(X_i) / X_{ij} \tag{7}$$

where X'_{ij} is the normalized value of X_{ij}, X_{ij} is the raw data for indicator i in j year, X_i represents all of the original data for indicator i, and $Max(X_i)$ obtains the maximum of indicator *i* by function during the study period, and $Min(X_i)$ obtains the minimum of indicator i by function during the study period.

5. RESULTS AND ANALYSIS

5.1. Entropy Change Analysis

The supportive entropy input showed a trend to remain stable during study period. The stressful entropy output was fluctuated within a slow upward trend (Table 3, Figure 1). For the stressful entropy output with the turning point in 2003 and 2008, the smallest value was in 2003, because the tourism industry was in a state of depression influenced by SARS. In addition, the global financial crisis and snow disaster of South China led to a smaller value in 2008. The burden on tourism destination ecosystems was decreased during those two years. However, the pressure of tourist destinations was increased with the recovery of the tourism industry. The burden of the Dunhuang tourism destination natural ecosystem was increased under the rapid development of tourism and increasing utilization of tourism resources, while the supportive entropy input experienced relatively slow growth. This shows that the pressure was increasing on the tourism destination socioeconomic ecosystem in some degree from 2008.

Table 3. Entropy production and total entropy change of the tourism destination socio-economic ecosystem in Dunhuang city on information entropy.

Year	Supportive Entropy Input	Stressful Entropy Output	Consumption Metabolism of Entropy	Regenerate Metabolism of Entropy	Entropy Flow	Entropy Production	Total Entropy Change
2000	0.7846	0.6370	0.6243	0.8002	−0.1476	−0.1759	−0.3234
2001	0.7925	0.6418	0.6504	0.7838	−0.1507	−0.1334	−0.2840
2002	0.7958	0.6398	0.6429	0.7864	−0.1559	−0.1435	−0.2994
2003	0.7974	0.6356	0.6330	0.7865	−0.1618	−0.1535	−0.3153
2004	0.7793	0.6607	0.6400	0.7793	−0.1186	−0.1393	−0.2579
2005	0.7980	0.6711	0.6505	0.7893	−0.1269	−0.1389	−0.2658
2006	0.7981	0.6789	0.6357	0.7987	−0.1192	−0.1630	−0.2821
2007	0.7969	0.6850	0.6352	0.8332	−0.1119	−0.1980	−0.3099
2008	0.7980	0.6659	0.6360	0.7271	−0.1321	−0.0911	−0.2233
2009	0.7991	0.6747	0.6010	0.7357	−0.1245	−0.1346	−0.2591
2010	0.7993	0.6831	0.5837	0.7549	−0.1162	−0.1713	−0.2875
2011	0.8095	0.6925	0.5953	0.7581	−0.1171	−0.1628	−0.2798
2012	0.8101	0.6914	0.5933	0.8556	−0.1187	−0.2623	−0.3810

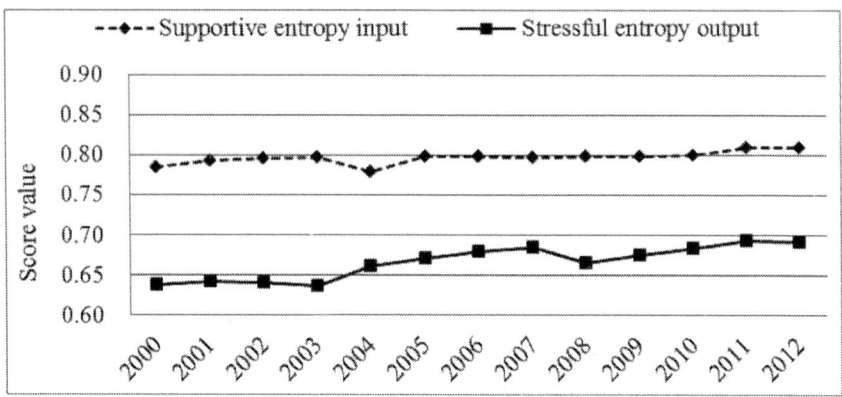

Figure 1. Score trends of the two entropy exchange types of the tourism destination.

The consumption metabolism of entropy showed a slowed trend down between 2000 and 2012, the regenerate metabolism of entropy fluctuated with a sharply upward trend. The turning point of regenerate metabolism of entropy was in 2007, 2008 and 2011, which first increased and then decreased with the turning point in 2007 and slowed down sharply. The minimum value was in 2008 and then showed a slowed upward trend. The turning point was a sharp upward trend in 2011 (Table 3, Figure 2). This indicates that the ecological security was improved, the potential of metabolism was better and the vitality was improved gradually in Dunhuang tourism destination from 2008.

The entropy flow showed fluctuation within a slow upward trend between 2000 and 2012, while the entropy production and total entropy change both sharply fluctuated. The turning points of entropy flow, entropy production and total entropy change were during the period of 2003 and 2008 because the tourism industry was mostly effected by the external environment, with the influences of SARS in 2003 and the global financial crisis and snow disaster of South China in 2008. The entropy flow fluctuated with a slow upward trend, which first decreased and then increased with the turning points in 2003 and 2008. The entropy production and total entropy change sharply fluctuated, first increasing and then decreasing with the turning

5. Results and Analysis

point in 2003. Also, it first increased and then decreased with the turning point in 2008 (Table 3). This indicates that the Dunhuang tourism destination ecosystem was orderly and healthy during the study period.

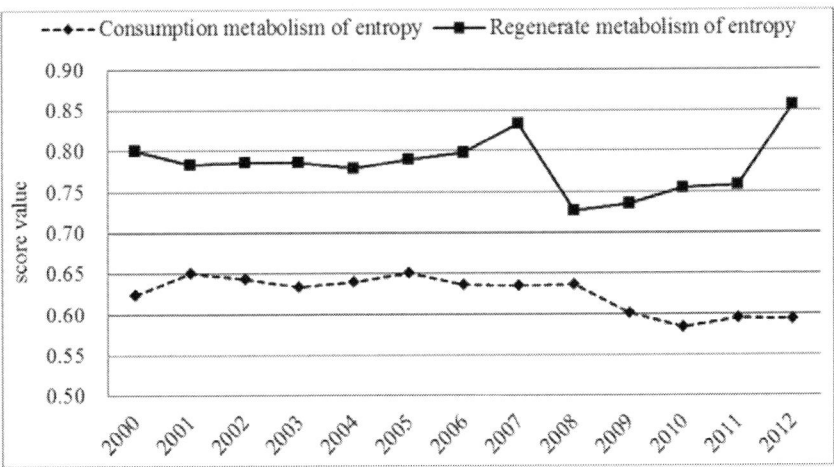

Figure 2. Score trends of the two entropy production types of tourism destination.

5.2. Analysis of Tourism Destination Ecosystem Sustainable Development Potential in Dunhuang

The values of supportive entropy input remained stable between 2000 and 2012. These values indicate the carrying capacity was relatively stable as a socio-economic ecosystem in Dunhuang. The values of stressful entropy output were decreased in 2003 and 2008, which showed that the pressure was increased on the nature ecosystem with the development of tourism (Figure 1). The values of consumption metabolism of entropy fluctuated with a slow downwards trend, and the values of regenerate metabolism of entropy fluctuated with a sharp upward trend. That indicated the metabolism of function was strengthened, which indicates some success in protecting the ecological environment and also its quality was improved (Figure 2). The values of sustainable development potential fluctuated with an upward trend (Figure 3). The lowest value was in 2008 and the highest was in 2012. This phenomenon may be related to external features of the tourism industry, which was in a status of trough influenced by the global financial crisis and snow disaster of Southern China in 2008. The stressful entropy output fluctuated with a downward trend; also the regenerate metabolism of entropy had the lowest value in 2008. The highest value is attributed to government attention and an improvement in ecological security, an increase in investment in anti-pollution projects as percentage of GDP and improved proportion of industrial solid waste treated and utilized.

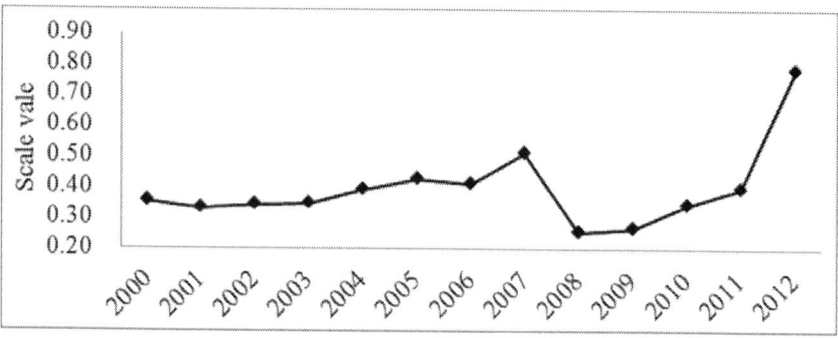

Figure 3. Score trends of tourism destination ecosystem sustainable development potential.

5.3. Analysis of Sustainable Development Measures Based on the Entropy Weights and Time Sequence Changes of the Indicators

The entropy weights of number of travel agencies, passenger-kilometers by highways and passenger-kilometers railways were largest among the supportive entropy input index in Dunhuang tourism destination ecosystem between 2000 and 2012 (Table 4). These indicate that significant increases of these three indexes had played an important role in strengthening the supportive entropy input system. However, the entropy weights of annual water supply were the smallest, and the entropy weights of passenger-km by civil aviation was smaller, showing some negative effects caused by the sharp decline of annual water supply and decrease of passenger-km by civil aviation in the Dunhuang tourism destination ecosystem. Those two indexes restricted the development of the supportive entropy input system. The shortage of water resources is a limiting factor for development of Dunhuang being located in the arid oasis area of northwest China. The transport passenger-kilometers focused on railways and highway, however, civil aviation supplemented the railway and highway with the development of the economy. The positive aspects of transport (railways, highway and civil aviation) and travel agencies should be improved for the supportive entropy input, as well as coordinating the relationship between water resources and to safeguard water supply by increasing effective use.

The entropy weights of number of visitor arrivals and water used by tourists were largest among the stressful entropy output index during the study period (Table 4). The value of those two indexes significantly increased the pressure of stressful entropy output index of Dunhuang tourism destination ecosystem. Those two indexes should be paid much attention, the number of visitor arrivals reasonably controlled and tourists guided on water use.

The entropy weights of industrial wastewater discharged and total wastewater discharged were largest among the consumption metabolism index of entropy between 2000 and 2012 (Table 4). Those indicate that the significant increases of wastewater discharged was strongly influenced by the consumption metabolism index of entropy and increasing the pressure on ecological environments of the tourism destination. The value of entropy weights for waste discharge by tourists was larger than that of carbon emissions, which indicates that the consumption metabolism of tourism destination is influenced more by increase of waste discharge by tourists than carbon emission by tourism. Given these circumstances, scientific controls should be put in place for the waste discharge of tourists.

The value of entropy weights for two indexes (proportion of industrial solid waste treated and utilized and investment in anti-pollution projects as percentage of GDP) were largest among regenerate metabolism index of entropy (Table 4). These show that those two indexes greatly impact the potential of regenerate metabolism system. Pollution should be controlled paying attention to the ecological security of the tourism destination, increasing investment in anti-pollution projects as a percentage of GDP and increasing proportion of industrial solid waste treated and utilized. The values of entropy weights were smallest in regenerate metabolism index of entropy, which include direct engaged persons in tourism industry as percentage of employees, rate of harmless garbage disposal and gardens per capita. These indicate improvement in the potential of regenerate metabolism system by increasing the direct engaged persons in the tourism industry as percentage of employees, improving the rate of harmless garbage disposal and increasing the gardens per capita.

Table 4. Information entropy and entropy weights of the sustainable development potential evaluation indicators for the tourism destination ecosystem in Dunhuang.

Indicator Type	Indicator	E_i	Q_i
Supportive entropy input index (A)	Number of travel agencies (A1)	0.9596	0.0295
	Number of direct engaged persons in tourism industry (A2)	0.9936	0.0047
	Number of star-rated hotels (A3)	0.9945	0.0040
	Number of beds in star-rated hotels (A4)	0.9941	0.0043
	Passenger-kilometers by highways (A5)	0.9669	0.0242
	Passenger-kilometers by railways (A6)	0.9674	0.0238
	Passenger-kilometers by civil aviation (A7)	0.9966	0.0025
	Annual water supply (A8)	0.9968	0.0023
Stressful entropy output index (B)	Number of visitor arrivals (B1)	0.9418	0.0426
	Transport expenditure as percentage of tourism expenditure (B2)	0.9919	0.0059
	sightseeing expenditure as percentage of tourism expenditure (B3)	0.9932	0.0050
	Hotels expenditure as percentage of tourism expenditure (B4)	0.9970	0.0022
	Catering expenditure as percentage of tourism expenditure (B5)	0.9906	0.0069
	Water used by tourists (B6)	0.9414	0.0429
Consumption metabolism index of entropy (C)	Total wastewater discharged (C1)	0.9153	0.0620
	Industrial wastewater discharged (C2)	0.8324	0.1226
	Emission of disulfide (C3)	0.9942	0.0043
	solid wastes discharged (C4)	0.9182	0.0598
	waste discharge by tourists (C5)	0.9414	0.0429
	Carbon emission by tourism (C6)	0.9627	0.0273
Regenerate metabolism index of entropy (D)	Number of training institutions for tourism (D1)	0.9683	0.0232
	Direct engaged persons in tourism industry as percentage of employees (D2)	0.9980	0.0014
	Tourism GDP (D3)	0.8609	0.1017
	Tourism GDP as percentage of GDP (D4)	0.9851	0.0109
	Investment in anti-pollution projects as percentage of GDP (D5)	0.8252	0.1279
	Proportion of industrial solid waste treated and utilized (D6)	0.7336	0.1948
	Rate of harmless garbage disposal (D7)	0.9972	0.0020
	Green coverage rate in developed areas (D8)	0.9799	0.0147
	Gardens per capita (D9)	0.9947	0.0039

6. CONCLUSIONS AND DISCUSSION

The analysis of the tourist destination ecosystem entropy change indicates an increase in the diversity and complexity of Dunhuang tourism destination's socio-economic ecosystem with the rapid development of the tourism industry; the demands placed on the natural ecosystem have increased. However, the pollution problems have been controlled, as shown by the overall upward trend for regenerate metabolism during the study period. The vitality of the tourism destination ecosystem was obviously strengthened from 2008. Based on the score of sustainable development potential for the tourism destination ecosystem between 2000 and 2012, the pressure on the natural ecosystem was increased, while the carrying capacity of its socio-economic system also strengthen. The regenerate metabolism system increased due to significant conservation achievements and development of the eco-environment in the Dunhuang tourism destination ecosystem. According to the entropy weight of this indicator and its impact on the sustainable development potential of Dunhuang tourism destination ecosystem, the countermeasures are as follows: Increase the potential of the supportive entropy input system by increasing the travel agencies and transportation; Reduce the pressure on the consumption metabolism system by decreasing the total wastewater discharged and industrial wastewater discharged; Enhance the potential of regenerate metabolism by focusing on the ecological security of the tourism destination, increase investment in anti-pollution projects as percentage of GDP and improve proportion of industrial solid waste treated and utilized.

The paper summarizes the former research results, and then further demonstrates by entropy change analysis, information entropy and negative entropy of dissipative structure system for evaluating the tourism destination ecosystem's sustainable development evolution feasibility. The numerical values show the orderly level of the tourism destination ecosystem demonstrating the system sustainable development potential. Combining the entropy weight of index and times series index will be more targeted for improving measures of Dunhuang tourism destination ecosystem's sustainable development. According to the data's availability, the article selects indexes focusing on supportive, stressful, consumed and regenerate indexes. The tourism destination ecosystem as a society-economics-environment artificial compound ecosystem, the tourists and local residents are significant participants and propellants of tourism sustainable development [27]. Their appreciation of tourism sustainable development plays an important role in system improvement; thus the indexes which are in line with their values should be chosen. On the other hand, using information entropy from the perspective of the development of tourism destination ecosystem evolution to analyze tourism destination sustainable development potential, it is beneficial to vertically analyze a single tourism destination. There are disadvantages in horizontally comparing tourism, and therefore this research must be improved.

ACKNOWLEDGMENTS

This research is supported by China Scholarship Council (CSC), The Fundamental Research Funds for the Central Universities (lzujbky-2013-m02), Natural Science Foundation of China (41471462, 41461119).

AUTHOR CONTRIBUTIONS

Huihui Feng, Xingpeng Chen, Peter Heck and Hong Miao designed the paper and all contributed to data collection and calculation.

CONFLICTS OF INTEREST

The authors declare no conflict of interest.

REFERENCES

1. General Assembly of the United Nations. The 19th Special Session for General Assembly of the United Nations resolutions. Available online: http://www.un.org/chinese/ga/spec/19/ar19_1.pdf (accessed on 19 September 1997).

References

2. Tang, C.; Zhong, L.; Cheng, S. A review on sustainable development for tourist destination. Prog. Geogr. 2013, 32, 984–992.

3. Butler, R. Sustainable tourism: A state-of-the-art review. Tour. Geogr. 1999, 1, 7–25.

4. Hardy, A.; Beeton, R.; Pearson, L. Sustainable tourism: An overview of the concept and its position in relation to conceptualizations of tourism. J. Sustain. Tour. 2002, 10, 475–496.

5. Lu, J.; Nepala, S. Sustainable tourism research: An analysis of papers published in the Journal of Sustainable Tourism. J. Sustain. Tour. 2009, 17, 5–16.

6. Buckley, R. Sustainable tourism: Research and reality. Ann. Tour. Res. 2012, 39, 528–546.

7. Liu, Z. Sustainable tourism development: A critique. J. Sustain. Tour. 2003, 11, 459–475.

8. Berno, T.; Bricker, K. Sustainable tourism development: The long road from theory to practice. Int. J. Econ. Dev. 2001, 3, 1–18.

9. Swarbrooke, J. Sustainable Tourism Management; Biddles Ltd., Guildford and King's Lyun: New York, NY, USA, 1999; pp. 4–5.

10. Bramwell, B.; Lane, B. Sustainable tourism: An evolving global approach. J. Sustain. Tour. 1993, 1, 1–5.

11. Smith, S. Tourism Analysis: A Handbook; Longman: London, UK, 1995; pp. 295–299.

12. Wight, P. Tools for sustainability analysis in planning and managing tourism and recreation in the destination. In Sustainable Tourism: A Geographical Perspective; Hall, C., Lew, A., Eds.; Addison Wesley Longman: New York, NY, USA, 1998; pp. 75–91.

13. Hunter, C. Sustainable tourism as an adaptive paradigm. Ann. Tour. Res. 1997, 24, 850–867.

14. Saarinen, J. Traditions of sustainability in tourism studies. Ann. Tour. Res. 2006, 33, 1121–1140.

15. Sharpley, R. Tourism and sustainable development: Exploring the theoretical divide. J. Sustain. Tour. 2000, 8, 1–19.

16. Gianna, M.; Laurie, M. There is no such thing as sustainable tourism: Re-conceptualizing tourism as a tool for sustainability. Sustainability 2014, 6, 2538–2561.

17. Aall, C. Sustainable tourism in practice: Promoting or perverting, the quest for a sustainable development. Sustainability 2014, 6, 2562–2583.

18. Saarinen, J. Critical sustainability: Setting the limits to growth and responsibility in tourism. Sustainability 2014, 6, 1–17.

19. Hunter, C. Aspects of the Sustainable Tourism Debate from a Natural Resources Perspective. In Sustainable Tourism: A Global Perspective; Harris, R., Griffin, T., Williams, P., Eds.; Butterworth-Heinemann: Oxford, UK, 2002; pp. 4–5.

20. Moyle, B.; McLennan, C.; Ruhanen, L.; Weiler, B. Tracking the concept of sustainability in Australian tourism policy and planning documents. J. Sustain. 2014, 22, 1037–1051.

21. Farrell, B.; Ward, L. Reconecptualizing tourism. Ann. Tour. Res. 2004, 31, 274–295.

22. Lansing, P.; Vries, P. Sustainable tourism: Ethical alternative or marketing ploy? J. Bus. Ethics 2007, 72, 77–85.

23. Delgado, A.; Saarinenc, J. Using indicators to assess sustainable tourism development: A review. Tour. Geogr. 2014, 16, 31–47.

24. Ko, T. Development of a tourism sustainability assessment procedure: A conceptual approach. Tour. Manag. Issue 2005, 26, 431–445.

25. Chris, H.; Sirakaya, E. Sustainability indicators for managing community tourism. Tour. Manag. 2006, 27, 1274–1289.

26. McElroy, J. Tourism penetration index in small Caribbean islands. Ann. Tour. Res. 1998, 25, 145–168.

27. McCool, S.; Moisey, R.; Nickerson, N.P. What should tourism sustain? The disconnect with industry perceptions of useful indicators. J. Travel Res. 2001, 40, 124–131.

28. Moore, S.; Polley, A. Defining indicators and standards for tourism impacts in protected areas: Cape Range National Park, Australia. Environ. Manag. 2007, 39, 291–300.

29. Castellani, V.; Sala, S. Sustainable performance index for tourism policy development. Tour. Manag. 2010, 31, 871–880.

30. Blancas, F.; Gonzalez, M.; Lozano-Oyola, M.; Pérez, F. The assessment of sustainable tourism: Application to Spanish coastal destinations. Ecol. Indic. 2010, 10, 484–492.

31. Twining, L.; Butler, R. Implementing STD on a small island: Development and use of sustainable tourism development indicators in Samoa. J. Sustain. Tour. 2002, 10, 363–387.

32. Roberts, S.; Tribe, J. Sustainability indicators for small tourism enterprises—An exploratory perspective. J. Sustain. Tour. 2008, 16, 575–594.

33. Schianetz, K.; Kavanagh, L. Sustainability indicators for tourism destinations: A complex adaptive systems approach using systemic indicator systems. J. Sustain. Tour. 2008, 16, 601–628.

34. Fernández, J.; Sánchez, R. Measuring tourism sustainability: Proposal for a composite index. Tour. Econ. 2009, 15, 277–296.

35. Blackstock, K.; White, V.; McCrum, G.; Scott, A.; Hunter, C. Measuring responsibility: An appraisal of a scottish national park's sustainable tourism indicators. J. Sustain. Tour. 2008, 16, 276–297.

36. Oyola, M.; Blancas, F.; González, M.; Caballero, R. Sustainable tourism indicators as planning tools in cultural destinations. Ecol. Indic. 2012, 18, 659–675.

37. Blancas, F.; Oyola, M.; González, M.; Guerrero, F.M.; Caballero, R. How to use sustainability indicators for tourism planning: The case of rural tourism in Andalusia (Spain). Sci. Total Environ. 2011, 412–413, 28–45.

38. Reddy, M. Sustainable tourism rapid indicators for less-developed islands: An economic perspective. Int. J. Tour. Res. 2008, 10, 557–576.

39. Miller, G. The development of indicators for sustainable tourism: Results of a Delphi survey of tourism researchers. Tour. Manag. 2001, 22, 351–362.

40. Wang, L. On the indicator system of sustainable development of tourism and the evaluating method. Tour. Tribune 2001, 16, 67–70.

41. Wan, Y. The method and indicates of evaluation on sustainable development for tourism. Stat. Decis. 2006, 2, 10–12.

42. Ahn, B.; Lee, B.; Shafer, C.S. Operationalizing sustainability in regional tourism planning: An application of the limits of acceptable change framework. Tour. Manag. 2002, 23, 1–15.

43. Hunter, C.; Shaw, J. The ecological footprint as a key indicator of sustainable tourism. Tour. Manag. 2007, 28, 46–57.

44. Gössling, S.; Hansson, C.; Hörstmeier, O.; Saggel, S. Ecological footprint analysis as a tool to assess tourism sustainability. Ecol. Econ. 2002, 43, 199–211.

45. Martín-Cejas, R.; Sánchez, P. Ecological footprint analysis of road transport related to tourism activity: The case for Lanzarote Island. Tour. Manag. 2010, 31, 98–103.

46. Mehdi, M.; Jerome, B. Ecotourism versus mass tourism: A comparison of environmental impacts based on ecological footprint analysis. Sustainability 2012, 4, 123–140.

47. Hunter, C. Sustainable tourism and the touristic ecological footprint. Environ. Dev. Sustain. 2002, 4, 7–20.

48. Yang, G.; Li, P. Touristic ecological footprint: A new yardstick to assess sustainability of tourism. Acta Ecol. Dinica 2005, 25, 1475–1480.

49. Castellani, V.; Sala, S. Ecological footprint and life cycle assessment in the sustainability assessment of tourism activities. Ecol. Indic. 2012, 16, 135–147.

References

50. Cui, F.; Liu, J.; Li, Q. Study of the theory and application of tourism bearing capacity index. Tour. Tribune 1998, 22, 41–44.

51. Xiao, J.; Yu, Q.; Liu, K.; Chen, D.; Chen, J.; Xiao, J. Evaluation of the ecological security of island tourist destination and island tourist sustainable development: A case study of Zhoushan Islands. Acta Geogr. Sin. 2011, 66, 842–852.

52. Salerno, F.; Viviano, G.; Manfredi, E.C.; Caroli, P.; Thakuri, S.; Tartari, G. Multiple Carrying Capacities from a management-oriented perspective to operationalize sustainable tourism in protected areas. J. Environ. 2013, 128, 116–125.

53. Zhong, L.; Deng, J.; Xiang, B. Tourism development and the tourism area life-cycle model: A case study of Zhangjiajie National Forest Park, China. Tour. Manag. 2008, 29, 841–856.

54. Yang, Y.; Luo, S.; Wang, X. Tourist destination life cycle early warning system research based on "Inflection Point" theory, China. Popul. Resour. Environ. 2009, 19, 110–113.

55. Rodríguez, J.; Parra-López, E.; Yanes-Estévez, V. The sustainability of island destinations: Tourism area life cycle and teleological perspectives: The case of Tenerife. Tour. Manag. 2008, 29, 53–65.

56. Yang, G. Targeting model of sustainable development in eco-tourism. Hum. Geogr. 2005, 20, 74–77.

57. Keitumetse, S. Cultural resources as sustainability enablers: Towards a community-based cultural heritage resources management (COBACHREM) model. Sustainability 2014, 6, 70–85.

58. Dina, R.; Geremia, G. A sustainable tourism paradigm: Opportunities and limits for forest landscape planning. Sustainability 2014, 6, 2379–2391.

59. Karatzoglou1, B.; Spilanis, I. Sustainable tourism in Greek islands: The integration of activity-based environmental management with a destination environmental scorecard based on the adaptive resource management paradigm. Bus. Strategy Environ. 2010, 19, 26–38.

60. Lu, M. Evaluation Pattern on the Sustainability of Tourist Destination; Central China Normal University: Wuhan, China, 2001.

61. Xuan, G.; Lu, L.; Zhang, J. Residents' perception of tourism impacts in coast resorts: The case study of Haikou and Sanya Cities, Hainan Province. Sciatica Geogr. Sonica 2002, 22, 741–746.

62. Su, Q.; Lin, B. Classification of residents in the tourist attractions based on attitudes and behaviors: A case study in Xidi, Zhouzhuang and Jiuhua Mountain. Geogr. Res. 2004, 23, 104–114.

63. Zhao, Y.; Li, D.; Huang, M. On the residents' perceptions and attitudes towards tourism development in tourist destinations overseas: A review. Tour. Tribune 2005, 20, 85–92.

64. Huang, Y.; Huang, Z. A study on the structural equation model and its application to tourist perception for agri-tourism destinations: Taking southwest minority areas as an example. Geogr. Res. 2008, 27, 1455–1465.

65. Choi, H.; Sirakaya, E. Measuring residents' attitude toward sustainable tourism: Development of sustainable tourism attitude scale. J. Travel Res. 2005, 43, 380–394.

66. Wearing, S.; Wearing, M. Understanding local power and interactional processes in sustainable tourism: Exploring village-tour operator relations on the Kokoda Track, Papua New Guinea. J. Sustain. 2010, 18, 61–76.

67. Haukeland, J.; Grue, B.; Veisten, K. Turning national parks into tourist attractions: Nature orientation and quest for facilities, Scandinavian. J. Hosp. Tour. 2010, 10, 248–271.

68. Lucia, S.; Nicholas, L.; Thapa, B. Visitor perspectives on sustainable tourism development in the pitons management area world heritage site. Environ. Dev. Sustain. 2010, 12, 839–857.

69. Cottrell, S.; Vaske, J.; Roemer, J.M. Resident satisfaction with sustainable tourism: The case of Frankenwald Nature Park, Germany. Tour. Manag. Perspect. 2013, 8, 42–48.

70. Sörensson, A.; Friedrichs, Y. An importance–performance analysis of sustainable tourism: A comparison between international and national tourists. J. Destin. Mark. Manag. 2013, 2, 14–21.

71. Dorcheh, S.; Mohamed, B. Local perception of tourism development: A conceptual framework for the sustainable cultural tourism. J. Manag. Sustain. 2013, 3, 31–39.

72. Cottrell, S.; Vaske, J. Resident perceptions of sustainable tourism in Chongdugou, China. Soc. Nat. Resour. Int. J. 2007, 20, 511–525.

73. Miller, D.; Merrilees, B. Sustainable urban tourism: Understanding and developing visitor pro-environmental behaviors. J. Sustain. Tour. 2014, 5, 1–21.

74. Chen, C.; Chen, S.; Hong, T. The destination competitiveness of Kinsmen's tourism industry: Exploring the interrelationships between tourist perceptions, service performance, customer satisfaction and sustainable tourism. J. Sustain. Tour. 2011, 19, 247–264.

75. Bimonte, S.; Punzo, L. Tourism, residents' attitudes and perceived carrying capacity with an experimental study in five Tuscan destinations. Int. J. Sustain. 2011, 14, 242–261.

76. Waligo, V.; Clarke, J.; Hawkins, R. Implementing sustainable tourism: A multi-stakeholder involvement management framework. Tour. Manag. 2013, 36, 342–353.

77. Jamal, T.; Stronza, A. Collaboration theory and tourism practice in protected areas: Stakeholders, structuring and sustainability. J. Sustain. Tour. 2009, 17, 169–189.

78. Vellecco, I.; Alessandra, M. Sustainability and tourism development in three Italian destinations: Stakeholders' opinions and behaviors. Serv. Ind. J. 2010, 30, 2201–2223.

79. Spenceley, A. Requirements for sustainable nature-based tourism in transfrontier conservation areas: A southern African Delphi consultation. Tour. Geogr. 2008, 10, 285–311.

80. Holden, A. Exploring stakeholders' perceptions of sustainable tourism development in the Annapurna Conservation Area: Issues and challenge. Tour. Hosp. Plan. Dev. 2010, 7, 337–351.

81. Timur, S.; Getz, D. Sustainable tourism development: How do destination stakeholders perceive sustainable Urban Tourism? Sustain. Dev. 2009, 17, 220–232.

82. Dabphet, S.; Scott, N.; Ruhanen, L. Applying diffusion theory to destination stakeholder understanding of sustainable tourism development: A case from Thailand. J. Sustain. Tour. 2012, 20, 1107–1124.

83. Hardy, A.; Beeton, R. Sustainable tourism or maintainable tourism: Managing resources for more than average outcomes. J. Sustain. Tour. 2001, 9, 168–192.

84. Larson, P.; Poudyal, N. Developing sustainable tourism through adaptive resource management: A case study of Machu Picchu. J. Sustain. Tour. 2012, 20, 917–938.

85. Dodds, R. Sustainable tourism and policy implementation: Lessons from the case of Calviá, Spain. Curr. Issues Tour. 2007, 10, 296–322.

86. Dodds, R.; Butler, R. Barriers to implementing sustainable tourism policy in mass tourism destinations. Int. Multidiscip. J. Tour. 2010, 5, 35–53.

87. Farsari, Y.; Butler, R.; Prastacos, P. Sustainable tourism policy for Mediterranean destinations: Issues and interrelationships. Int. J. Tour. 2007, 1, 58–78.

88. Pigram, J. Sustainable tourism—Policy considerations. J. Tour. Stud. 1990, 1, 2–9.

89. Solstrand, M. Marine angling tourism in Norway and Iceland: Finding balance in management policy for sustainability. Nat. Resour. Forum 2013, 37, 113–126.

90. Xu, H.; Sofield, T. Sustainability in Chinese development tourism policies. Curr. Issues Tour. 2013, 10, 30–38.

References

91. Michael, C. Local Government and Tourism Public Policy: A Case of the Hurunui District, New Zealand; Lincoln University: Lincoln, UK, 2013.

92. Yasarataa, M.; Altinay, L.; Burns, P.; Okumus, F. Politics and sustainable tourism development—Can they co-exist? Voices from North Cyprus. Tour. Manag. 2010, 31, 345–356.

93. Whitford, M.; Ruhanen, L. Australian indigenous tourism policy: Practical and sustainable policies? J. Sustain. Tour. 2010, 18, 475–496.

94. Halla, C. Policy learning and policy failure in sustainable tourism governance: From first- and second-order to third-order change? J. Sustain. Tour. 2011, 19, 649–671.

95. Bahaire, T.; White, M. The application of geographical information systems (GIS) in sustainable tourism planning: A review. J. Sustain. Tour. 1999, 7, 159–174.

96. Boers, B.; Cottrell, S. Sustainable tourism infrastructure planning: A GIS-supported approach. Tour. Geogr. 2007, 9, 1–21.

97. Bunruamkaew, K.; Murayama, Y. Site suitability evaluation for ecotourism using GIS & AHP: A case study of Surat Thani province, Thailand. Procedia Soc. Behav. Sci. 2011, 21, 269–278.

98. White, V.; McCrum, G.; Blackstock, K.L.; Scott, A. Indicators and sustainable tourism: Literature review. Tour. Manag. 2006, 27, 1274–1289.

99. Ivan, K.; Mikulić, J. Research note: Measuring tourism sustainability: An empirical comparison of different weighting procedures used in modelling composite indicators. Tour. Econ. 2014, 20, 429–437.

100. Mikulića, J.; Kožićb, I.; Krešić, D. Weighting indicators of tourism sustainability: A critical note. Ecol. Indic. 2015, 48, 312–314.

101. Delgado, T.A.; Palomeque, F.L. Measuring sustainable tourism at the municipal level. Ann. Tour. Res. 2014, 49, 122–137.

102. Coccossis, H.; Mexa, A. The Challenge of Tourism Carrying Capacity Assessment: Theory and Practice; Ashgate, Basingstoke: Hampshire, UK, 2004; pp. 277–288.

103. Simón, F.; Narangajavana, Y.; Marqués, D.P. Carrying capacity in the tourism industry: A case study of Hengistbury Head. Tour. Manag. 2004, 25, 275–283.

104. Juradoa, E.; Tejadab, M.T.; García, F.A.; González, J.C.; Macías, R.C.; Peña, J.D.; Gutiérrez, F.F.; Fernández, G.G.; Gallego, M.L.; García, G.M.; et al. Carrying capacity assessment for tourist destinations. Methodology for the creation of synthetic indicators applied in a coastal area. Tour. Manag. 2012, 33, 1337–1346.

105. Green, H.; Hunter, C.; Moore, B. The environmental impact assessment of tourism development. In Perspectives on Tourism Policy; Johnson, P., Thomas, B., Eds.; Thomson Learning: Boston, MA, USA, 1992; pp. 29–47.

106. Green, H.; Hunter, C.; Moore, B. Assessing the environmental impact of tourism development: Use of the Delphi technique. Tour. Manag. 1990, 11, 111–120.

107. Peng, J.; Wu, J.; Jiang, Y.-Y.; Ye, M.-T. Shortcomings of applying ecological footprints to the ecological assessment of regional sustainable development. Acta Ecol. Sin. 2006, 26, 2716–2722.

108. Haywood, K. Can the tourist-area life cycle be made operational? Tour. Manag. 1986, 7, 154–167.

109. Guinee, J.; Heijungs, R.; Huppes, G.; Zamagni, A.; Masoni, P.; Buonamici, R.; Ekvall, T.; Rydberg, T. Life cycle assessment: Past, present, and future. Environ. Sci. Technol. 2011, 45, 90–96.

110. McCool, S. Planning for sustainable nature dependent tourism development: The Limits of Acceptable Change system. Tour. Recreat. Res. 1994, 19, 51–55.

111. Avdimiotis, S.; Mavrodontis, T.; Dermetzopoulos, A.S.; Riavoglou, K. GIS applications as a tool for tourism planning and education: A case study of Chalkidiki. Tourism (Zagreb) 2006, 54, 405–413.

112. Schianetz, K.; Kavanagh, L.; Lockington, D. Concepts and tools for comprehensive sustainability assessments for tourism destinations: A comparative review. J. Sustain. Tour. 2007, 15, 369–389.

References

113. Lacitignola, D.; Petrosillo, I.; Cataldi, M.; Zurlini, G. Modelling socio-ecological tourism-based systems for sustainability. Ecol. Model. 2007, 206, 191–204.

114. Bailey, K. Social entropy theory: An overview. Syst. Pract. 1990, 3, 365–382.

115. Swanson, G.A.; Kenneth, D.B. Social Entropy Theory, Macro Accounting, and Entropy Related Measures. 2006. Available online: http://www.isssbrasil.usp.br/isssbrasil/pdfs/2006-247.pdf (accessed on 30 July 2014).

116. Stepanic, J.; Stefancic, H.; Zebec, M.S.; Perackovic, K. Approach to a quantitative description of social systems based on thermodynamic formalism. Entropy 2000, 2, 98–105.

117. Wilson, A. Entropy in urban and regional modelling: Retrospect and prospect. Geogr. Anal. 2010, 42, 364–394.

118. Cabral, P.; Augusto, G.; Tewolde, M.; Araya, Y. Entropy in urban systems. Entropy 2013, 15, 5223–5236.

119. Leiper, N. Industrial entropy in tourism systems. Ann. Tour. Res. 1993, 20, 221–226.

120. Sabkbar, H.; Ahmad, A.; Karimian, T. Spatial tourism planning in Isfahan by entropy model. Agrochimica 2014, 58, 67–82.

121. Hao, C. A Tentative analysis of entropy and dissipative structure in tourist destination ecological system. J. Taiyuan Norm. Univ. (Soc. Sci. Ed.) 2010, 9, 66–68.

122. Zhao, L.; Zhang, W. Applied research of dissipative structure theory on tolerance threshold of tourism system-on the basis of tourism system entropy principle. Tour. Forum 2010, 3, 10–16.

123. Qian, Y. Application of the dissipative structure theory in the traveling environment management. J. Anhui Agric. Sci. 2006, 34, 5006–5007.

124. Ministry of Agriculture of the People's Republic of China. This Is the Time to Be Excavated of Dunhuang Culture: The Report for Culture Industry of Dunhuang, Gansu Province. Available online:

http://www.agri.gov.cn/dfv20/GS/ncwh/dtzx/201306/t20130603_3480954.htm (accessed on 3 June 2013).

125. Statistics Report for Economic and Social Development of Dunhuang City in 2012. Available online: http:// zwgk.dunhuang. gov. cn/ReadNews.asp?NewsID=2367 (accessed on 28 July 2014).

126. It is Urgent Time to Combating Desertification in Dunhuang by Expert's Opinions. Available online: http:// news. xinhuanet.com/local/2010-09/11/c_12542051.htm (accessed on 14 September 2010).

127. Xing, X. Evolution equation for physical entropy and information entropy. Sci. China (Ser. A) 2001, 31, 78–84.

128. Li, X. The entropy in Socio-economic system. J. Syst. Dialectics 1996, 10, 84–87.

129. Shannon, C. A mathematical theory of communication (1–2). Bell Syst. Tech. J. 1948, 27, 623–656.

130. Zhang, Y.; Yang, Z.; Li, W. Measurement and evaluation of interactive relationships in urban complex ecosystem. Acta Ecol. Sin. 2005, 25, 1734–1740.

131. Lin, Z.; Xia, B. Analysis of sustainable development ability of the urban ecosystem in Guangzhou City in the perspective of entropy. Acta Geogr. Sin. 2013, 68, 45–57.

132. Leife, R. Understanding organizational transformation using a dissipative structure model. Hum. Relat. 1989, 42, 899–916.

133. Kondoh, Y. Function for dissipative dynamic operators and the attractor of the dissipative structure. Phys. Rev. E 1993, 48, 2975–2979.

134. Dooley, K. A complex adaptive systems model of organization change nonlinear dynamics. Psychol. Life Sci. 1997, 1, 69–97.

135. Zhan, K.; Sheng, X. Prigogine and Theory of Dissipative Structure; Shanxi Science & Technology Press: Xi'an, China, 1998.

136. Lu, L.; Bao, J. The course and mechanism of evolution about Qiandao Lake based on the theory of dissipative structure. Acta Geogr. Sin. 2010, 65, 755–768.

137. Valentin, A.; Spangenberg, J. A guide to community sustainability indicators. Environ. Impact Assess. Rev. 2000, 20, 381–392.

138. Dai, Y. Sustainable development indexes systems in tourist cities. Yunnan Geogr. Environ. Res. 2006, 18, 35–39.

139. Indicators of Sustainable Development for Tourism Destinations: A Guidebook. Available online: http://www.e-unwto.org/content/ x53g07/ fulltext?p=a6a03d7c67d24258bf5c4ae9e96b6530&pi=0#section=890050 &page=2&locus=40 (accessed on 1 October 2004).

140. European Tourism Indicator System Toolkit for Sustainable Destinations. Available online: http://ec.europa.eu/ enterprise/ sectors/ tourism/sustainable-tourism/indicators/ documents_indicators/ eu_ toolkit_indicators_en.pdf (accessed on 10 February 2013).

141. Niu, Y. The study on index system of sustainable tourism. Chn. Pop. Resour. Environ. 2002, 12, 42–45.

142. Zhang, J.; Singh, V. Theory and Applications of Information Entropy; China Water Power Press: Beijing, China, 2012; pp. 79–80.

143. Yang, X. Evaluation research of sustainable development of tourism based on the information entropy and AHP. Sci. Technol. Eng. 2008, 22, 6176–7004.

144. Jiang, C.; Zhang, Y. Study on the mechanism of evolution about island resorts system based on the theory of dissipative structure: A case study of Meizhou Island. Chin. Agric. Sci. Bull. 2013, 29, 213–220.

145. Di, Q.; Han, X. Sustainable development ability of China's marine ecosystem in the perspective of entropy. Sci. Geogr. Sin. 2014, 34, 6664–6671.

146. Chen, Y.; Liu, J. An index of equilibrium of urban land-use structure and information dimension of urban form. Geogr. Res. 2001, 20, 146–152.

147. Dunhuang City Bureau of Statistics. 10th Five-Year Statistical Yearbook of Dunhuang City; Wang, W., Ed.; Juelun Culture Communication Co., Ltd.: Xiamen, China, 2006; pp. 252–305.

148. Dunhuang City Bureau of Statistics. 11th Five-Year Statistical Yearbook of Dunhuang City; Ren, B., Ed.; Sanmenxia Culture Communication Co., Ltd.: Sanmenxia, China, 2011; pp. 258–309.

149. Dunhuang City Bureau of Statistics. 2011 Statistical Yearbook of Dunhuang City; Dunhuang City Bureau of Statistics: Dunhuang, China, 2012.

150. Dunhuang Municipal Environmental Protection Bureau. Environmental Quality Bulletin of Dunhuang City (2006–2012); Dunhuang Municipal Environmental Protection Bureau: Dunhuang, China, 2013.

CHAPTER 2

Comparative Study of Functions Affecting the Behavioral Patterns of Tourists in Iran and America's Tourism Marketing Plan Utilizing SWOT Model

Issa Ebrahimzadeh, Monir Yari

Faculty of Geography and Environmental Planning, University of Sistan and Baluchestan, Zahedan, Iran.

ABSTRACT

Behavioral patterns of tourists are designed aiming to present a scientific framework of how the variables are interacted and factors influencing consumer behavior can help decision makers and market activists in predicting and controlling consumer behavior. This study aims to explore the functions influencing tourist's behavioral patterns in Iran and America's tourism marketing planning which has comparatively been studied. The research methodology is analytical-descriptive and it has been studied and analyzed utilizing SWOT technique. Analytical results of this study show that America in proportion with culture, potentials and operational functions in Iran can be in some cases exploited.

KEYWORDS

Tourism; Tourists Behavioral Patterns; Marketing; SWOT Technique; Iran; America

1. INTRODUCTION

Tourism development, particularly for developing countries to get away from problems such as high unemployment, limited foreign exchange and single-product economy is very important. Iran's economy which is highly dependent on the incomes obtained from oil exchange, its macro-economic variables are affected to strong fluctuations over time, in order to diversify sources of economic growth and foreign exchange earnings as well as creating new job opportunities in the country, tourism development is highly important in [1].

The researches performed by World Tourism Organization and other ones show that marketing is essential for developing this industry in any country or region' however, people must be known the tourism potentials of that specific region to attract tourists to a region [2]. However, the point is that the majority of tourists attractions in Iran are not only are not well known not only in Iran, but in abroad. Therefore, it seems that marketing weaknesses and lack of appropriate advertising with tourist's needs and tourism markets and also lack of designing tourists' behavioral patterns is one of the factors that can be associated with Iran's underdevelopment tourism industry. Although the studies indicate that only marketing cannot be led to sustainable development of tourism industry for the attractions of a region without existing infrastructure facilities and services, the necessity and significance of marketing planning is to the extent that it is sometimes considered in the field of tourism literature as a main pillar of developing this industry in tourism plans. Therefore, designing and developing macroeconomics and also strategic approaches for tourism development in the framework of "tourism marketing planning process" is so critical [3].

2. RESEARCH THEORETICAL APPROACH

2.1. Tourist's Behavioral Patterns

Behavioral, conceptual and motivational differences have great impact on personal decisions; individuals' behavior depends on their conception of the world and is influenced by their subjective perception toward a location or a travel agency which are determined by factors such as childhood, family experiences and working personality. Tourism factors include supply and demand and the study of tourist's behavior is taken into account the supply dimension, as most researchers focused their attention to research attractions (supply). It is essential that tourism enterprises identify their demographic and geographical characteristics of their consumers [4]. Tourism behavioral studies will be taken shape in three categories namely before travel, during travel and after travel.

- Before the travel category includes the study of complex affecting issues on travel decision making and the intention to visit a tourist destination. For example, the destination image is influenced by people's features, psychology, previous experiences and motivation.

- By during the travel, it means the tourist behavior and their perception of the quality of facilities and the amount of money they spent.

- After the travel includes satisfaction and fulfillment of tourists and the intention to return again as well as the tourist's reception with the host community.

Tourists behaviors (Figure 1) includes human's reaction against needs and their demands to decide to have a travel, then the satisfaction or dissatisfaction of tourists with tourist facilities, services and attractions during travel and ultimately their understanding of tourism destinations, persistency, the decision to return or not to return to the tourist destination and travel reflection. Meanwhile, tourist's behavioral patterns consist of categorizing the tourists in similar and convergent groups based on their behavior in three stages of before the travel, during the travel and after the travel [5].

2.2. Market and Tourism Marketing

Market is a set of actual and potential purchasers of a product; a potential market is the customers that have the desire and interest to buy goods or services. In order to potential customers become actual customers, they should both have access to product and have enough income. Therefore, an accessible market is the set of customers who have the interest and power to buy products and have access to the regarded product. The concept of market finally brings us to the concept of marketing. Marketing means working with markets to provide ex changes with the aim of supplying human's needs and desires [6]. Meanwhile, the concept of marketing can be defined in a simple term; that is coordinating the product with the needs and demands of the market and people and what they want to buy should be offered and suggested [7]. Marketing services is certainly complex, for they do not have the capability to store and stock services and for this respect its marketing differs from marketing products and its supplying method to markets is also different [8].

Marketing is necessary and essential for sustaining the existence of tourism industry. In fact, the potential tourists can be attracted to the purposes intended for through marketing methods. Therefore, developing strategic marketing programs can effectively contribute in achieving the objectives of tourism projects [3]. Despite the fact that marketing discussions are extensive, its position in tourism industry management and development is essential [9]. In fact, market segmentation is a process of identifying major differences in the characteristics of the buyer to divide the market into two or more groups with a specific goal, selecting the part of the market that the institution will serve with and also designing the product and marketing programs in such a way that it satisfies the specific features of the targeted group [10]. Tourism marketing aims to identify and predict the tourist's needs and providing facilities to meet their needs and making them informed and creating motivation within them to visit the destination again [11]. Given the specific characteristics of tourism products, the role of marketing in this industry has particular importance than other industries and effective use of tourism marketing tools is essential for a country or region, for the potential tourists can be supplied with information about what a specific region can provide through marketing and encourage them to visit that region [12].

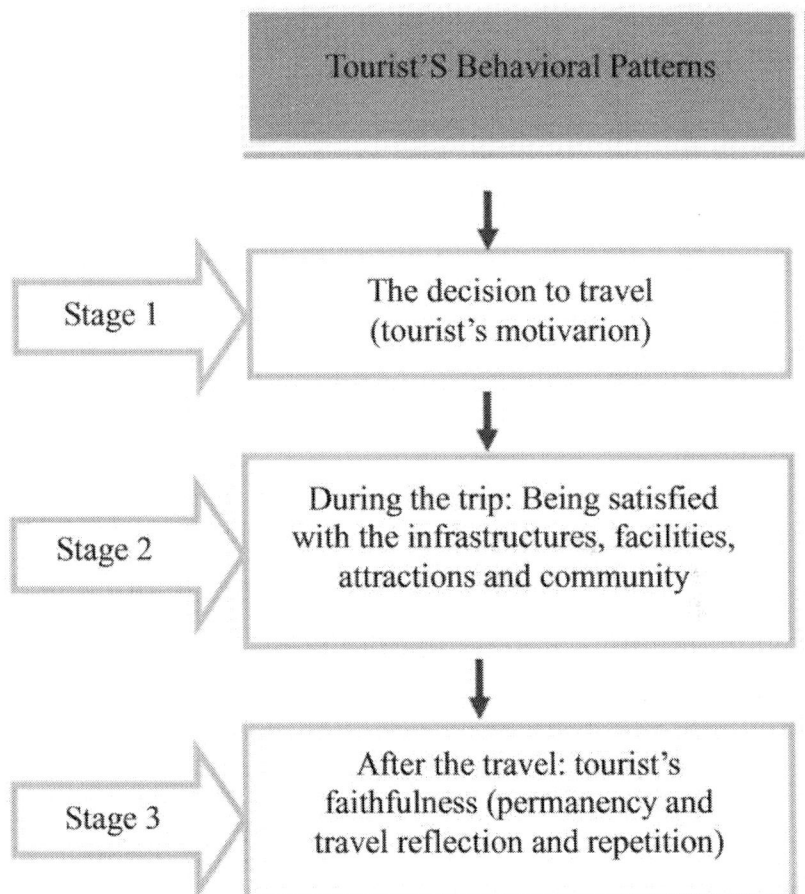

Figure 1. Tourist's behavioral patterns in three stages of decision to travel, during the travel and after the travel [5].

3. EXAMPLES OF COMPREHENSIVE BEHAVIORAL PATTERNS OF TOURISTS IN AMERICA

Tourist's behavioral pattern and American's lifestyle: Stanford Research Institute devised an improved method for America's market segmentation which is called VALS2, which is based on the concept of self-justification and customer-provided resources, and its justification is based on how customers search for and provide goods and services; those which satisfy them and show their personal identity. Therefore, they are prone to purchase influenced by the principles, conditions and their actions [13]. Its most significant segmentation is called VALS2 which stands for Values and Lifestyle which people within are divided into 8 groups based on how they spend their time and money as follows:

- Self-flourishing or self-discovery: customers who have high income and are successful, active, modern and responsible. These customers are looking for the best products.
- Prosperous: mature professional and responsible customers who have higher education, and their information about global issues is good and are ready to accept new ideas and social changes. They prefer products which are durable and have appropriate performance and value.
- Successful: people who are successful and professional who obtain their satisfaction from their family and business. They are looking for products that provide them prestige which indicates their success.
- Experience-driven: these consumers are young people with 25 years mean age; they have lots of energy spent for exercise and social activities. A large part of their income is spent on clothing, food, music, cinema and other new products with an emphasis on being modern.
- Believers: they are conservative and traditional people which welcome familiar and well-accustomed products as well as are middle-income.
- Hard-working: people with low financial resources and feeling of insecurity who are seeking for other's approval. These customers are looking for buying fashionable products to show themselves as those who have high financial resources.

3. Examples of Comprehensive Behavioral Patterns of Tourists in America

- Constructive: they are pragmatic, self-sufficient-traditional and family-oriented people. Their income is low and they are seeking for products that can be applied.
- Activist: these consumers are in the lowest income levels. They are elderly, retired and passive people; their average age is 61 years old and they are loyal to their favorite brands [6].

Motivational Behavioral Pattern of American Tourists: Maslow presented for the first time the pyramid of human needs with psychological and human motivation approach in his valuable works entitled "Motivation and Personality" in 1954 (Figure 2) which has taken into consideration a pattern of human behaviors. In fact, Maslow maintains that it is possible that human motivations and behaviors in the field of tourism suggest his/her needs. These needs are often manifested behind human's activities and personal or hidden matters [14].

The above-mentioned theory contributes the consumption behavior in tourism industry [15]. Also, the differences and its influences in tourist's behaviors and relationships with physical and cultural milieus as well as people during travel and in destination has caused that the categories are useful in such a way that it offers a simple classification of complex phenomena, allowing them to be understood. Much of this work was done during the 1970s and early 1980s and part of it was reviewed by Lowick and Fen Rongeno and Blaert (1992). The especially began discussion about Gallup Organization which was representative of American Express Travel Services. The aim of the above study was traveller's classification and by making initial talks from 6500 volunteers' respondents in United States, Germany, Great Britain and Japan, the world's four major economic forces, almost 4000 people were among the travellers from the total of respondents. The evaluation in overall showed that almost 13,000 people have been traveling over the past 12 months. According to this study, the tourists who decided to travel in different ways regardless of origin and destination have been divided into five groups as follows:

Figure 2. Hierarchy of needs according to Maslow's ideas (1954).

- Adventurers: they are independent and reckless and trying to create new activities, see new people and also become familiar with different cultures. They are generally more educated and more affluent than members of other groups. Travelling has a man role in life for these people. On the other hand, adventurers are young men who 44% of the members of the group are in the age range of 11 - 34 years old.
- Heartsick: these people are affected by too much worries and concerns and do not have confidence in decisions related to travelling and are generally scared of flying. These people are less educated and affluent than others. This group travel less than other four groups and if they initiate to have a travel, they will tend to domestic travels. Most of these people are women and a major part of them are elderly and nearly more than half of them have 50 years old.
- Whimsical people: these people are living with travelling imagination and supposition and hence they give importance for travelling during life. Despite many studies and discussions they made about new routes and destinations, their business travel is much less impressive than their thoughts and beliefs and often they have a tendency to rest relaxation

than adventure and seeking for incidence. Such people can be among low income and less educated people who most of them are 50 year women or older than it. When these people travel to new places, they often carry maps and guide books with themselves.

- Thrifty people: travelling provides an opportunity for relaxation and leisure for thrifty people and it is an experience from the perspective of travelling that gives meaning to their life. Travelling costs does not matter for them and do not think of the notion that they should pay extra money for the services and facilities provided for. Such people have a moderate income and have not culturally been advanced. It can be said that they travel twice a year; even they do not have the ability to do so. These people are more women than men and are usually a little older than their travelling companions.
- Indulgent and wasteful people: the financial condition of such people is usually much better than other travellers and they want to pay more money for more convenience and better services during the travel. They tend to stay in great hotels compared to others and are more interested in indolence. They support the adventurers during travelling and the number of men and women in such group is the same [3].

4. EXAMPLES OF COMPREHENSIVE BEHAVIORAL PATTERNS OF TOURISTS IN IRAN

In order to achieve a comprehensive and sustainable development as well as replace alternative resources of revenue instead of oil resources in today Iran, using all the facilities and capabilities in our country are needed. In this regard, developing the tourism industry believed to the third dynamic and growing economic phenomenon after oil and automobile industry in the world by the economists is also considered as a basic requirement in Iran. Hence, reviewing the obstacles in the development of this industry in our country is very necessary. In this regard, the tourism marketing weaknesses can be one of the obstacles to tourism development in Iran. Meanwhile, the lack of conducting basic and applied research in the behavioral patterns of Iranian tourists as a prerequisite for tourism marketing is considered the most constraints of tourism development in Iran. Little research has been done in recent years in this field in a sporadic fashion as follows: in the study conducted by Ranjbarian & Zahedi [16], the effects of repeated travels to Isfahan on the foreign tourist's satisfaction were explored and the results of their study indicate that the satisfaction level of tourists who repeat the travel to Isfahan was much lower than those who experience such a travelling for the first time. In the study performed by Kazemi M. [17], Zahedan citizen's understanding on Chabahar tourism development had has been analyzed. His findings indicate that inasmuch as product marketing of tourism in any country or region depends on buyer's interests and perceptions and with regard to the comprehensive tourism plan of Sistan and Baluchestan, Iran in intra-provisional tourism development strategy, now the motivation of Zahedan citizens to travelling to Chabahar are cultural, commercial, physical activities, gaining peace, natural, luxurious and emotional motivations. Reza Jaravandi & Forghani [18] conducted a study entitled "comparison of travel motives among both youths and adults (a case study of Shiraz travellers and came to this conclusion that there is no significant difference between different age groups in travelling to Dubai. His studied motivation were divided into five groups among which the impact of age and life cycle are considered less effective and he maintained that its cause is to how advertise in that he represents travelling to Dubai as a trip to the whole family and no distinction is considered in age or gender in advertising, as the motivation all family members are more influenced by family and familial conditions than influenced by variables as cultural sub-groups or reference groups. Alilou et al. [19] compared in another study the personality traits of a group of Iranian tourists with a non-tourist group. The findings of this study showed that there is a significance difference between tourists and non-tourists in seeking emotion, extraversion and openness against experience. The tourist group achieved higher scores than the non-tourist ones in personality traits of extraversion, openness against experience. The extroverts are gregarious and plebiscites who prefer large groups and gatherings; as a result the extroverts are more willing to travel.

5. RESEARCH METHODOLOGY

Research methodology in this study is analytical-descriptive which describes and quantitatively and qualitatively analyze behavioral patterns and influential functions upon it in the tourism market section and the necessity of its development as the main pillar of developing this industry in tourism goals are studied. To collect required information and data, documentary and library studies have been used and the behavioral patterns in the two countries of Iran and America were analyzed. Then SWOT model was used to analyze findings and finally guidelines and strategies appropriate to developing influencing behavioral patterns as one if the pillars of marketing in Iran and America have been presented. The analysis of affecting factors facing tourism marketing using the gradation factors (Tables 1-4) [20].

5.1 Research Locational Scope

According to research purposes, which comparatively study the influential functions on tourist's behavioral patterns in the two developed and developing countries, the research locational scope has been selected in both Iran and America (Figure 3). United States of America as an area of 9,363,364 square kilometers is located in North American continent. This country gained independence from England on 4/7/1776. The government type id federal republic with two legislative houses. English is the official language of this country and its capital is Washington, and the Islamic Republic of Iran as an area of 1,648,195 square kilometers is located in Asia. Iran has an ancient history and its government type is Islamic republic with one legislative house. Persian is the official language of this country and its capital is Tehran [21].

6. RESEARCH FINDINGS ANALYSIS

Today, the functions affecting Iranian and American tourist's behavioral patterns benefitted from SWOT analysis are considered as a new tool for performance analysis and functional gaps status will be used by strategic designers and evaluators. Therefore, this method is used in this study. In fact, this model is considered a conceptual framework for systemic analysis which makes possible the evaluation and comparison of bottle-necks, threats, harmful aspects, opportunities, demands and external environment conditions associated with strategic strengths and weaknesses (Table 5). The combination and incorporation of these factors with each other is in fact the base of devising four types of strategies, includeing 1) combining the strengths and opportunities (SO), 2) combining the strengths and threats (ST), 3) combining the weaknesses and opportunities (WO), 4) combining the weaknesses and threats (WT) [22].

Table 1. Results of external factors analysis affecting marketing tourism of America.

External environment (threats and opportunities)	weight	Grade	Weighted score
O1—creating a balance between tourism firms, tourists and society's advantages	0.01	3	0.03
O2—estimating the tourism markets opportunities	0.04	2	0.08
O3—government's support against competitors	0.07	5	0.35
O4—the growing process of America's tourism destination acceptance through domestic and foreign markets	0.04	4	0.14
O5—clear goals and proper planning of firms to achieve the objectives	0.03	4	0.12
O6—the possibility of competition in international tourism markets	0.05	4	0.20
O7—creating new ideas in marketing	0.03	1	0.03
O8—tourism development as one of the factors of economic growth	0.05	4	0.20
O9—developing promoting activities in the field of advertising	0.05	1	0.05
O10—identifying strengths and weaknesses of one's own firms and those of competitors	0.03	3	0.09
O11—formulaing the costs	0.02	2	0.04
O12—increasing the country's exchange incomes	0.08	4	0.32
O13—advertising in order to expand the use of country's aviation services	0.04	4	0.16
O14—creating motives in potential tourists to travel to America	0.04	3	0.12
O15—gaining loyal customers in domestic and foreign markets			
T1—tourism system leads strongly people towards dependence to materials	0.01	2	0.02
T2—water pollution (people are invaded by advertising)	0.01	2	0.02
T3—dividing the tourism markets into smaller sectors with less profitability	0.05	3	0.15
T4—lack of ability to implement all strategies through tourism organization	0.03	1	0.03
T5—taking into consideration general solutions in tourism markets	0.04	3	0.12
T6—global markets segmentation with general variables without taking into consideration the subcultures	0.05	3	0.08
T7—the competitor's progress in other developing countries	0.04	4	0.16
T8—recission in America	0.08	5	0.40
T9—demonetizing the tourist's demands with changing the tourist's imaginations from the adopted style	0.02	4	0.08
T10—removing the small companies through big ones	0.05	5	0.25
T11—using unfair competitive practices for damaging or destructing other firms	0.02	4	0.16
Total	1	-	3.43

Source: [The analysis of research findings by authors, 2010].

6. Research Findings Analysis

Table 2. Results of internal factors analysis affecting marketing tourism of America.

Internal environment (weaknesses and strengths)	weight	grade	Weighted score
S1—existing tourism markets with the highest compliance with what are existed and tourism facilities	0.07	4	0.28
S2—adapting the type of advertisements with the needs and demands of actual and potential tourists	0.03	4	0.12
S3—estimating the possible demand in each sector of tourism markets	0.03	3	0.09
S4—enough identification of the status quo of tourism markets	0.02	3	0.06
S5—clarifying the customers in each sector of the tourism market	0.05	2	0.10
S6—identifying the needs and demands of tourists	0.04	1	0.04
S7—coordination between demand and supply of products and tourism services	0.02	1	0.02
S8—paying attention to new needs of tourists	0.01	2	0.02
S9—evaluating the criteria of market segmentation	0.04	1	0.04
S10—increased right of customer's choice	0.03	2	0.06
S10—increased right of customer's choice	0.03	2	0.06
S11—maximizing the tourist's consumption	0.02	3	0.06
S12—existing various courses related to tourism and marketing	0.03	1	0.03
S13—consecrating adequate funding to marketing programs	0.08	5	0.40
S14—existing much research in the field of market and marketing studies	0.02	4	0.08
S15—existing statistical bases and appropriate tourism information	0.05	5	0.25
S16—a complete and comprehensive information regarding attractions and tourists locations	0.04	4	0.16
S17—existing skillful personnel in marketing	0.05	4	0.20
S18—existing tourism offices in other countries	0.02	4	0.08
W1—increasing prices in some market segments	0.04	4	0.16
W2—exaggeration in portraying the destination features	0.02	4	0.08
W3—changing the structures of life cycle in America	0.02	1	0.02
W4—much migrations to America	0.02	1	0.02
W5—changing the population structure in America	0.03	3	0.09
W6—misuse of state officials	0.08	4	0.32
W7—high cost of over-distributing goods and services	0.05	3	0.15
W8—high cost of advertisement	0.05	4	0.20
W9—lack of attention to ethical issues in marketing	0.04	4	0.16
Total	1	-	3.34

Source: [The analysis of research findings by authors, 2010].

Table 3. Results of the analysis of the internal factors affecting tourism marketing in Iran.

External Environment	Weight	Grading	Weighted score
O1—predicting demand in various market assessments	0.04	3	0.12
O2—identifying the goals and motivations of tourists in the form of behavioral patterns	0.05	2	0.10
O3—being aware of potential tourists	0.04	2	0.08
O4—identifying the expected level and beyond the expectations of tourists	0.03	1	0.03
O5—understanding how to supply goods and services to different market segments	0.03	3	0.09
O6—promoting the level of tourist's loyalty to the company's regarded products and services	0.05	2	0.10
O7—existing the areas for the future development of tourism in Iran	0.07	4	0.28
O8—meeting the needs and demands of internal tourists to prevent from the outgoing exchange from the country	0.09	3	0.27
O9—identifying the needs and demands of tourists in domestic and foreign tourism markets	0.04	4	0.16
O10—increased rights for tourists to have a choice	0.05	2	0.10
O11—finding tourism markets with the highest compliance with tourism attractions and facilities	0.06	4	0.24
O12—job creation in tourism sector	0.08	5	0.40
T1—the superficiality of some classification of behavioral patterns	0.06	2	0.12
T2—lack of repeating surveys from tourists	0.07	3	0.21
T3—distributing survey forms in an inappropriate time and place	0.02	3	0.06
T4—lack of meeting tourist's needs and demands of one market sector in good atmosphere	0.05	5	0.25
T5—poor research in tourism market and marketing	0.08	4	0.32
T6—lack of appropriate tourist acceptance culture in the country	0.04	4	0.16
T7—lack of coordination of tourism-related sectors	0.05	2	0.10
Total	1	-	3.3

Source: [The analysis of research findings by authors, 2010].

Table 4. Results of analysis of external factors on tourism marketing in Iran.

External Environment	Weight	Grading	Weighted score
O1—predicting demand in various market assessments	0.04	3	0.12
O2—identifying the goals and motivations of tourists in the form of behavioral patterns	0.05	2	0.10
O3—being aware of potential tourists	0.04	2	0.08
O4—identifying the expected level and beyond the expectations of tourists	0.03	1	0.03
O5—understanding how to supply goods and services to different market segments	0.03	3	0.09
O6—promoting the level of tourist's loyalty to the company's regarded products and services	0.05	2	0.10
O7—existing the areas for the future development of tourism in Iran	0.07	4	0.28
O8—meeting the needs and demands of internal tourists to prevent from the outgoing exchange from the country	0.09	3	0.27
O9—identifying the needs and demands of tourists in domestic and foreign tourism markets	0.04	4	0.16
O10—increased rights for tourists to have a choice	0.05	2	0.10
O11—finding tourism markets with the highest compliance with tourism attractions and facilities	0.06	4	0.24
O12—job creation in tourism sector	0.08	5	0.40
T1—the superficiality of some classification of behavioral patterns	0.06	2	0.12
T2—lack of repeating surveys from tourists	0.07	3	0.21
T3—distributing survey forms in an inappropriate time and place	0.02	3	0.06
T4—lack of meeting tourist's needs and demands of one market sector in good atmosphere	0.05	5	0.25
T5—poor research in tourism market and marketing	0.08	4	0.32
T6—lack of appropriate tourist acceptance culture in the country	0.04	4	0.16
T7—lack of coordination of tourism-related sectors	0.05	2	0.10
Total	1	-	3.3

Source: [The analysis of research findings by authors, 2010].

Source: [The analysis of research findings by authors, 2010].

6. Research Findings Analysis

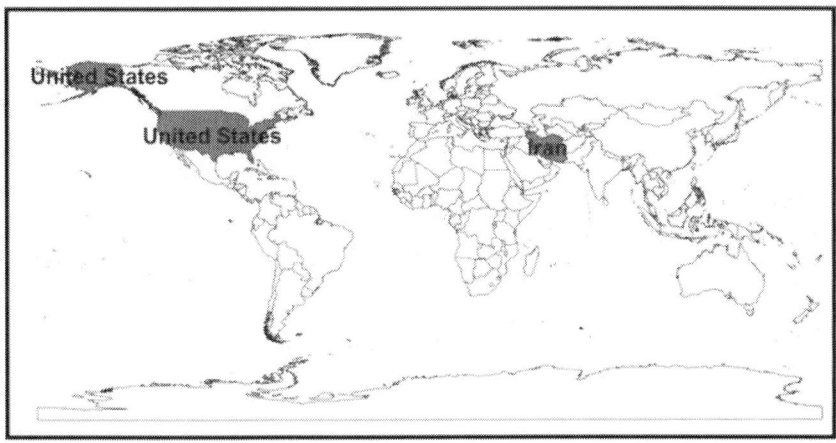

Figure 3. Iran and America location in world map (authors, 2010).

6.1. Analysis of Internal and External Factors Affecting Tourism Marketing Behavioral Patterns in AMERICA

The results achieved from analyzing influential functions on tourism behavioral patterns in America, as noted in Table 1, indicate that the most important opportunities from the perspective of those polled (students of Tourism Industry Planning), include protection against competitors with a score of 0.35, increasing foreign exchange earning with a score of 0.32, tourism development as one of the factors of economic growth and the possibility of competitiveness in international tourism markets, each with a score of 0.20. At the same time, of the most important threats are recession in America with a score of 0.40, outfighting of small organizations through mammoth ones and preventing the entry of new companies with a score of 0.25, using unfair competitive marketing practices for damaging or destruction of other companies and the progress of competitors in other developed countries, each with a score of 0.16. The weighted score of external environment in marketing tourism of America is 3.43 which indicate its high influence.

Now, according to the results obtained in Table 2, the most important strengths from the perspective of those polled (students of tourism industry planning) include sufficient funding t marketing programs with a score of 0.35, existing tourism markets with the highest compliance with what is owned and tourism facilities with a score of 0.28, existing statistical bases and appropriate tourism information with a score of 0.25, existing skillful personnel in marketing with a score of 0.20. The most important weaknesses include misuse of state officials with a score of 0.32, the high cost of advertising with a score of 0.20, lack of attention to ethical issues in marketing and high prices in some market segments, each with a score of 0.16. The weighted score in tourism marketing of America is 3.34 which indicate its high influence.

6.2 Developing Affective Strategies on Tourist's Behavioral Patterns in Order to Optimize Marketing in America

At this stage, interfering with any of the other factors and by analyzing how the strengths and weaknesses are influencing and being influenced from each other as well as threats and opportunities facing the tourism market in America, a final functional analysis is made by offering effective strategies on the tourist's behavioral patterns in order to optimize the tourism marketing. In this way, these functions and its strategic planning are presented in the framework of four competitive basic strategy, namely competitive, diversity, defensive and revisionist strategies as follows:

Table 5. Quadruplet strategies of SWOT matrix and how to determine it.

Weaknesses (W) (are listed) Subjects	Strengths (S) (are listed) Subjects	SWOT Matrix
Opportunities (O) (are listed) Subjects	SO strategies (max-max). By making benefits of strengths, the opportunities are utilized.	WO strategies (min-max) By making benefits of weaknesses, weaknesses are removed.
Threats (T) (are listed) Subjects	ST strategies (max-min), strengths are used to avoid threats.	WT strategies (min-min), weaknesses are reduced and threats are avoided.

Source: [22].

6. Research Findings Analysis

Competitive strategies affecting tourist's behavioral patterns in America's marketing:

So1—designing conceptual diagrams: these diagrams are designed in two parts; in part one the expected level and beyond expectations in every part of the market are scored and in the second part, the tourists give scores to their satisfaction about facilities on their patterns which by comparing the two parts, the deficiencies are known in the part of tourism and some plans are devised to eliminate it;

So2—presenting behavioral patterns with new variables by the companies and organizations to find new markets;

So3—the participation of same level enterprises in order to take advantage of new opportunities; in this state, companies are able to reach their specific targets by combining capital, managerial capabilities and marketing resources;

So4—Finding the most appropriate media with regard to tourist's behavioral patterns; considering the fact that which media the parts of the market deals with, the media appropriate to that market can be selected for advertising;

So5—product development and tourism services with the current product development, services, creativity and innovation to create new products and services.

Diversity strategies (ST) affecting tourist's behavioral patterns in America marketing:

St1—focusing on submarkets in other countries;

St2—research regarding other countries' domestic markets;

St3—changing the criteria and segmentation standards of the market according to new demands;

St4—developing ethical marketing policy; these policies should include measures of advertising, giving services to customers in diverse parts of the market, pricing and competitive ethical criteria;

St5—appointing prizes and supporting new ideas in the field of marketing and competition

Defensive strategies (WT) affecting tourist's behavioral patterns in America marketing:

Wt1—reviewin and assessing advertising and progressive tax on the cost of advertising;

Wt2—development of observer sector interfering with each factor on each other in the whole country's marketing system;

Wt3—taking advantage of focused marketing; whenever the market is faced with restrictions, instead of trying to look for a small proportion of a great market, the company tries to create a relatively small portion of one or more submarkets.

Revisionist strategies (WO) affecting on tourist's behavioral patterns in America's marketing:

Wo1—redicing costs in order to fit the prices of goods and services in market sectors;

Wo2—reviewing the consumer's demands and tourism organizations needs and long-term interests of consumers.

6.3. Analysis of Internal and External Factors Affecting the Behavioral Patterns of Tourism Marketing in Iran

The results obtained from analyzing functions affecting the behavioral patterns of tourism marketing in Iran, as noted in Table 3, suggests that the most important strengths from the perspectives of those polled (students of tourism industry planning) include a variety of different landscapes and climatic conditions with a score of 0.40, existing tangible and intangible historical and cultural heritages with a score of 0.28 and Iran's geographical position in the Middle East and Central Asia with a score of 0.27; meanwhile their most important weaknesses are infrastructures and tourism facilities

6. Research Findings Analysis

weaknesses with a score of 0.45, lack of statistical and information base for the tourists with a score of 0.28 and lack a complete and comprehensive information regarding tourism attractions with a score of 0.20. The weighted score of internal environment in tourism marketing in Iran is 3.13 which indicate its high influence.

Now, according to the results obtained from Table 4, the most important opportunities from the perspectives of those polled (students of tourism industry planning) affecting tourism marketing is job creation in the field of tourism with score 0.40, existing the field of future development of tourism in the area with the score 0.28, satisfying the needs and demands of foreign tourists to avoid outgoing foreign exchange from the country with the score 0.27; in addition, the most important threats are weaknesses in market research and tourism marketing with the score 0.32, lack of meeting the needs and demands of tourists of one market segment in an appropriate environment with the score 0.25 and superficiality of some behavioral patterns division with the score 0.12.

The weighted score of external environment in Iran's tourism marketing is 3.3 which show its high effectiveness.

Developing Strategies Affecting Tourist's Behavioral Patterns to Optimize Marketing in Iran's Tourism

Now basic competitive, diversity, defensive and revisionist strategies and their functions and strategic planning in Iran is presented as follows:

Competitive strategies (SO) affecting tourist's behavioural patterns in Iran's marketing:

So1—analyzing the tourist's behavioral patterns in Middle East and Central Asia and Persian Gulf countries as targeted markets towards planning to achieve tourism markets appropriate to the country's tourism attractions;

So2—Organizing the tourism advertisement according to tourist's needs and demands and tourism destinations attractions;

So3—Creating advanced courses of management and tourism marketing for youths in order to utilize and employ a trained group in the positions related to marketing in the country and supporting new ideas in tourism marketing system.

Diversity strategies (ST) affecting tourist's behavioural patterns in Iran's marketing:

St1—designing products and presenting services appropriate to tourist's needs and demands of different cities and ethnics;

St2—using secondary data, including the ones other people and organizations or that organization itself have collected for other purposes such as sociological, psychological and geographical information which can be utilized in Iran's targeted tourism marketing.

Defensive strategies (WT) affecting tourist's behavioural pattern in Iran's marketing:

Wt1—selecting efficient managers for each segment of the market to establish intersection cooperation;

Wt2—using advertisement in math media to make publicize the tourism culture among people;

Wt3—constricting behavioral patterns with variables consistent with Iran's tourism marketing;

Wt4—designing new patterns with repeated surveys due to changes in variables such as changes in population structure of internal and external tourism markets;

Wt5—utilizing and employing tourism experts out of successful marketing companies in developed countries.

Revisionist strategies (WO) affecting tourist's behavioral patterns in Iran's marketing:

Wo1—application of copycat marketing and taking advantage of successful marketing patterns in developed countries appropriate to targeted markets in Iran's marketing;

Wo2—developing and designing local patterns appropriate to market variables to correctly segment the tourism market;

Wo3—participating of same level organizations for taking more advantage of new opportunities; in such case, the mentioned organizations can reach their goals with the combination of capital, managerial abilities and marketing resources.

7. CONCLUSIONS

Tourism markets can be flourished primarily through identifying behavioral patterns and attracts the potential tourists to purposes intended. On the other hand, the needs and demands of tourist can be identified well with a customer-focused marketing and marketers will be consciously make a plan for attracting tourists and therefore the level of tourist's satisfaction will be increased. Travelling repetition will be increased by increasing the level of satisfaction and demand will be maintained in a desirable level and hence tourists are indirectly involved in tourism planning.

Reviewing the analysis of the factors influencing tourist's behaviors in America's market division showed that tourism market in this country was vastly developed and marketing science is closely linked to marketing practice through supporting the valuable studies done in behavioral patterns and the external environment with the weighted score of 3.43 is more influencing than its internal environment with the weighted score of 3.34. However, the studies and researches in Islamic Republic of Iran associated with tourism marketing and the factors affecting tourist's behavioral patterns is in its starting point and then requires being supported and more publicly and personally financed. Nevertheless, the external environment with the weighted score of 3.3 is more effective than internal environment with the weighted score of 3.13 in this country. Now, a fundamental question is raised indicating that given the America's successes in tourism marketing and its difference with Iran, is America's strategies can be utilized for improving and promoting Iran's tourism marketing as well as the strategies recommended for Iran?

According to the studies performed by the writers in this regard, it can be said that America's tourism development strategies will be applied in Iran when it is put in the socio-cultural, economic and potential contexts of Iran's tourism; otherwise, this industry may be led to instability due to developing inappropriate tourism markets. In other words, application of copycat marketing and utilizing successful marketing patterns in developed countries according to Iran's targeted tourism markets will be effective by providing and developing indigenous patterns appropriate to Iran's market variables in order to properly exploit the tourism market.

REFERENCES

1. S. Tayebi, R. Babaki and A. Jabari, "Study of the Relationship between Tourism Development and Economic Growth in Iran (1959-2004)," Bulletin of Humanities and Social Sciences Especially in Economics, Vol. 26, No. 7, 2007, pp. 83-110.
2. N. Naserpour, "Studying and Explaining the Obstacles Facing Tourism Industry in Lorestan Province and Presenting Appropriate Strategies in this Regard," MA Thesis, Mazandaran University, Babolsar, 2003.
3. R. Heydari, "Principles of Tourism Industry Planning," SAMT Publications, Tehran, 2008.
4. S. Amirtahmaseb, "The Typology of Inbound Tourists Visiting Iran," Master Thesis, Department of Business Administration and Social Sciences, Division of Industrial Marketing and E-Commerce, Lulea University of Technology, Lulea, 2007.
5. M. Yari, "Behavioral Patterns of Khoramabad Pre-Province Tourists," MA Thesis, University of Sistan and Baluchestan, Zahedan, 2011.

REFERENCES

6. H. Esmaelpour, "Principles of Marketing Management," Negah Danesh Publications, Tehran, 2005.
7. R. Doswell, "Tourism Management," Cultural Research Office Publications, Tehran, 2010.
8. M. Ketabchi, "Principles of Marketing of Travel and Marketing Services," Faras Publications, Tehran, 2004.
9. M. Kazemi, "Analysis of Zahedan Citizen's Perceptions in Chabahar Tourism Development," Journal of Geography and Development, Vol. 6, No. 12, 2006, pp. 81-100.
10. K. Kamali and M. Dadkhah, "Networking and Network Management (Scientific-Practical)," Shahre Ashoub Publications, Tehran, 2005.
11. K. G. Brown, "Island Tourism Marketing Music and Culture," International Journal of Culture Tourism and Hospitality, Vol. 3, No. 1, 2009, pp. 25-32.
12. P. Laimer and W. Juergen, "Portfolio Analysis as a Strategic Tool for Tourism Policy," Tourism Review, Vol. 64, No. 1, 2009, pp. 17-31.
13. O. C. Walker, J. Mullins and H. Boyd, "Marketing Strategy: A Decision Focused Approach," 7th Edition, McGraw-Hill/Irwin, 2010.
14. R. Lankovar, "Tourism Sociology and Travelling," Shahid Beheshti University, Tehran, 2002.
15. L. Lumsdon, "Tourism Marketing," Thomson Business Press, London, 1997.
16. B. Ranjbarian and M. Zahedi, "Study of the Effects of Repeating Traveling to Isfahan on Foreign Tourist's Satisfaction," Journal of Geography and Development, Vol. 9, No. 9, 2007, pp. 65-78.
17. M. Kazemi, "Tourism Management," SAMT Publications, Tehran, 2008.
18. R. Jarvandi and N. Forghani, "Comparison of Travelling Motives Among Two Young and Elderly Generations (Case Study of Shiraz City Travellers)," Journal of Youth Studies, Culture and Society, 2nd Edition, Vol. 2, 2009, pp. 123-143.
19. M. Alilou, S. H. Razmi and F. Nemati Sogoulitapeh, "Comparison of Group Personality Traits of Iranian Tourists with Non-Tourist Groups," Studies on Behavioral Sciences, Vol. 1, No. 7, 2009, pp. 55-62.

20. Ebrahimzadeh and A. Aghasizadeh, "Analyzing the Factors Affecting Tourism Development in Chabahar Coastal Area, Using SWOT Strategic Model," Journal of Urban and Regional Researches and Studies, Vol. 1, No. 1, 2009, pp. 107-128.
21. Research Unit and Cosmography Compilation, "Cosmography Comprehensive Atlas," Gitashenasi Publications, Tehran, 2005.
22. Ebrahimzadeh, "Land Preparation and Environmental Planning in South East of Iran," Etelaat Publications, Tehran, 2010.

CHAPTER 3

Determinants of Destination Management System (DMS) and Tourism Industry Assessment of Madagascar

Dare Aurelien, Jing Zhao

School of Economics and Management, China University of Geosciences, Wuhan, China

ABSTRACT

Tourism is one of the largest industries in the world. It is particularly significant in island economies, but also benefits significantly a range of developing and developed countries across the world. The sector is compound largely because of the interdependency between the universal production, largely managed in the established world, and the purposes around the world for which it assembles visitors. Information and communication technologies (ICTs) and tourism are two of the most dynamic drives of our worldwide economy. In the Madagascar, the tourism sector is the third largest provider of currency. Madagascar has a remarkable array of biodiversity, natural beauty and cultural resources to support tourism. Unexpectedly, of the 200,000 visitors the island per year, only about 60,000 come expressly for tourism, the rest traveling for other reasons but which could include some tourism activity. Madagascar has the potential to welcome many more tourists if the sector's growth is well planned in abroad, multi-sectorial way focusing on economic aspects, infrastructure and environmental and social concerns, particularly for community participation. This report sets out a program for equitable development of the sector and evaluates the opportunities for growth and the barriers that currently block progress and some suggestions that how to cover this era to maintain its stander of tourism sector along with. This paper describes the tourism of Madagascar, some lack of management, condition of the present tourism sector and at the end there is a conclusion of this study and recommendations on the base of present study.

KEYWORDS

Tourism, Madagascar, Economics, DMS

1. INTRODUCTION

Information and communication technologies (ICTs) and tourism are two of the most dynamic drives of our worldwide economy. As many authors have claimed, tourism must be treated as an information-intensive industry [1] [2] . Travel and tourism can be defined as an information business; because information is one of the most important factors which support actions in the tourism field such as the service industry. The technical revolution has facilitated the commercialization of tourism produces [3] . The World Tourism Organization (WTO) is trying to identify the ICT as having a role to play in tourism in 2007. E-tourism is definite as the act of choosing, organizing and booking holidays from the Internet. According to the investigations of an e-Marketer expert firm, e-tourism has become the most byproduct on the internet during the year, with turnovers exceeding 1000 billion.

1. Introduction

Figure 1. Map of Madagascar.

1.1 Madagascar

Madagascar, officially the Republic of Madagascar and previously known as the Malagasy Republic, is an island country in the Indian Ocean, off the coast of Southeast Africa. The nation comprises the island of Madagascar (the fourth-largest island in the world), as well as numerous smaller peripheral islands. Madagascar is a biodiversity hotspot; over 90 percent of its wildlife is found nowhere else on Earth. The island's dissimilar ecosystems and exclusive wildlife are exposed by the violation of the rapidly growing human population and other environmental threats. In 2012, the population of Madagascar was estimated at just over 22 million, 90 percent of whom live on less than two dollars per day. Malagasy and French are both official languages of the state. The majority of the population adheres to traditional beliefs, Christianity, or an amalgamation of both (Figure 1).

2. TOURISM INDUSTRY MADAGASCAR

For Madagascar, the tourism sector is the third largest provider of currency. The total revenue is estimated to 366651047.85 USD in 2013 (Ministry of Tourism of Madagascar, 2013) against 261353189.21 USD in 2012. In 2013, the number of international tourists is totaled to 196,375 with an annual increase of 20%. The tourism industry is composed of 1500 establishments employing more than 50,000 people (Malagasy Institute of Statistics, 2012). The island dominates the niche market of ecotourism with a natural capital 90% endemic [4] . The prospective is enormous but the country is still under exploited. The Government is currently developing the tourism sector as a priority. Tourism development necessarily goes through massive use of ICT. The use of the internet sector is growing over the past decades. Today almost 50% of players in tourism have a website, and at the same time 70% of the Malagasy territory is covered by digital networks. In 2003, the number of Internet subscribers was approximately 60,000 (300% growth compared to 2001) with 49,800 Internet users in the capital Antananarivo and the rest is located in the other five provinces [5] . 80% of Internet users are private traders and the 35% are involved in activities directly related in tourism. Travel representatives only rely on information; virtual dealings are still in its start; only 35% of communications in tourism come from the internet. Actors, especially the private operators (travel agencies, hotels, tour operators, tourist facilities, etc.) are conscious that the addition and use of ICT in the tourism industry is a precondition for its growth as the development of the access to the country as well as the tourism organization [6].

2.1. Tourism as a Development Tool

Tourism can be a effective development tool, creating economic development, expanding the economy, contributing to poverty improvement and also creating backward and forward relations to other manufacture and service sectors. In Madagascar, where rural deficiency is extensive and where the poor put pressure on the natural resource base, tourism could produce positive externalities. First, because the resources cover throughout the island, tourism generates pockets of economic growth in areas that have no substitute sources of income and occupation. In remote regions, mostly, tourism helps to ease poverty by expanding income sources. Second, tourism, properly accomplished, can help to preserve the environment, whether for ecotourism or for resort-based tourism. Madagascar's natural resources of flora and fauna and its coastal zone are among its most important but fragile economic assets. The assessment of Madagascar's assets for tourism undertaken for this report suggests that the current small size of the sector reflects substantial unrealized potential. In the past, tourism has been considered at worst as a residual to conservation, or at best a way of partially funding conservation. But tourism is complex and requires its own analysis, mainly as it is one of the major in the world and rapidly joining into a few large players. More needs to be done to build an active partnership between business and preservation, in credit of the fact that a sound business plan for tourism, an actual environmental plan, and a framework for social presence are equally strengthening and that absenteeism of one may put the others in risk.

2.2. Virtual Benefit

Madagascar's extraordinary natural assets, land-based and marine, make it a target destination for a diverse range of tourists. As one of the world's few mega-biodiversity countries, 95% of its animal and plant species are endemic. Among these are 32 species of lemurs, which are the main attraction for many tourists. As the world's fourth largest island, with nearly

5000 km of coastline and a continental shelf that is equal to 20% of its land area, Madagascar has world-class resort assets. These are enhanced by its marine and coastal biological diversity, which is greater than in any other Western Indian Ocean country, and, in particular, by its spectacular marine mega-fauna.

1) Because most tourism to Madagascar is natural resource based, the terms ecotourism, nature tourism and tourism are used interchangeably. Although a distinction is made between ecotourism and resort tourism, the latter is also primarily based on marine natural resources and its wildlife.

2) The variety of the country's assets is reflected in its World Heritage Site: Stingy Bemaraha Strict Nature Reserve in the west, which consists of a spectacular karst geological formation. The 16 National Parks, as well as other protected areas, cover 3% of Madagascar's land surface and are distributed throughout the island. Four marine parks are clustered around Mananara Nord, a Biosphere Reserve, on the northeast coast. Private reserves also dot the island and are visited by tourists. Because of its high endemism, Madagascar's ecotourism assets are unique and place it among the prime ecotourism destinations in the world. Its sun, sea and sand assets are outstanding but compete with the other better-known and better-developed resort destinations in the Indian Ocean. Madagascar also has potential as a cruise ship destination.

2.3. The Advantage Base

Madagascar's natural resources and, therefore, its asset base for tourism are much studied and have received considerable international financing and technical assistance for their conservation. Data about the tourism sector proper, on the demand side particularly, are, however, poor. Information is unreliable about numbers of bonafide tourists as opposed to foreign visitors, their expenditures, average length of stay, distribution around the country, and their socio-economic characteristics. This information is essential to: Know the current size of the sector and calculate benefits from tourism. Understand well what type of accommodation should be built where and which related services and products are essential Target promotion and marketing to existing and potential segments of demand and countries of tourist origin.

2.4. Tourism and the Surroundings

Tourism, carefully managed, is a tool for environmental protection and for financing conservation. Many "willingness to pay" studies indicate that tourists can be tapped to support environmental or cultural protection either through entrance fees, departure or other taxes, and voluntary contributions. To enhance the island's image, as well as help preserve the natural resource base, the larger accommodation units and, particularly those in sensitive ecological areas, should begin to aim for hotel accreditation that signifies that the unit has met clear environmental standards and is also a valuable marketing tool.

While economic cost benefit analysis may capture all streams of revenues and costs, where assets are public goods no quantification may be available. Yet tourism assets are scarce and generate "rents" and much dissention between countries and developers can be traced to how best distribute these rents between the travel industry, tourists and the host country. Clearly, a charge for use of these assets is desirable if the asset is to be used sustainably. Many direct and indirect tools are available but it is important that tourism operators as well as government policy makers and environmental officials take heed of this important issue. Measures for ensuring that resources collected are managed responsibly include arrangements with NGOs and other providers.

2.5. Sector Management

The Ministry of Tourism has capable people in its leadership echelons, but a low budget that renders it ineffective. The Ministry has also not succeeded in convincing other branches of government of tourism's significance and contribution to the economy, i.e., in classifying tourism as an export industry or attracting additional resources for infrastructure and promotion. The Association National pour la Gestation des Aires Protégées (ANGAP) has been an effective enough manager of the national parks and reserves, has built a solid reputation and works effectively with NGOs but faces funding shortfalls in the short run. Given the critical role of the national parks and reserves for conservation and for tourism, ANGAP urgently requires resources to enable it to continue its operations. GEF is working with ANGAP to create a trust fund and to increase its revenue generation from park fees.

2.6. Future Development

Tourism to Madagascar has recently been affected by acts of terrorism worldwide (September 11, Moscow, Bali and Nairobi) and by political upheavals at home. The WTO suggests that tourism is resilient and will bounce back-and indeed some countries in Asia saw growth in tourism in 2002. Mauritius for example, has been very successful in 2002 selling itself as a "safe haven" and its tourism grew. Indeed, in spite of the political hiatus in Madagascar in 2002, occupancies in higher income hotels seem to have held up; smaller operations have seen a decline. The emerging pattern in major markets is one of taking more frequent, shorter vacations, with long- haul destinations suffering by comparison with short haul trips (WTO). It is time for caution in extrapolating past trends but in the absence of crises, tourism can increase its Contribution to the economy. In the absence of targeted interventions for the sector, tourism can accommodate some Expansion in current numbers until capacity constraints kick in. Capacity is defined, principally, as the availability of quality accommodation and services and, where relevant, seats on internal and international flights. Without a more detailed survey of the many destinations visited by tourists, the exact timing of when such constraints will restrict tourist numbers cannot be forecast. It is a remarkable tribute to Madagascar's assets and its people, that despite lack of competitiveness in many facets of the sector, most tourists express high satisfaction with their visit. The number of return tourists also confirms the level of satisfaction. Madagascar has sufficient and varied assets that can be brought on stream successively over time to cater to different segments and niche markets, as well as income groups, of the tourism market. The resort market is the most competitive segment of demand worldwide and Madagascar competes with destinations in the Caribbean, South Pacific, and other "island paradises", as well as with the rest of the Indian Ocean. A challenge for Madagascar is to maintain its preeminent position as an ecotourism and adventure destination and simultaneously build up its image for other segments of demand, such as the resort market. It should also explore options for creating regional tours as suggested above with other islands in the Indian Ocean and neighboring countries on the continent.

2.7. A Future Approach for Tourism

Tourism may be an efficient development tool for Madagascar because it can be a Catalyst for economic growth, much of it in rural areas, and can create a number of positive externalities to alleviate poverty and protect the environment. As several other countries have done, Madagascar intends to complete master plans 4 for its tourism zones (some are already underway) and, with the support of GATO AG (a German apex organization for tourism), develop a "concept for tourism" in order to provide short- and long-term frameworks for tourism development. These studies should examine tourism's potential externalities and integrate tourism within the macro-economic framework to create linkages to other productive sectors.

The study should also identify poles for development, ways to eliminate constraints and barriers to growth, encourage sustainability through physical planning and community participation, as well as on an analysis of demand and proposals for appropriate tourist accommodation and services. Local communities are to be included in a participatory process at an early stage and the study should also address the policy framework, financing of tourism, the "greening" of the island, pro-poor tourism, and creating clusters of high-quality accommodation and services to upgrade the product. The study is to be completed by the end of 2003.

3. TOURISM AND ECONOMY OF MADAGASCAR

Tourism is one of the largest industries in the world. It is particularly significant in island Economies, but also benefits significantly a range of developing and developed countries across the world. The sector is compound largely because of the interdependency between the universal production, largely managed in the established world, and the purposes around the world for which it assembles visitors. An urbane supply chain and rapid alliance in the nineties, in airlines, hotel chains and tour operators/travel agents describe the industry, which makes concentrated use of information organization in its operations and of information expertise in questions and marketing. Within an individual country, tourism is also complex because of its cross-sectorial relations and the connections that are obligatory for its sustainability between the domestic and foreign private sector and several branches of central and local government, as well as with local communities visited by tourists. The sector has the possible to attract asset in a range of housings and services, to generate investments in substructure that also advantage other sectors and communities, and to stimulate demand in other sectors. In spite of a high possible for tourism, tourism in Madagascar is underdeveloped

4. ECONOMIC IMPACT OF TOURISM

Tourism produces considerable economic movement in other production and service sectors. The structure sector is clearly impacted during structure of hotel and tourism-related facilities, including substructure. Hotel and other types of accommodation produce economic activity through regressive and onward relations to agriculture and fishing and developed. Tourist services and tourists themselves, through individual expenditure in and outside the tourist accommodation, create a call for transport, banking, insurance, telecommunications, medical, security and retail services, and, particularly, handcrafts and other souvenirs. Tourism clues to the formation of businesses connected to water, mountain and exploration sports and other recreation activities, as well as every facet of travel and transport. Personal services related to spa treatments and traditional beauty services are also expanded by the presence of tourists. In 1993, the United Nations called on all countries to develop a National Satellite. Account for Travel and Tourism (TSA) to provide a reliable measure of its true influence to the national economy.

The Dominican Republic, with assistance from WTO and UNDP, undertook an experimental TSA in 1993, updated in 1998, designed to be a model for small island nations. The data for 1996 indicate that tourism expenditure as a percentage of GDP amounted to 20.5%. The comment of the Director of National Accounts of the 36 tourism to GDP and to export revenues is considerably higher in TSA than in conventional calculations that use only estimated direct tourist expenditures. Traditionally, tourism is included in sectorial GDP in Madagascar as "Trade, Hotels and Restaurants". By lumping trade with tourism, the contribution of neither can be well understood. In Madagascar, ecotourism or adventure tourism is the largest segment and some argue for foregoing the "resort market" (beach or mountain based). This requires closer analysis; for many countries find that their product line is made up of a variety of segments. It does not mean that Madagascar should forego the resort market in favor of ecotourism quite the contrary, Madagascar needs an internationally recognized brand name hotel or two to put it on the tourism map. In many successful tourism countries the tourism product mix is quite varied in Senegal where the Government argues only for "high end" tourists, in fact there are many backpackers, assumed to be low spenders; similarly, Mauritius, although known for its superlative five star hotels, is seeing a rapid growth of (often unregulated) villa and self-catering accommodation. There needs to be careful analysis of the Product mix within Madagascar and attention should also be given to creating regional circuits in the Indian Ocean community

4.1. Employment

Tourism generated some 17,564 jobs in hotels and restaurants, of which 3554 in travel and entertainment establishments in 2001 which increased with the passage of time, as shown in Table 1.

4.2. Taxes

Government revenues are produced through numerous taxes, including sales; value added (VAT), room, airport or departure, corporate income, payroll, social security, and property taxes. Revenues are also resulting from import duties and from aircraft arrival fees and voyage and boat docking fees. The particular amount of taxes elevated from tourism is difficult to approximation because of the dispersal of tourism expenses and the variety of goods and services used by tourism. In Madagascar, tourism generated tax revenues estimated at MF 62.5 billion in 1999 including VAT and income taxes. Whether these are income taxes on hotel and restaurant employees only or also include tour operators, transport services related to tourism, etc., needs to be clarified with the Government.

5. ANALYTICAL APPROACH

The information base to assume the type of economic analysis that would deliver a better understanding of tourism's involvement to the economy is not in place in Madagascar. Policy preparation for the sector needs to be based on a better understanding of the costs and benefits of different types of tourism development. The Government would need backing with this enterprise. Whether Madagascar develops a major tourism destination will be contingent on the quality of sector management. Tourism is fundamentally a private sector movement but is highly reliant on public sector sustenance. Tourism cuts transversely many sectors and there is frequently little communications between them. Famous among the ministries whose actions are related to tourism, in addition to the Ministry of Tourism, are, for example, the Ministries of Finance, Land, Culture, Public Works, Agriculture, Labor and Commerce. To be successful, tourism requires coordination and complementarity be- tween the Government and the private sector, civil society in general and, in particular, with local communities that are specifically impacted by tourism, where NGOs can enable the process. Only an ongoing dialogue between the public and private sectors can lead to effective tourism management. Despite some development, that dialogue is not yet in place. A strengthening of public and private sector institutions, joint with a strengthening of organization mechanisms among them, should facilitate the process but specific measures should be put in place as soon as possible to validate the discourse.

5. Analytical Approach

Table 1. Tourism employment.

	1996	1997	1998	1999	2000	2001
Hotels & Restaurants	11,103	11,318	11,655	12,640	13,628	14,010
Travel & Entertainment	2604	2661	2708	2934	3231	3554
Total	13,707	13,979	14,363	15,574	16,859	17,564

Source: Ministry of Tourism.

Destination Management is the only tool to make the betterment in tourism sector, Destination Management Organization will be a leading and most significant body to manage a destination developing into a "total destination management system". The DMS will become the final source of strategic aptitude and grow as a communication center for a destination. The DMSs community will be required after by Government organizations and large establishments for information and leadership [7] . DMS should not only help to hand pre-trip, post arrival information requests, but also assimilate an availability and booking service too [8] [9] . DMS could increase visitor traffic, attract the right market segment with the provision of an accurate and up to date comprehensive electronic database [10] . Madagascar has an inspiring array of biodiversity, natural beauty and cultural resources to provision tourism. The nation includes the island of Madagascar, as well as many smaller marginal islands. For Madagascar, the tourism sector is the third largest earner of currency. The potential is huge but the country is still under exploited. The Government is currently developing the tourism sector as an importance. Due to the fact that tourism has been familiar as one of the key sector to drive economic development and that the usual typical has not carried about sensitive socio-economic growth, a number of attitudes have been scheduled to improve tourism economic associations.

In Table 2 and Figure 2 we can see the number of people arrive in Madagascar from the world to see the beauty and attractions of Madagascar. In table we can see history of people from 1995-2012. In these years we scan see ups and down of people which were travelling to Madagascar. In should increase but in some years its decrease the reason may the lack of government management in this sector or unavailability of DMS within the country, one reason may be the high rate of air fare, so Government should look into these issues to make the tourism sector more efficient, which can increase the economy of the country.

6. CONCLUSIONS AND SUGGESTIONS

As we can say Madagascar is a paradise on earth due to its beauty and abundant attractive scenery. The Government has a big portion of GDP which gets from this sector, but as per my conclusion, it can increase more if we try to develop the strong DMS in the country along with strong ICT system. The research shows that DMO within Madagascar does appreciate the importance of understanding DMS and those they can deliver the destination with diffident benefit and long-term benefits. An important number of DMS are not delegating DMS regardless of the advantages can be expanded. The DMS still prefers to purchase the system and few advances to build their own DMS, lack of subsidy and scientific proficiency are seen as two of the major problems.

The World Bank's existing program in Madagascar is assisting to remove constraints to development. With greater planned focus, it could raise the outreach of tourism to rural areas and help relieve poverty. Because tourism is cross-sectoral nearly any donor-assisted-assisted activity could have an important impression on tourism, as, for example, all organization projects, including water, power, telecommunications, as well as health and education, if a focus on tourism is involved as part of the significant sector approach. The national transport plan will help to improve those roads, secondary airports and, possibly, railways, ports and river navigation that would directly benefit the growth and/or expansion of tourism. The rural roads project could be involved in opening up isolated areas with tourism possible. New projects in Madagascar should consider the impacts on and potential benefits to tourism. Policy discussions with the Government should comprise the need to progress the business environment for tourism and, particularly, to adopt measures that will enable new investors to follow-through on their asset plans. An agricultural sector evaluation should reflect

6. Conclusions and Suggestions

tourist plea. Public expenditure reviews should analyze the taxes derived from tourism and government expenditures for the tourism sector, including for elevation and training, which should be funded through a partnership with the private sector. Poverty missions should review the potential of tourism to produce income in specific rural areas and environmental assignments should assessment the potential of tourism for rising revenues to preserve natural resources used for tourism.

Table 2. International tourism, number of arrival in Madagascar.

Sr. No	Year	No. of people	Sr. No	Year	No. of people
1	1995	75,000	10	2004	229,000
2	1996	83,000	11	2005	277,000
3	1997	101,000	12	2006	312,000
4	1998	121,000	13	2007	344,000
5	1999	138,000	14	2008	375,000
6	2000	160,000	15	2009	163,000
7	2001	170,000	16	2010	196,000
8	2002	62,000	17	2011	225,000
9	2003	139,000	18	2012	256,000

Source: http://data.worldbank.org/indicator/ST.INT.ARVL.

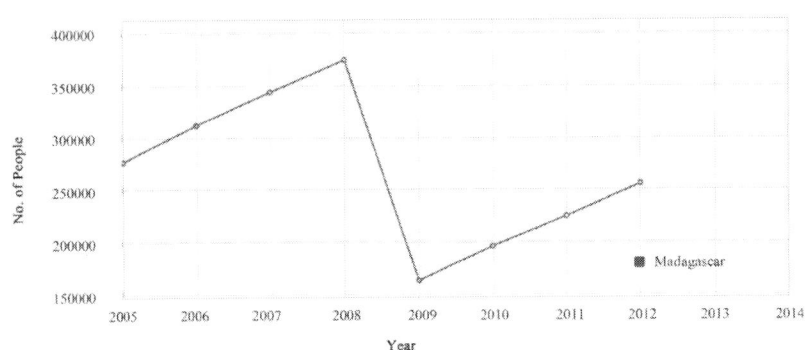

Figure 2. Graph number of people arrival in Madagascar.

DMS can help the traveler, but also growth the economy of any country, a country like Madagascar, which is a under developing country, not only need to create these types of system itself, but the Government should also inspire DMS system in the country for the foreigner investors. The vendors of DMO are also responsible to create a confirmatory environment for the tourists, develop the market request and driven tourism packages, enlarge the service quality and product growth of local contractors and improve the environmental performance of private businesses. To support maintainable tourism development, promote responsible travel and give motivations to travelers to understand and protect the local culture and environment they are staying. Government of Madagascar should take some serious and suitable steps for the betterment of this sector because SWOT analyses that the weakness and threats of this sector are high. Apart from it, a large number of Madagascar people attach with this occupation, so by the betterment of this sector, the Government can promote the economy of the country as well as the life of people. A sustainable and developed DMS are in need of time for this sector within country.

1) Tourism is highly modest in some markets, mainly the international beach resort market; In any event, the country will need to launch a product line that offers assorted involvements. There is also the chance to create regional markets with nearby islands and countries on the African landmass and thus produce a produce line with very high value additional for the country.

2) The Government identifies the need to advance the statistical base for tourism but is unnatural by lack of funding. These costs may persuade the Government to find the resources to assist with this important task of improving the tourism database.

3) Policy design for the sector needs to be based on a better considerate of the costs and benefits of diverse types of tourism and policies should include an considerate of how to stimulate the regressive and forward relations to other creation and service sectors. A study of the linkages between tourism and agriculture would help understand the demand from the tourism sector for agricultural produce; a study of handicrafts would help strengthen linkages between tourism and that sector and promote higher value added from tourism.

4) Air travel to Madagascar is ultra-expensive, inflexible, and limits the number of tourists who can afford to travel as well as the main source markets from which tourists come. Part of this is due to the distances

but part is also due to high fares. An expansion of demand from European countries other than France and from other supplier markets will require better air access, cheaper fares and direct marketing. Moreover, the international airport in Antananarivo needs upgrading to permit the landing of larger planes and for greater passenger comfort. It would be interesting to review existing bilateral agreements 70 in Africa and to open several airports in Madagascar to international flights and thus diversify the supply of air services.

5) The Government should focus on reform measures to reduce transactions costs for investors and machinists.
6) The lack of training and talents is a major weakness in all activities and all sectors, including the handicrafts sector. To enlarge Madagascar's tourist markets, a significant effort must be made to teach languages in schools and to local people who have regular interaction with tourists.

REFERENCES

1. Poon, A. (1993) Tourism, Technology and Competitive Strategies. CAB International, New York.
2. Sheldon, P. (1997) Tourism Information Technology. CAB International, New York.
3. Sheldon, P.J. (1993) Destination Information Systems. Annals of Tourism Research, 20, 633-649.
4. Crompton, D.E. (1999) An Action Plan for Sustainable Tourism Development in Panama. Tourism Development and Marketing Plan for Panama. Nathan Associates, Government of Panama. IADB.
5. http://www.omert.org/plan-du-site/
6. Crompton, D.E. (1996) Annexes III (Coastal Zone Management) and VI (Tourism) "Belize Environmental Report" No. 15543-BEL. World Bank. Latin America and the Caribbean Region ODA.
7. Varghese, B. (2013) Intervention of Destination Management Organization's in Tourist Destinations for Branding, Image Building and Competitiveness—A Conducive Model for Karnataka. International Journal of Investment and Management, 2, 50-56.

8. Buhalis, D. (1997) Information Technology as a Strategic Tool for Economic, Social, Cultural and Environmental Benefits Enhancement of Tourism at Destination Regions. Progress in Tourism and Hospitality Research, 3, 71-93.
9. Frew, A.J. and O'Connor, P. (1998) A Comparative Examination of the Implementation of Destination Marketing System Strategies: Scotland and Ireland. In: Buhalis, D. and Schertler, W., Eds., Information and Communication Technologies in Tourism1998, Springer, Wien, New York, 258-267.
10. Sheldon, P.J. (1993) Destination Information Systems. Annals of Tourism Research, 20, 633-649.

CHAPTER 4

Destination management in New Zealand: Structures and functions

Douglas G. Pearce

School of Management, Victoria University of Wellington, P O Box 600, Wellington, New Zealand

ABSTRACT

This article seeks to explain why structural differences occur with destination management in New Zealand and to examine the benefits and disadvantages of particular structures. Drawing on the analysis of a wide range of documents and in-depth interviews with practitioners, these issues are analyzed by taking a functional approach, one which views functions as the building blocks of destination management, takes into account a range of associated functions and considers the various organizational structures through which these are delivered. No one model of destination management exists. Rather, the country's administrative regime permits a mix of statutory and discretionary functions to be carried out under a range of different structures which have been adopted by local governments and the tourism sector in each destination depending on local or regional circumstances. Different views prevail as to which functions are important, whether they should be carried out by specialized organizations or units or brought together in multi-functional bodies.

KEYWORDS

Destination management; Destination marketing; DMOs; Local government; Planning; Regional tourism organizations

1. INTRODUCTION

Considerable variation occurs throughout New Zealand in the organizational and inter-organizational framework for destination management and the ways and extent to which aspects of destination management are undertaken. In some instances all or most destination management functions are carried out by territorial local authorities (city and district councils); in others, a more complex set of structural arrangements are found in which the role of councils is complemented by the work of regional tourist organizations (RTOs), economic development agencies and macro-regional marketing alliances. The destination management landscape is dynamic: from time to time the respective functions of councils and RTOs may be expanded or pared back; RTOs periodically move inside and outside of council structures; occasionally they disappear altogether. This variation and state of flux suggests there is an ongoing tension regarding destination management as each destination seeks to find the most appropriate structure. This article seeks to explain why these structural differences arise and to examine the benefits and disadvantages of particular structures. These issues are analyzed by taking a functional approach, one which views functions as the building blocks of destination management, takes into account a range of associated functions and considers the various organizational structures through which these are delivered. In doing so, it moves beyond the New Zealand case to address broader issues about the nature of destination management and how this is organized.

2. LITERATURE REVIEW

2.1. Structures and functions

Longjit and Pearce (2013) present a conceptual framework of destination management based on three inter-related features of management: purpose or goals, activities or functions and structures or organization. Activities or functions are undertaken to achieve a particular purpose or set of goals. Where multiple functions are needed to achieve this they are generally differentiated by specialization tasks and then integrated horizontally or vertically in some organizational structure (Hodge, Anthony, & Gales, 2003) or inter-organizational framework (Pearce, 1992). Organizations may be mono-functional and specialize in a particular function (e.g. marketing, planning, development, visitor servicing) or multi-functional and carry out several (Pearce, 1992 and ROS Development and Planning, 2008). Functions can be classified in other ways. Twyoniak (1998), (cited by Marsat, Guerra, & Lepinay, 2010), proposes a threefold typology of the competences of an organization:

- -elementary or operational competences;
- -intermediate level or functional competences (R & D, production, marketing…); and
- -upper level: interfunctional (e.g. quality control) or general competences (coordination, decision-making, *incitation*…).

Functions might also be distinguished in terms of whether they are statutory or discretionary, that is, whether organizations are mandated, or not, to perform a particular function or are free to do so if they choose (d'Angella, De Carlo & Sainaghi, 2010; Dredge et al., 2011, Gerbaux and Marcelpoil, 2006 and Pearce, 1992). Further distinction might be made between enabling and regulatory functions, between those which encourage some activity or development and those which control or restrict it (Ruhanen, 2013 and Simmons and Shone, 2002). The relationships between the units or organizations responsible for particular functions may be formal and tightly structured or informal and loose (Ruekert & Walker, 1987). These relationships may be dyadic, networked or take some other form (Longjit & Pearce, 2013).

Approaches to destination management, whether in terms of research or practice, also depend on one's concept of a destination. This study follows Pearce (2014a, p. 149) who conceptualized a destination as 'a dynamic, geographically based mode of production which provides interdependent and complementary products to tourists and transforms the spaces and places in which this production occurs'. In particular, attention is given to the interdependent and complementary nature of destination management functions and to the geographical or territorial context in which destinations are embedded, two key features which influence destination management structures.

What functions then does destination management involve? This depends in large part on the particular stream of literature considered: core destination management studies, more specific papers relating to DMOs (destination marketing or management organizations) and the related literature on the role of local government in tourism. Each of these streams includes multiple functions, with differing weight given to them.

2.2. Core destination management functions

Table 1 summarizes the key functions of destination management drawn from core studies in this field, that is those which focus specifically on destination management. The table should be regarded as indicative as some variation in terminology occurs from one study to another. Moreover, a specific function such as destination marketing can incorporate a range of activities. The functions are listed in the order in which they recur. Recurrence is an ordering device; it provides an indication of the frequency with which the different functions are cited but does not necessarily imply the relative importance of these. Multiple listings of particular studies indicate the different functions mentioned in the more comprehensive definitions or concepts.

2. Literature review

Table 1. Destination managemen functions.

Functions	Authors
Destination marketing, branding and positioning	Aberg (2014), Anderson (2000), ATRN (n.d.), Crouch & Ritchie (1999), Harrill (2005), Jamieson (2006), Laesser and Beritelli (2013), Longjit and Pearce (2013), Morrison (2013), Pearce and Schänzel (2013), Pearce (2014b), Pechlaner, Herntrei, and Kofink (2009), Ryglová (2008) and WTO (2007)
Destination planning, monitoring and evaluation	Aberg (2014), ATRN (n.d.), Crouch and Ritchie (1999), Dwyer and Kim (2003), Fuchs and Weiermair (2004) and Jamieson (2006), Laesser and Beritelli (2013), Morrison (2013), Pearce (2014b), Ryglová (2008) and WTO (2007)
Product development	ATRN (n.d.), Ivanis (2011), Jamieson (2006), Laesser and Beritelli (2013), Morrison (2013), Pearce (2014b), Pechlaner et al. (2009), Risteski, Kocevski, and Arnaudove (2012) and WTO (2007)
Research, information management and knowledge-building	Anderson (2000), ATRN (n.d.), Crouch and Ritchie (1999), Harrill (2005), Morrison (2013), Pavlovich (2003), Pearce (2014b) and WTO (2007)
Resource stewardship, environmental management	Crouch and Ritchie (1999), Dwyer and Kim (2003), Fuchs and Weiermair (2004), Longjit and Pearce (2013), Pavlovich (2003), Risteski et al. (2012) and Ryglová (2008)
Visitor management, managing the visitor experience, adventure risk management, safety management	Anderson (2000), Crouch and Ritchie (1999), Longjit and Pearce (2013), Pavlovich (2003), Pearce and Schänzel (2013) and Risteski et al (2012)
Relationship building	Crouch and Ritchie (1999), Dwyer and Kim (2003), Jamieson (2006), Morrison (2013) and WTO (2007)
Human resource development, training	Dwyer and Kim (2003), Longjit and Pearce (2013), Pearce (2014b) and WTO (2007)
Organizational responsibility, leadership and partnership	Anderson (2000), Crouch and Ritchie (1999), Jamieson (2006), Risteski et al. (2012)
Specific decisions and actions	Hawkins (2004) and Sainaghi (2006)
Lobbying	Jamieson (2006)
Service coordination	Laesser and Beritelli (2013)
Information provision	Laesser and Beritelli (2013)
Regulating and channeling tourism pressure	Pearce and Schänzel (2013)
Managing phases in the life cycle of a district (e.g. relaunch or start-up)	Laws (1995)
Managing particular problems (e.g. carrying capacity)	Sainaghi (2006)
Business support	WTO (2007)

In this core literature, destination management is generally portrayed as an over-arching process or approach which addresses the need to manage the diverse facets of a destination. It is most commonly expressed in terms of the upper level process of coordinating and integrating the management of supply and demand, functions and resources or a process which involves the collaboration, cooperation and interrelationships of relevant agencies or stakeholders. As Table 1 shows, there is general agreement that destination management involves multiple functions. The number and type of these vary from study to study but most frequently relate to destination marketing, positioning and branding; destination planning, monitoring and evaluation; product development; resource stewardship and environmental management; research, information management and knowledge-building; and various aspects of visitor management. Other functions are listed by only one or two studies, for example, lobbying (Laesser & Beritelli, 2013) or information provision (Pearce & Schänzel, 2013). There is relatively little debate in these studies over which functions are relevant or not. The studies cited here are primarily normative; they generally indicate what functions should be undertaken rather than those which are actually being carried out in a particular destination.

2.3. DMO functions

A second stream of literature relates to the functions of DMOs. The issues here mainly revolve around whether the term DMO refers to destination *marketing* organizations (the initial application), to more recent usage as destination *management* organizations, to destination *marketing and management* organizations or to some other related organization (e.g. *ente turístico*, ROS Development and Planning, 2008). This is not simply a question of semantics but also a question of the extent to which the title reflects the basic functions undertaken by the organization. Table 2 depicts the functions of DMOs (variously described) using the same procedure as for Table 1 but focusing on more specific reference to these organizations. Destination marketing is the most frequently cited function of DMOs, followed by relationship building and coordination, where DMOs are seen to play a key role, and product development. However, fewer studies see destination planning and resource stewardship as a function of DMOs. Destination management now appears as a separate category but not a frequently cited nor over-arching one.

Again, many of these studies are normative in nature, expressing what DMOs should or might do. The comprehensive set of external destination marketing and internal destination development activities put forward by Presenza, Sheehan, and Ritchie (2005), for example, constitutes a list of "possible" activities developed in their attempt to model the roles of destination management organizations. Empirical studies from North America have drawn attention to the difficulties faced by destination marketing organizations attempting to expand their role to include various destination management functions (Getz et al., 1998 and Gretzel et al., 2006). Getz et al. (1998) identified a planning policy gap between destination marketing and destination development in Canada due to the constraints faced by CVBs, such as lack of direct control over products or issues of competition should they develop their own products and the more political nature of involvement in the supply side compared to marketing. The functions of Spanish DMOs have evolved over time: an initial focus on providing information was complemented in the 1960s and 1970s with marketing and organizing small events, and in the 1980s with planning (ROS Development and Planning, 2008). That report also notes a more recent trend in large cities towards specialization. Pike and Page (2014, p. 205) go so far as to argue:

'using the term Destination Management Organization as a generic descriptor is unhelpful in adding clarity and purpose to the discussion of the DMOs' role *because it confuses the perceived need for management with the largely marketing function they actually undertake*...a DMO is an organization responsible for marketing of an identifiable tourism destination with an explicit geographical boundary'. (emphasis added)

In other words, for Pike and Page destination marketing is a completely separate phenomenon from destination management, rather than one of the components of the latter as the core destination management literature proposes.

Table 2. DMO functions.

Functions	Authors
Destination marketing, branding and positioning	Baggio (2008), Bieger et al. (2009), Bornhorst, Ritchie, Bornhorst, and Sheehan (2010), Bramwell and Rawding (1994), Crouch and Ritchie (1999), Elbe et al. (2009), Gretzel et al. (2006), Heath and Wall (1992), Osmankovic, Kenjic, and Zrnic (2010), Pechlaner, Volgger, and Herntrei (2012), Presenza et al. (2005), Prideaux and Cooper (2002), ROS Development and Planning (2008), Sheehan, Ritchie, and Hudon (2007) and Socher (2006)
Relationship building/coordination/facilitation	Bornhorst et al. (2010), Crouch and Ritchie (1999), Elbe et al. (2009), Heath and Wall (1992), Presenza et al. (2005), Prideaux and Cooper (2002), Sheehan et al. (2007) and WTO (2007)
Product development/development activities	Baggio (2008), Bieger et al. (2009), Bornhorst et al. (2010), Osmankovic et al. (2010), Pechlaner et al. (2012), Socher (2006) and WTO (2007)
Destination planning, strategy formulation monitoring and evaluation	Baggio (2008), Bornhorst et al. (2010), Heath and Wall (1992), Pechlaner et al. (2012), Jenkins et al. (2011) and WTO (2007)
Resource stewardship, environmental management	Bornhorst et al. (2010), Presenza et al. (2005) and ROS Development and Planning (2008)
Human resource development, training	Presenza et al. (2005), ROS Development and Planning (2008) and WTO (2007)
Destination management	Bornhorst et al. (2010), Crouch and Ritchie (1999) and Gretzel et al. (2006)
Quality assurance	Osmankovic et al. (2010) and ROS Development and Planning (2008)
Information provision and reservations	Bieger et al. (2009), ROS Development and Planning (2008) and WTO (2007)
Research, information management and knowledge-building	Osmankovic et al. (2010) and Presenza et al. (2005)
Visitor management, managing the visitor experience	Bornhorst et al. (2010) and Presenza et al. (2005)
Service provision, coordination	Bornhorst et al. (2010) and ROS Development and Planning (2008); Prideaux and Cooper (2002) and ROS Development and Planning (2008)
Business support	Baggio (2008) and Bornhorst et al. (2010)
Policy making or enforcement	WTO (2007)
Destination and site operations	Presenza et al. (2005)
Crisis management	Presenza et al. (2005)
Assistance with accessing finance	Bornhorst et al. (2010)
Enhance well-being of destination residents	ROS Development and Planning (2008)
Animation	

2.4. Tourism and local government

The broader question which arises when comparing the differences between Table 1 and Table 2 is who, if anyone, is carrying out functions deemed to be necessary for destination management if they are not undertaken by a destination marketing/management organization? Some of the answers to this question are found in a third stream of literature, that on tourism and local government. This stream is rather diverse but raises a range of critical points. Foremost amongst these is that it widens the debate away from a narrow sectoral view of destination management to a broader consideration of how destinations are managed in terms of the territories in which they are embedded (Barrado Timón, 2004; Marsart et al., 2010; Gerbaux & Marcelpoil, 2006). That is, tourism is generally but one activity which occurs in any place and as a consequence destination management must be considered alongside that of other sectors (e.g. residential, commerce, transport, other services, manufacturing...) and the place as a whole. Such a perspective gives prominence to other functions. In particular, the emphasis shifts to spatial planning, policy making, provison of infrastructure and utilities and management of public good assets such as parks, museums and galleries. Statutory responsibility for these functions vary by jurisdiction but generally belongs to some form of local government (Dredge et al., 2011 and Jenkins et al., 2011). Local government (*communes*) in France also has statutory responsibility for the *offices de tourisme* (information provision) and for management of ski lifts, though the latter is frequently delegated to a concessionaire (Escadafal, 2007, Gerbaux and Marcelpoil, 2006 and Marsat et al., 2010). Other organizations such as regional planning agencies or departments responsible for managing coastal or protected natural areas may also be involved alongside local authorities in aspects of destination management (Longjit & Pearce, 2013).

Studies in urban tourism in particular have drawn attention to the way in which and extent to which planning and policy making for tourism is integrated, or not, with broader urban plans, policies and management (Evans, 2000 and Hinch, 1996). Other research on tourism and local government has examined the linkages between functions. Bramwell and

Rawding (1994, p. 431) suggested that a trend in industrial cities in Britain towards the creation of corporate city marketing organizations might integrate tourism promotion with other city marketing policies and objectives but make it harder to 'coordinate marketing with the policies and practice in the economic, social, development and planning spheres'. Conversely, Palmer (1996) found a greater degree of inter-functional coordination within local authorities where public-private tourism development companies had been established.

Rather than being framed as studies of destination management, much of this stream of research focuses on particular functions such as marketing and planning or on an issue such as sustainability. Dredge et al. (2011), for example, portray destination management as one of the three pillars of sustainable tourism management alongside destination development and destination marketing. Marsat et al. (2010) is one of the few studies which takes a comprehensive, multi-functional view of destination management (*management stratégique de destination touristique*) and considers the links between local government functions and territorial management.

2.5. Destination management structures

As 2.2, 2.3 and 2.4 have shown, the core destination management literature identifies a range of destination management functions (Table 1) whereas the research streams dealing with DMOs (Table 2) and local government each focus on a smaller set of functions which they commonly undertake. However, both the core destination management studies and the organizational literature emphasize the need for multiple functions to be coordinated and integrated for 'the better management of the destination' (WTO, 2007) and to achieve an organization's goals (Hodge, Anthony, & Gales, 2003). The relative lack of integration between the three research streams outlined in 2.2, 2.3 and 2.4 is paralleled by the paucity of research which examines the organizational structures needed to deliver multiple functions when a broad view of destination management is taken. Studies which deal with organizational or inter-organizational structures and relationships often do not address the functions being carried out (Bodega,

Cioccarelli, & Denicolai, 2004) or focus on particular functions such as destination marketing (Beritelli et al., 2014, Elbe et al., 2009 and Prideaux and Cooper, 2002).

Pike and Page (2014) note that the perceived need for the management of destinations might be met by linkages between destination marketing organizations and territorial local authorities before concentrating on the former. Some studies though do go further in linking structure and function. Functions vary with scale from local organizations through to national and international ones (Bieger et al., 2009, Pearce, 1992 and Spyriadis et al., 2013). Spyriadis et al., 2013, p. 82) observe that: 'At a regional level, dual structures are also present, where regional or local government departments constitute the senior policy-making element with subsidiary bodies to take charge of operations, which in the majority of cases is external marketing'. Socher (2000) argues that because marketing and product development exhibit different forms of market failure, different forms of organization with different funding mechanisms are needed to deliver these functions.

The ROS Development and Planning (2008) study of the Spanish situation assesses the advantages and disadvantages of various organizational models for destination management. The comprehensive model (*modelo integral*) enables all functions to be brought together in one organization and may be especially suitable for small and medium destinations. However, difficulties may be experienced carrying out the functions, staff may lack sufficient specialized knowledge and the risk of political interference increases as functions associated with public goods are included. On the other hand, organizations specializing in one or more specific functions may be more efficient, they can focus on particular goals with suitably qualified staff and budgetary control may be easier but such specialization may also make itdifficult to obtain an overall vision for the destination and new opportunities may be neglected as being unprofitable. Opportunities to include the private sector vary across the two models.

2.6. Overview of functions and structures

In summary, comprehensive destination management requires multiple functions to be undertaken in a coordinated and integrated manner. However, analysis of such coordination and integration is complicated by the various inter-related ways in which functions can be conceptualized and structured. Functions can be viewed in terms of the actual nature of the activity (Table 1 and Table 2) and whether they constitute elementary, intermediate or upper level competences (Twyoniak, 1998), are statutory or discretionary, and enabling or regulating. In addition, these functions may be carried out by mono-functional organizations which specialize in performing a particular task or by multi-functional ones which undertake several. Organizations may focus on tourism alone, such as DMOs, or have much broader responsibilities, as is the case with local government. Coordination and integration can thus refer to bringing different tourism functions together (e.g. tourism planning and destination marketing) and/or linking tourism functions with those of other sectors (e.g. incorporating planning for tourism in territorial planning or combining destination marketing with place marketing).

Structurally, these considerations raise a set of inter-related questions:

- What different functional structures exist, what are the functions performed and what factors influence these?
- What are the perceived advantages and disadvantages of different functional structures?
- Are particular functions best carried out together by some multi-functional organization or distributed amongst a set of different organizations or organizational units?
- How is functional coordination and integration achieved?
- These questions are addressed in the New Zealand context.

3. METHOD

As the literature review has shown, different functions of destination management are likely to be undertaken by different organizations. In particular, studies of DMOs and local government tend to report and emphasize different functions and structures (2.3 and 2.4). A major challenge therefore in attempting to take a more comprehensive approach to destination management is to select examples that cover a range of situations and to collect information that enables a good understanding of both organizational structures and functions performed by the relevant organizations or agencies to be understood. In New Zealand there are 30 RTOs (regional tourism organizations, the New Zealand form of destination marketing organizations) and 67 territorial local authorities (TLAs): 11 city councils, 50 district councils and 6 unitary councils. Other studies on related topics have focused on case studies of particular regions (Simmons and Shone, 2002 and Zahra, 2011) or are based on a nation-wide survey of TLAs on a specific theme (Connell, Page, & Bentley, 2009). The former provide some depth but through their limited focus do not allow any commonalities or differences to be identified; the latter reports national patterns but does not flesh these out or illustrate them with particular examples. This study adopts an intermediate course; it examines local government and RTO participation in destination management in 14 selected destinations based on the compilation and analysis of documentation and interviews.

As the RTOs are the most visible tourism body at the regional or local destination level the selection of destinations began with an analysis of their websites to identify those which had prepared some form of tourism strategy or plan or indicated some other destination management related activity. Discussions with the national RTO body (RTONZ) also assisted with this selection by suggesting RTO managers who might be particularly knowledgeable about the theme. With this initial selection made, the websites of the relevant councils were then scanned to identify any documents potentially relating to destination management. The final selection included a range of destinations: large urban centres (Auckland and Wellington), middle order cities (Dunedin and Hamilton); major destinations (Rotorua and Queenstown) and more rural districts (e.g. Central Otago, Coromandel Thames).

The RTO documents compiled from their websites or during field visits consisted of tourism plans, strategies, annual reports and statements of intent while those from the council included statutory and non-statutory plans and economic development strategies. Analysis of these documents enabled an assessment of:

- the structure and functions of the RTOs and their formal links with local government;
- the extent to which there is explicit recognition of destination management in these documents and the form this takes;
- the alignment of any tourism/destination specific policies, plans or practices with broader ones of local government; and
- the way in which tourism/destination issues are incorporated in the broader policies, plans and practices of local government when no tourism/destination specific ones were found.

The compilation and analysis of documents were complemented by twenty-five in-depth semi-structured interviews conducted in late 2013 with CEOs, managers, planners and business development advisors from RTOs, councils and economic development agencies (EDAs – some of which are responsible for RTOs). These interviews provided additional information and insights from those directly involved. The interviews were guided by a series of open questions relating to their concept of destination management, how their organizations were structured, what functions they undertook and how they considered these, what links they had with other organizations and how these were viewed. The interviews lasted on average an hour, were audio-recorded and transcribed. The transcribed responses were compiled into a single file, read and re-read to identify and code key themes which were then sorted using NVivo and systematically reviewed. For reasons of space, the respondents' views on destination management *per se* and a more detailed analysis of functional matters will be reported elsewhere.

This focus on a diverse selection of destinations using these two sources of information has enabled a balanced approach to destination management to be taken in which information obtained from formal documents is complemented by insights from those involved. The interviews, for example, were particularly useful for providing insights into the way in which the formal structural linkages outlined in the various documents are complemented by more informal working relationships. This approach has required some trade-off between depth and breadth – not all RTOs and TLAs were included, nor were the roles of regional councils and of the Department of Conservation, the national agency responsible for protected areas, examined. While the sample is varied, the selection process will have included destinations with more active destination management functions and may therefore overstate the amount of destination management being undertaken throughout the country as a whole.

4. RESULTS: DESTINATION MANAGEMENT IN NEW ZEALAND

Taking 'a lead role in destination management by forming partnerships with key stakeholders' was one of four strategic aims put forward in the 2003 *Local Government Tourism Strategy* (Local Government New Zealand, 2003, p. 11). In elaborating on this aim the strategy identified multiple functions of destination management, the central role of territorial local authorities(TLAs) and the involvement of other agencies:

In association with the Department of Conservation, local government is the cornerstone of destination management in New Zealand. Destination management is essentially about cross agency co-operation in areas of planning, provision and functioning of information centres, management of infrastructure, Regional Tourism Organisation[RTO]/private sector marketing activities and site management

These functions need to be seen in the context of the broader purpose and role of local authorities in New Zealand. Local government is subject to national legislation which has been frequently amended with an emphasis this century on greater consultation, accountability and efficiency. Under the Local Government Act 2002, TLAs were 'to play a broad role in promoting the social, economic, environmental, and cultural well-being of their communities... [through] the prudent use and stewardship of community resources'. Amendments to the act in 2012 encouraged councils to focus on their core services, with their purpose being restated as 'to meet the current and future needs for good quality local infrastructure, local public services, and performance of regulatory functions' in a cost-effective way. The 2002 Act required local authorities to publish and update a 10-year Long-Term Council Community Plan (LTCCP), subsequently amended to Long-term Plan (LTP). Douglas (2007, p. 13) concluded that: 'Through the LTCCP process there is at last acceptance that Local Government strategic planning ...is the most senior function of local government'. Winefield (2007), however, was more circumspect in his assessment, noting a range of challenges to strategic planning by councils, including consistency with district plans. Under the Resource Management Act (1991) local authorities must prepare a district plan which sets out how 'the effect of using, developing and protecting the district's natural and physical resources will be managed in the future'. These statutory documents may be guided by other non-statutory strategies and policies.

4.1. RTOs, marketing alliances and i-SITES

Considerable variation occurs throughout New Zealand in the organizational and inter-organizational framework for destination management (Fig. 1 and Fig. 2) and the ways and extent to which different functions are undertaken. RTOs vary in size and structure but their prime function is destination marketing. RTOs may exist within the TLA (Fig. 1a), being located within such units or departments as Economics and Regulatory (Rotorua) or Community Services (Central Otago). In other instances RTOs may be established at arm's length as Council Controlled Organizations (CCO) (e.g. Positively Wellington Tourism, Tourism Dunedin, Destination Great Lakes Taupo) or some form of not for profit entity such as a trust (e.g. Tourism Bay of Plenty) which are governed by

4. Results: destination management in New Zealand

independently appointed directors (Fig. 1b). These RTOs are funded primarily, if not exclusively, by councils to undertake functions specified in a statement of intent. For example, the role of Destination Great Lakes Taupo (DGLT), is:

... to ensure that the greater Taupo region is marketed as a visitor destination so as to maximize the long-term benefits of the Taupo regional economy. Its specific functions are to develop, implement and promote strategies for tourism as a wider Taupo region. (DGLT, 2013, p. 1)

The objectives for the DGLT are spelled out mainly in terms of marketing; the main 'destination management' focus is to 'build a mandate to be in the long term infra-structure planning and deliberations for the region'. The DGLT sits physically within 'The Hub', alongside other agencies 'working for the Taupo District' (the EDA, an event organizing agency, an arts trust and a towncentre group). The scope of Tourism Bay of Plenty's activities covers destination marketing, management, leadership and development. The 'destination management' focus is on advocacy for infrastructure and facilitating the provision of readily accessible tourism information (Tourism Bay of Plenty, 2013).

In other destinations the RTO may form an integral part of an economic development agency (EDA) which has a wider brief than just promoting tourism and which itself might be located inside or outside a TLA (Fig. 1c and d). EDAs take different forms. In the southern part of the country, Venture Southland was established in 2001 as a joint initiative of Invercargill City, Southland District and Gore District Councils to promote an integrated region-wide approach to economic development, community development and destination marketing. Its mission is 'To actively work with groups and organizations to identify opportunities and to facilitate the development of projects and initiatives that will enhance the prosperity and quality of life of Southland communities' (Venture Southland, 2014). Tourism, events and community development come under a group manager. Venture Southland's budget has a tourism/destination marketing section where the emphasis is on destination marketing but provision is also made for other activities including industry management and attractions development. The Southland RTO is complemented by Destination Fiordland.

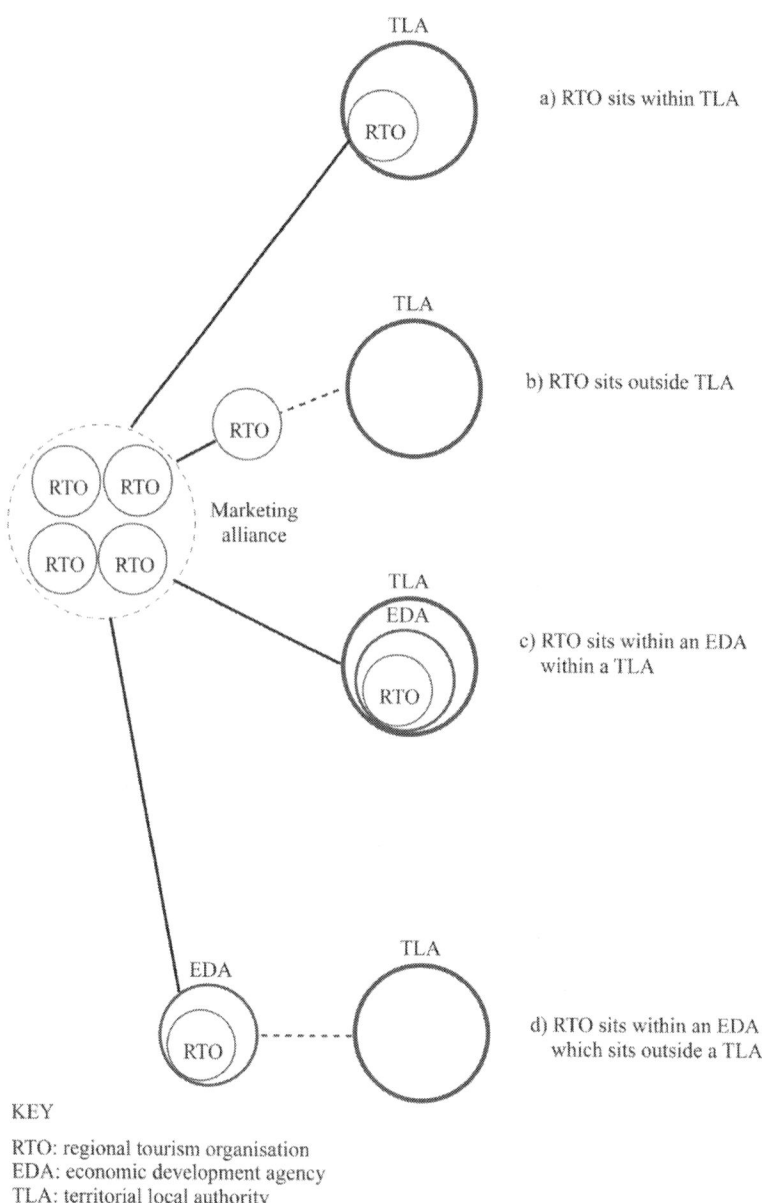

KEY

RTO: regional tourism organisation
EDA: economic development agency
TLA: territorial local authority

Figure 1. Main destination management structures in New Zealand.

4. Results: destination management in New Zealand

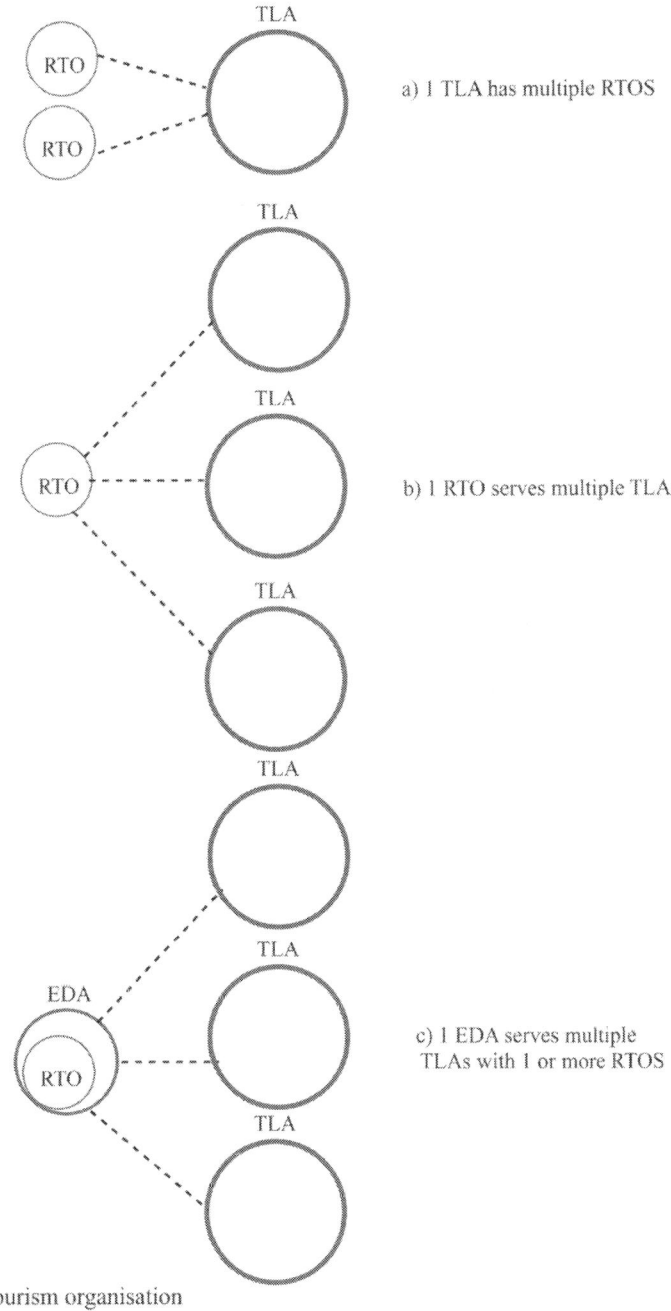

Figure 2. Destination management structures in New Zealand with multiple TLAs and RTOs.

Auckland Tourism, Events and Economic Development (ATEED) was formed in 2010 as a CCO following amalgamation of the metropolitan area's local authorities into a new 'super city'. ATEED's vision is to 'Improve New Zealand's economic prosperity by leading the successful transformation of Auckland's economy' (ATEED, 2013, p. 5). Tourism falls under Destination and Marketing, one of ATEED's eight business units. The unit is charged with 'growing the visitor economy' by delivering on the *Auckland Visitor Plan December 2011* (ATEED, 2011). The plan's emphasis is on destination marketing but other activities focus on enhancing the city's visitor proposition through a range of supply side actions. In the 2014/15 Statement of Intent (ATEED, 2014) these include measures to increase the capacity for growth and attraction development and support. In 2013/14 ATEED had an overall budget of NZ$51 million compared with Venture Southland's total budget of $5million (of which NZ1.37 million for tourism). The creation and competitive impact of ATEED appears to have stimulated interest in the EDA model. Early in 2014 Enterprise Dunedin was established as the single marketing agency for Dunedin by bringing together the RTO (Tourism Dunedin), the i-SITE and the council's Economic Development Unit. Later in the year proposals were put forward to create a similar EDA in the capital, the Wellington Regional Economic Development Agency.

RTOs may also take other forms. The two RTOs in the Queenstown Lakes District – Destination Queenstown and Lake Wanaka Tourism – are incorporated societies. They are funded primarily from a tourism levy collected on their behalf by the district council on the basis that tourism is the 'primary industry providing benefits to all local businesses' (Destination Queenstown, 2013, p. 6). Membership consists mainly of local businesses. There is a council representative on the boards. Hamilton & Waikato Tourism is a subsidiary of the regional airport company which in turn is owned by five councils. This structure follows closure of an earlier RTO serving the Waikato region (Zahra, 2011) and was prompted by the need for a lead tourism marketing organization for the airport and Tourism New Zealand to work with in promoting the region.

4. Results: destination management in New Zealand

For reasons of geography, local politics and the size and nature of the destination, a TLA may have more than one RTO (e.g. Queenstown Lakes District Council) and an EDA or RTO may serve more than one TLA (Fig. 2). Destination Coromandel, for example, was established to provide the Hauraki and Thames Coromandel District Councils 'with a vehicle to lead, manage and market tourism for the two districts under an umbrella brand, *The Coromandel*' (Destination Coromandel Trust Inc., 2012, p. 3). Tourism Bay of Plenty covers Tauranga and the Western Bay of Plenty. Hamilton & Waikato Tourism acts as the RTO for seven TLAs in the region. In the case of Southland, Venture Southland serves three TLAs and there are two RTOs.

RTOs may also come together with others to form regional marketing alliances (Fig. 1) to more effectively fund and carry out international marketing campaigns in association with Tourism New Zealand. These are based either on a macro region or a touring route. For example, in addition to joint initiatives between Southland RTO and Destination Fiordland, Southland is represented in:

- Waitaki/Dunedin/Clutha/Southland IMA (International Marketing Alliance);
- Southern Lakes IMA (Wanaka, Queenstown, Fiordland RTOs);
- Southern Scenic Route (Dunedin, Clutha, Southland, Fiordland and Queenstown RTOs); and
- SOUTH- collective marketing of the South Island targeted at the Chinese and Australian markets led by Christchurch International Airport with the South Island RTOs.

Similar alliances exist in the North Island, such as Explore Central North Island and the Classic New Zealand Wine Trail.

Additional structural and functional complexity arises from the various ways i-SITEs are operated. Over 80 i-SITEs form part a nation-wide network of accredited visitor information centres which provide independent destination information and book and sell travel products. In New Zealand they remain an important means of distribution to domestic and independent international visitors who make many 'at destination' decisions

(Pearce & Tan, 2004). Fourteen of the RTOs, within or outside council, operate one or more i-SITEs. Others are run by councils or operate independently. Where an RTO serves multiple TLAs, councils usually run i-SITES separately.

4.2. Local government

Although some of these structural arrangements are complex, RTOs, marketing alliances and i-SITES are the more visible agents of aspects of destination management, being responsible primarily for destination marketing and information provision. In contrast, other functions carried out by TLAs are often less evident, but not necessarily less important, as they are frequently undertaken as part of the broader functions of local government and not framed specifically in terms of destination management. Although all the councils have similar responsibilities under the Local Government Act 2002 and the RMA, they vary widely in the ways and extent to which they engage directly or indirectly with tourism and destination management. The neighboring districts of Central Otago and Queenstown Lakes provide two contrasting approaches to destination management. The former is a largely rural district in which significant tourism development is a relatively recent development whereas the latter is made up primarily of the long-established destinations of Queenstown and Wanaka.

The Central Otago tourism strategy (CODC, 2007, p. 5) is one of the few to have destination management – defined as being 'essentially about communities and cross-agency co-operation in all areas to both capitalize on and maintain what is special in this place'- as its central thrust and for this to be embedded widely and explicitly in the district council's plans and activities (CODC, 2012a, CODC, 2012b, CODC, 2013a, CODC, 2013b and CODC, 2013c). These also have a strong community focus. The linkages between tourism and community interests is reinforced by the structure of the council in which the RTO, the i-SITEs and responsibility for regional identity sit within the Community Services department. Amongst the outcomes set out in the *2012–2022 Long Term Plan* (CODC, 2012a, p.

4. Results: destination management in New Zealand

12) reference is made to 'A tourism industry that is well managed, which focuses on our natural environment and heritage with marketing plans that reflect this...' Challenges as well as opportunities were identified in the district's heritage strategy (CODC, 2012b):

The challenge of managing tourism so that it does not undermine heritage values is real, but it is somewhat alleviated by the destination management approach taken in Central Otago. In community plans, communities identify their opportunities and aspirations and weigh them against the values they want to retain in relation to lifestyle to determine goals and an action plan for the future. Promoting tourism takes coordination, a level of expertise and requires funding.

The regional brand – 'A World of Difference' – offers scope for tourism to be marketed jointly with the region's wines and horticultural products.

In contrast, destination management is not mentioned explicitly in the Queenstown Lakes District plans and strategies that were examined, even though, or perhaps because, tourism dominates the district's economy and the council's planning and service delivery is in effect directed at managing the two major destinations of Queenstown and Wanaka and the smaller settlement of Arrowtown. The town centre strategies for Queenstown and Wanaka emphasize the need to provide for residents as well as tourists (QLDC, 2009a and QLDC, 2009b). More generally, there is an implicit understanding that the district depends on tourism and that this must be managed. The District Plan (QLDC, 2013, sec. 4, p. 6), for instance, states:

The District relies in large part for its social and economic wellbeing on the quality of the landscape image and environment and has included provisions in the District Plan to avoid development which would detract from the general landscape image and values.

Various documents clearly signal that the district is in growth mode. The district's growth management strategy (QLDC, 2007, p. 2) was:

prepared on the basis that the Council will not (and cannot) stop growth from occurring in the District. However the Council will act with determination to manage the quality, location and type of growth to help ensure that new activities add to the economic, social, cultural and environmental wellbeing of the District ...

In her introduction to the 2014/15 Annual Plan (QLDC, 2014, p. 1) the mayor observed:

We live in one of the fastest growing districts in the country with a huge visiting population. Our challenge is to deliver better service at less cost. This plan sets out our activities for 2014/15. Many of them are routine – maintaining safe roads; delivering good quality drinking water; collecting rubbish and recycling; treating wastewater; providing parks, pools and libraries.

Provision of such services, while routine, is nonetheless crucial for the destinations to function successfully. At the same time, the plan also records the council's decision to approve the development of a major new convention centre for Queenstown which would necessitate a NZ$32.5 million council contribution.

Elsewhere tourism is accorded varying levels of importance in economic development plans (e.g. Dunedin City Council, 2013, Rotorua District Council, 2011, TCDC, 2014 and Waipa District Council, 2012) but is generally treated in a rather fragmented fashion in the various statutory and non-statutory plans and strategies which guide council activities. Reference may be made to tourism and to the needs of visitors as well as residents in terms of service or infrastructural needs but the measures outlined and steps taken tend not to be drawn together in a concerted fashion which suggests the councils have a comprehensive, coherent approach to destination management. This is not altogether surprising as the councils' responsibilities go beyond tourism to managing their cities and districts as a whole. These issues are examined in more detail for the three cities of Dunedin, Rotorua and Hamilton in a related paper (Pearce, in press).

Variation also occurs in the extent to which explicit reference is made to the RTOs and i-SITES in these wider plans and activities. The *2013/14 Annual Plan* (CODC, 2013a) for Central Otago, for example, comments on the critical role that the council's i-SITES play in connecting visitors with tourism operators. The economic development action plan for the Thames Coromandel District (TCDC, 2014, p. 12) draws attention to the restructuring of their RTO 'to focus on delivering more effective and successful visitor marketing' (explicit reference to destination management was dropped), increased funding for the RTO and i-SITES and closer partnerships with them (e.g. joint promotions aimed at encouraging holiday home owners to spend more time in the region). ATEED's most recent statement of intent (ATEED, 2014) underlines the need for its priorities to be aligned with the strategic direction of the higher level Auckland Plan and Economic Development Strategy.

Other practical, operational aspects of destination management are less visible in these documents and may seem routine but they emerge from the interviews as important for the successful functioning of the destinations. These include such basic matters as provision of signage, public toilets and parking; dealing with road closures and sale of liquor bans during major sporting events; and assisting operators and developers in terms of compliance and planning regulations. The recent upsurge in freedom camping has generated a variety of problems which have been addressed through attempts to limit overnight camping on public land through by-laws or by re-directing this demand to more appropriate sites through information provision, targeted promotion or the development of new facilities.

4.3. Structures and functions

These variations in how destination management is structured and undertaken in New Zealand show that the national administrative regime permits such variation and that different structures are perceived to have advantages and disadvantages in different destinations. In terms of the first point, under the Local Government Act 2002, the RMA and other legislation, local government has statutory responsibility to carry out a range of general functions that include those relating to aspects of destination management, notably provision of local infrastructure and services, spatial planning and environmental protection. Although some of these supply-side functions may be contracted out, they remain the responsibility of local government. Despite the changes in 2012 to the Local Government Act (2002), local government is still able to engage in a wide range of economic and community activities. Councils continue to have a considerable amount of discretion in terms of the manner and extent to which they engage in destination management, whether in terms of making explicit provision for tourism in their statutory functions or in the discretionary ones with regard to economic development, destination marketing and information provision (QLDC, 2014 and TCDC, 2014). It is this mix of statutory and discretionary functions which permits variation in functional structures.

The resulting variation in turn reflects differing circumstances and differing views on the advantages and disadvantages of bringing different functions together or having them in separate organizations or organizational units specializing in particular functions. As the relevant statutory functions are performed by divisions of council, the key structural issue which arises is which discretionary functions are best carried out within council and which are best performed through a CCO or some other arm's length organization or independent entity having one or more specialized functions. In practice this relates to where, structurally, the RTOs and, to a lesser extent, the i-SITES are located, what form they take, what functions they perform and how these are linked to the statutory functions of council? The RTO, council and EDA interviewees advanced a range of advantages and disadvantages associated with different structures with cases being made both for specialized or multi-functional organizations and for locating RTOs within and outside of council. These two issues – specialization/multi-functionality and location inside/outside of council – are inter-related.

Proponents of specialization, particularly of having RTOs concentrate on destination marketing, emphasized the benefits of clarity of purpose and a tighter focus which enabled them to get on with the job of drawing more tourists to the destination and having them stay longer. Some also expressed the view that destination marketing should not be limited to just attracting tourists but should also incorporate the broader aspects of place marketing.

4. Results: destination management in New Zealand

Those who favoured multi-functional organizations or units acknowledged the inter-dependence of functions and stressed the benefits of enhanced functional interaction; that is, having two or more functions carried out by the same body increases the synergy between them and results in more effective destination management. Various forms and levels of inter-functionality were identified: marketing/information provision; marketing and product development; enabling and regulating. Opposing views were occasionally presented.

Some RTO respondents saw a close association between marketing the destination externally and providing information at the destination: '[The i-SITES are] a really important component to our team for understanding what's going on at the ground level with our visitors'. Others underlined a clear separation between the two functions: 'Our [the RTO's] job is to get people to ... [the destination] and their job [the i-SITES] is to keep them there'. Another manager saw pros and cons with his RTO running the destination's i-SITES: 'It can largely be quite a big distraction for our business. However, I think it's a necessary link ... We attract them to the region and we also want to make sure they're looked after well when they get to the region'. Resourcing is also often an important consideration in this debate; the cost of staffing i-SITES can be high and reduce the budget available for marketing. Opportunities for product development to be strengthened by marketing may be enhanced when both functions are undertaken within the same organization, such as an EDA with a broader economic development brief.

One of the strongest advocates for a multi-functional structure and approach came from Rotorua where the RTO is located within the Economics and Regulatory division of the district council. Such a structure, it was argued, brought greater problem solving ability to destination management issues:

[being in this group] really forces a fusion between destination marketing and management because... in my team are the people that are responsible for growing the economy... but in that same team... are the people responsible for regulating the environment through central government legislation, by-laws... all the things that could constrict development. So by putting both into the

same group, we're really making each other… fight to find solutions…. We're having a lot more discussion where maybe in silos people wouldn't have really had that perspective.

The Rotorua interviewee also acknowledged that clarity of purpose across other units of council, such as those responsible for convention, events and management of council owned venues, although complementary, could become blurred and complicated due to each unit having different goals. The relationships between regulation and economic development were illustrated by a manager in another TLA in a similar way:

… the district plan is a key enabler and we've worked to rezone a whole lot of land and make events easier to happen and change all those rules around events, so it does have a big economic development consideration because that is actually the blocker or enabler.

Earlier, with regard to the West Coast, Simmons and Shone (2002) raised the issue of conflicts of interest which might arise between a council's economic and regulatory functions and suggested that separation of these may be needed to maintain a system of checks and balances.

As these latter examples show, the question of specialization and multiple functions is closely related to which destination management functions are performed inside or outside of council. Again, views vary. Where the RTO is located within the council structure, benefits are seen to lie in being able to form closer connections with the broader functions of councils, to have greater access to decision-makers and thereby have a greater ability to effect change. In the case of Rotorua this was raised in terms of tourism, regulation and economic development; in Central Otago it has enabled close links between tourism and community development. Proponents for having the RTO, and thus destination marketing, as a separate entity outside of council stress the advantage of having a skill's based board and being at arm's length from the 'meddling' of local politicians; of being more agile and better able to respond to changing market conditions; and of having staff with more flair and different skills and mind-sets than other council staff. Others would contest these latter points. Having the RTO outside of council may also facilitate operating a RTO at a larger regional scale, as in the Waikato, the Coromandel and the Bay of Plenty. However, whether inside or outside, RTOs do come together in a range of macro-regional marketing alliances.

In Twyoniak's (1998) terms the functions discussed here are largely operational or intermediate level ones. It is less clear what upper level interfunctional destination management actually takes place, at least formally. Destination management is a designated responsibility of one of the advisors in Dunedin City Council's Economic Development Unit, a role which involves coordination and communication between council units and with external stakeholders. The recent establishment of Enterprise Dunedin has also meant that tourism marketing in the city is now more closely linked with other aspects of place marketing there. In small TLAs oversight of destination management may result from an individual having several responsibilities. The explicit and embedded destination management policies and practices in Central Otago are undoubtedly due to the manager of community services also being the deputy CEO, the general manager of the RTO and a person with strong views on the need for such an approach. The business development facilitator of another small TLA claimed 'it really depends on the expertise and the drive of the bloke who sits in this seat'.

Elsewhere, much interfunctional interaction appears to be occurring informally between individuals in different organizations or units in the manner of open social systems discussed by Ruekert and Walker (1987, p. 2) whereby:

- 'Behavior among members of the social system is motivated by both individual and collective interests.
- Interdependent processes emerge because of the specialization and division of labor.'

Interviewees reported frequent contact between council and RTO managers and staff. CEOs of RTOs are regularly invited to comment on destination management issues. One asserted that ' The two things [destination marketing & management] are distinct but it is essential they don't operate separately. You don't have to be in each other's camp but must be connected.' Physical co-location, such as in Taupo's Hub, facilitates such interaction. In stressing the advantages of bringing functions together, one EDA manager argued 'when you separate out roles, then people become very protective and they spend most of their time protecting their patch rather than actually aligning and working with others'. Tension was evident in some destinations but how much of this was due to functional differences, especially between destination marketing and other activities, or the personalities of the individuals involved was less clear, especially with small entities such as many RTOs. Several interviewees expressed the view that structure was less important than interpersonal relationships between those carrying out different tasks and that most structures could be made to work if those involved had a mind to do so. Others made the point that funding was a fundamental issue, that even if appropriate structures were in place and there was a meeting of minds without adequate resources to carry out the various functions then any destination management would be limited.

5. CONCLUSIONS

This study of destination management in New Zealand has clearly shown that no single model exists. Rather, the country's administrative regime permits a mix of statutory and discretionary functions to be carried out under a range of different structures which have been adopted by local governments and the tourism sector in each destination depending on local or regional circumstances. As the variety of structural arrangements for RTOs, i-SITEs and macro-regional alliances shows, this flexibility is particularly evident in dealing with demand-side functions of marketing and information provision where administrative boundaries have not proven to be insuperable barriers in many places to increasing the scale of activity or having multiple local entities when this is thought to be appropriate. Likewise, the growing move towards EDAs is breaking down internal barriers and leading to greater integration of tourism in city marketing. The extent to which statutory functions incorporate other aspects of destination management, either explicitly or implicitly, varies considerably, however,

5. Conclusions

depending on the significance of tourism in the region and the degree to which local government chooses to engage with the sector. Different views prevail as to which functions are important, whether they should be carried out by specialized organizations or units or brought together in multi-functional bodies. In these respects the situation in New Zealand is not dissimilar to that in Spain (ROS Development and Planning, 2008). There is not a lot of evidence to show that that destination management is being carried out as an explicit higher level function with a concerted effort to integrate all relevant functions in the coordinated fashion encouraged by the WTO (2007). However, in many destinations there appears to be a fair amount of informal, practical interaction between individuals responsible for different functions.

Issues of specialization and multi-functionality also apply to research on destination management functions and structures. Although the largely, normative core management literature emphasizes multiple functions and views destination management as an over-arching activity, much empirical research in this field has focused either on particular functions, such as destination marketing or planning, or on specific organizations such as DMOs (Table 2) and local government (Section 2.3). Such specialized research is useful but if functions are regarded as interdependent and a broad approach to destination management is to be taken (Table 1), then such work needs to be complemented by studies such as this which span across functions and associated structures to provide a more holistic view.

In particular, future research might focus on the inter-related questions of functional interdependence and structural effectiveness at a destination level. How, for example, are marketing and information provision inter-connected at the scale of the destination, how does spatial planning affect product development and what are the synergies produced when these are all brought together? This applies not only to the interdependence of functions across the tourism sector but also to the way in which these functions mesh with broader territorial management functions, for example by incorporating tourism explicitly in district plans and economic development strategies rather than by treating it separately. Assessing the effectiveness of these different functions and how they are performed under different structures is critical but challenging. At present, many RTO KPIs in New Zealand are commonly expressed in terms of increased visitor numbers, expenditure or length of stay but taking a broader view of destination management will also require more explicit measures of the quality of the visitor experience as well as taking into account the impact of destination management on other stakeholders such as local residents, tourism providers and other businesses. In terms of relationships and coordination (Table 2), consideration must be given to the roles of formal structures and informal relationships in bringing about effective interfunctionality, that is, is it the intra- or inter-organizational structure which is the most important factor here, the informal inter-personal interactions between groups or individuals, or some blend of both? How do structures influence interaction and thus interfunctionality?Much remains to be done. Many opportunties and challenges clearly exist for extending research on destination management and for managing destinations.

ACKNOWLEDGMENT

Fig. 1 and Fig. 2 were drawn by Jo Heitger. This research was supported by the Victoria Research Trust under Research grant 8-113873-2309.

REFERENCES

1. Aberg, K. G. (2014). The importance of being local – prioritizing knowledge in recruitment for destination development. *Tourism Review*, 69(3), 229–243.

2. Anderson, D. (2000). Destination management. In: J. Jafari (Ed.), *Encyclopedia of Tourism* (p. 146) New York: Routledge.

3. ARTN n.d. The guide to best practice destination management. Australian Regional Tourism Network.

References

4. ATEED. (2011). *Auckland's visitor plan 2011-2021*. Auckland: ATEED.

5. ATEED. (2013). *Statement of intent for Auckland Tourism, Events and Economic Development Ltd (ATEED) 2013-2014*. Auckland: ATEED.

6. ATEED. (2014). *Statement of intent for Auckland Tourism, Events and Economic Development Ltd (ATEED)* (pp. 2014–2015) Auckland: ATEED.

7. Baggio, R. (2008). Symptoms of complexity in a tourism system. *Tourism Analysis, 13*(1), 1–20.

8. Barrado Timón, D. A. (2004). El concepto de destino turístico: una aproximación geográfico-territorial. *Estudios Turísticos, 160,* 45–68.

9. Beritelli, P., Bieger, T., & Laesser (2014). The new frontiers of destination manage- ment: applying variable geometry as a function-based approach. *Journal of Travel Research, 53*(4), 403–417.

10. Bieger, T., Beritelli, P., & Laesser, C. (2009). Size matters! Increasing DMO effective- ness and extending tourism destination boundaries. *Tourism, 57*(3), 309–327.

11. Bodega, D., Cioccarelli, G., et al. (2004). New inter-organizational forms: evolution of relationship structures in mountain tourism. *Tourism Review, 59*(3), 13–19.

12. Bornhorst, T., Ritchie, Bornhorst, T., J. R. B., & Sheehan, L. (2010). Determinants of tourism success for DMOs & destinations: an empirical examination of stakeholders' perspectives. *Tourism Management, 31*(5), 572–589.

13. Bramwell, B., & Rawding, L. (1994). Tourism marketing organizations in industrial cities: organizations, objectives and urban governance. *Tourism Management, 15* (6), 425–434.

14. CODC. (2007). *Central Otago tourism strategy 2007-2012*. Alexandra: Central Otago District Council.

15. CODC. (2013). *2012-2013 annual report*. Alexandra: Central Otago District Council. CODC. (2012). *2012-2022 long term plan*. Alexandra: Central Otago District Council. CODC. (2012). *Towards better heritage outcomes for Central Otago*. Alexandra: Central Otago District Council.

16. CODC. (2013). *Economic and business development strategy 2013-2016*. Alexandra: Central Otago District Council.

17. CODC. (2013). *2013/14 annual plan*. Alexandra: Central Otago District Council. Connell, J., Page, S. J., & Bentley, T. (2009). Towards sustainable tourism planning in

18. New Zealand: monitoring local government planning under the Resource Management Act. *Tourism Management, 30*, 867–877.

19. Crouch, G. I., & Ritchie, J. R. B. (1999). Tourism, competitiveness and societal prosperity. *Journal of Business Research, 44*, 137–152.

20. d'Angella, F., De Carlo, M., & Sainaghi, R. (2010). Archetypes of destination governance: a comparison of international destinations. *Tourism Review, 65* (4), 61–73.

21. Destination Coromandel Trust Inc. (2012). *Statement of intent for 2012-2013*. Thames: Destination Coromandel Trust Inc.

22. Destination Queenstown. (2013). *Annual report Destination Queenstown Incorpo- rated (01 July 2012-30 June 2013)*. Queenstown: Destination Queenstown Incorporated.

23. DGLT. (2013). Destination Great Lake Taupo statement of intent 2013/2014. *Taupo District Council Meeting, 30(July), 2013*.

24. Douglass, M. (2007). LTCCPs on the first round. *Planning Quarterly*, 12–13 (March). Dredge, D., Ford, E.-J., & Whitford, M. (2011). Managing local tourism: building sustainable tourism management practices across local government divides. *Tourism and Hospitality Research, 12*(2), 101–116.

25. Dunedin City Council. (2013). *Dunedin's economic development strategy by Dunedin for Dunedin and beyond* (pp. *2013–2023*)Dunedin: Dunedin City Council.

26. Dwyer, L., & Kim, C. (2003). Destination competitiveness: determinants and indicators. *Current Issues in Tourism, 6*(5), 369–414.

27. Elbe, J., Hallén, L., & Axelsson, B. (2009). The destination-management organisation and the integrative destination-marketing process. *International Journal of Tourism Research, 11*(3), 283–296.

28. Escadafal, A. (2007). Attractivité des destinations touristiques: quelle strategies d'organization territorial en France? *Téoros, 26*(2), 27–32.

29. Evans, G. (2000). Planning for urban tourism: a critique of borough development plans and tourism policy in London. *International Journal of Tourism Research, 2*, 307 -306.

30. Fuchs, M., & Weiermair, K. (2004). Destination benchmarking: an indicator- system's potential for exploring guest satisfaction. *Journal of Travel Research, 42*(3), 212–225.

31. Gerbaux, F., & Marcelpoil, E. (2006). Gouvernance des stations de montagne en France: les spécificités du partenariat public-privé. *Revue de Géographie Alpine, 94*(1), 9–19.

32. Getz, D., Anderson, D., & Sheehan, L. (1998). Roles, issues, and strategies for convention and visitors' bureaux in destination planning and product devel- opment. *Tourism Management, 19*(4), 331–340.

33. Gretzel, U., Fesenmaier, D. R., Formica, S., & O'Leary, J. T. (2006). Searching for the future: challenges faced by destination marketing organizations. *Journal of Travel Research, 45*(2), 116–126.

34. Fundamentals of destination management and marketing. In: R. Harrill (Ed.), Lansing: MI:The Education Institute of the American Hotel & Lodging Association.

35. Hawkins, D. E. (2004). Sustainable tourism competitiveness clusters: application to World Heritage sites network development in Indonesia. *Asia Pacific Journal of Tourism Research, 9*(3), 293–307.

36. Heath, E., & Wall, G. (1992). *Marketing tourism destinations: a strategic planning approach.* New York: Wiley.

37. Hinch, T. D. (1996). Urban tourism: perspectives on sustainability. *Journal of Sustainable Tourism, 4*(2), 95–110.

38. Hodge, B. J., Anthony, W. P., & Gales, L. M. (2003). *Organization theory: a strategic approach.* Upper Saddle River, New Jersey: Prentice Hall.

39. Ivaniš, M. (2011). General model of small entrepreneurship development in tourism destination in Croatia. *Tourism & Hospitality Management, 17*(2), 231–250.

40. Community destination management in developing economies. In: W. Jamieson (Ed.), New York: Haworth Hospitality Press.

41. Jenkins, J., Dredge, D., & Taplin, J. (2011). Destination planning and policy: process and practice. In: Y. Wang, & A. Pizam (Eds.), *Destination marketing and management: theories and applications* (pp. 21–38). Wallingford: CABI.

42. Laesser, C., & Beritelli, P. (2013). St Gallen consensus on destination management. *Journal of Destination Marketing & Management, 2,* 46–49.

43. Laws, E. (1995). *Tourist destination management: issues, analysis and policies.* London: Routledge.

References

44. Local Government New Zealand. (2003). *Postcards from home:the local government tourism strategy*. Wellington: Local Government New Zealand.

45. Longjit, C., & Pearce, D. G. (2013). Managing a mature coastal destination: Pattaya, Thailand. *Journal of Destination Marketing & Management, 2*(3), 165–175.

46. Marsat, J.-B., Guerra, F., & Lepinay, T. (2010). *Management stratégigue du destination touristique et management territorial: le cas du Massif du Sancy. Colloque Joint ASRDLF-AISRE: Identite, Qualite et Competitivite Territoriale* (pp. 1–25)Aosta: Association de Science Regionale de la Langue Francaise.

47. Morrison, A. (2013). *Marketing and managing tourism destinations*. London and New York: Routledge.

48. Osmankovic, J., Kenjic, V., & Zrnic, R. (2010). Destination management: consensus for competitiveness. *Proceedings of Tourism & Hospitality Management 2010 Conference*, 513–525.

49. Palmer, A. (1996). Linking external and internal relationship building in networks of public and private sector organizations. *International Journal of Public Sector Management, 9*(3), 51–60.

50. Pavlovich, K. (2003). The evolution and transformation of a tourism destination network: the Waitomo Caves, New Zealand. *Tourism Management, 24*(2), 203–216.

51. Pearce, D. G. (1992). *Tourist organizations*. Harlow: Longman.

52. Pearce, D. G. (2014a). Towards an integrative conceptual framework of destinations. *Journal of Travel Research, 53*(2), 141–153.

53. Pearce, D. G. (2014b). Gestión de destinos turísticos en hinterlands costeros y urbanos. *Revista Geográfica de Valparaíso, 48*, 57–73.

54. Pearce (in press). Urban management, destination management and urban destina- tion management: a comparative review with issues

and examples from New Zealand. *International Journal of Tourism Cities*, 1(1).

55. Pearce, D. G., & Schanzel, H. A. (2013). Destination management: the tourists' perspective. *Journal of Destination Marketing & Management*, 2(3), 137–145.

56. Pearce, D. G., & Tan, R. (2004). Distribution channels for heritage and cultural tourism in New Zealand. *Asia Pacific Journal of Tourism Research*, 9(3), 225–237. Pechlaner, H., Herntrei, M., & Kofink, L. (2009). Growth strategies in mature destinations: linking spatial planning with product development. *Tourism*, 57(3), 285–307.

57. Pechlaner, H., Volgger, M., & Herntrei, M. (2012). Destination management organizations as interface between destination governance and corporate governance. *Anatolia*, 23(2), 151–168.

58. Pike, S., & Page, S. J. (2014). Destination marketing organizations and destination marketing: a narrative analysis of the literature. *Tourism Management*, 40, 202–227.

59. Presenza, A., Sheehan, L., & Ritchie, J. R. B. (2005). Towards a model of the roles and activities of destination management organizations. *Journal of Hospitality, Tourism and Leisure Science*, 3, 1–16.

60. Prideaux, B., & Cooper, C. (2002). Marketing and destination growth: a symbiotic relationship or simple coincidence? *Journal of Vacation Marketing*, 9(1), 35–51. QLDC. (2007). *Queenstown Lakes District Council growth management strategy*: final. Queenstown: Queenstown Lakes District Council.

61. QLDC. (2009). *Wanaka town centre strategy*. Queenstown: Queenstown Lakes District Council.

62. QLDC. (2009). *Queenstown town centre strategy*. Queenstown: Queenstown Lakes District Council.

References

63. QLDC. (2013). *District Plan*. Queenstown: Queenstown Lakes District Council. QLDC. (2014). *Queenstown Lakes District Council 2014/15 Annual Plan*. Queenstown: Queenstown Lakes District Council.

64. Risteski, M., Kocevski, J., & Arnaudove, K. (2012). Spatial planning and sustainable development as basis for developing competitive tourist destinations. *Procedia- Social and Behavioral Sciences, 44*, 375–386.

65. ROS Development and Planning. (2008). *Modelos de gestión turístico local: principios y practicas*. Madrid: Federación Española de Municipios y Provincias/Secretaría General de Turismo.

66. Rotorua District Council. (2011). *Rotorua sustainable economic growth strategy*. Rotorua: Rotorua District Council.

67. Ruekert, R. W., & Walker, O. C. (1987). Marketing's interaction with other functional units: a conceptual framework and empirical evidence. *Journal of Marketing, 51*, 1–19.

68. Ruhanen, L. (2013). Local government: facilitator or inhibitor of sustainable tourism development. *Journal of Sustainable Tourism, 21*(1), 80–98.

69. Ryglová, K. (2008). Destination management. *Agricultural Economics, 54*(9), 440–448.

70. Sainaghi, R. (2006). From contents to processes: versus a dynamic destination management model (DDMM). *Tourism Management, 27*(5), 1053–1063.

71. Sheehan, L., Ritchie, J. R. B., & Hudon, S. (2007). The destination promotion triad: Understanding asymmetric stakeholder interdependencies among the city, hotels, and DMO. *Journal of Travel Research, 46*(1), 64–74.

72. Simmons, D. G., & Shone, M. C. (2002). Roles and responsibilities in tourism: managing the potential tensions caused by tourism on the West Coast. *Planning Quarterly, 145*(20-22), 28.

73. Socher, K. (2000). Reforming destination management organizations and financing. *Tourist Review, 2*, 39–44.

74. Spyriadis, T., Fletcher, J., & Fyall, A. (2013). Destination management organizational structures. In: C. Costa, E. Panyik, & D. Buhalis (Eds.), *Trends in European tourism planning and organisation* (pp. 77–91). Bristol: Channel View.

75. TCDC. (2014). *Future coromandel: Thames-Coromandel District economic development plan. Summary 2014-2018*. Thames: Thames-Coromandel District.

76. Tourism Bay of Plenty. (2013). *Tourism Bay of Plenty annual report year ending*. Tauranga: Tourism Bay of Plenty (*30th June 2013*).

77. Twyoniak, S. (1998). Le modèle des ressources et des compétences: un nouveau paradigme pour le management stratégique? In: H. Laroche, & J.-P. Nioche (Eds.), *Repenser la stratégie* (pp. 166–204). Paris: Vuibert. *Business plan 2014-15*. Invercargill: Venture Southland.

78. Waipa District Council. (2012). *Economic development strategy*. Waipa District Council.

79. Winefield, P. (2007). Strategic planningin local government – an oxymoron. *Planning Quarterly*, 5–7 (March).

80. WTO. (2007). *A practical guide to tourism destination management*. Madrid: World Tourism Organization.

81. Zahra, A. L. (2011). Rethinking regional tourism governance: the principle of subsidiarity. *Journal of Sustainable Tourism, 19*(4-5), 535–552.

CHAPTER 5

Progress and prospects for event tourism research

Donald Getz[a,], Stephen J. Page[b,]

[a] The University of Calgary, Canada
[b] Faculty of Management, Bournemouth University, UK

ABSTRACT

This paper examines event tourism as a field of study and area of professional practice updating the previous review article published in 2008. In this substantially extended review, a deeper analysis of the field's evolution and development is presented, charting the growth of the literature, focussing both chronologically and thematically A framework for understanding and creating knowledge about events and tourism is presented, forming the basis which signposts established research themes and concepts and outlines future directions for research. In addition, the review article focuses on constraining and propelling forces, ontological advances, contributions from key journals, and emerging themes and issues. It also presents a roadmap for research activity in event tourism.

KEYWORDS

Event tourism; Trends; Research; Theory; Ontology

1. INTRODUCTION

The field of event studies, reviewed by Getz, 2012a and Getz, 2012b depicts the expanding field of event management and the wider social science contribution to this interdisciplinary area of study, heralded as a major success story in terms of its educational provision within higher education, its expansion of research activity and its contribution to tourism development within the commercial arena. Within the context of tourism and the tourism system (Leiper, 1990), events comprise a key element in both the origin area (i.e. events are an important motivator of tourism) as well as within the destination area (i.e. events feature prominently in the development and marketing plans of most destinations). Events are both animators of destination attractiveness but more fundamentally as key marketing propositions in the promotion of places given the increasingly global competitiveness to attract visitor spending. To use Leiper's analogy of the tourism system, events have become a core element of the destination system where accommodation, attractions, transport and ancillary services have been utilised or specifically developed (e.g. the provision of infrastructure for mega events) to enhance the destination offer thereby expanding the tourism potential and capacity of destinations beyond a narrow focus on leisure-based tourism (e.g. holidays). Recent research (e.g. Connell, Page, & Meyer, 2015) also demonstrates the critical relationship that exists between events as a bridge between the market for visitor attractions created by tourists and the use of events to fill the gap left in the off-peak season by a seasonal drop in tourism demand, as residents and domestic visitors provide a substitutable form of demand stimulated by events. In this respect, events have a wider remit than destination-related tourism although the focus of this article is primarily on the destination-related issues of event tourism and the studies associated with this area.

Interestingly, place marketing, often referred to as boosterism (where events are used to 'boost' visitor numbers and appeal) has emerged as a key feature associated with events to develop a unique selling proposition that differentiates the destination from the competition. With its nineteenth century origins (Pike and Page 2014) event-led place marketing and

1. Introduction

development initially promoted in the USA (Ford & Peeper, 2007) has continuously utilised conventions and events to achieve key tourism and other visitation objectives linked to place development, although the analysis of its wider contribution to tourism development as both a field of study and area of critical research is a more recent outcome of the evolution of event studies (Getz, 2008). In the previous review of event tourism, Getz (2008) outlined many of the principal themes around the growth of event research and subsequent studies (e.g., Getz, 2012a and Getz, 2012b) expand upon the nature of the contributing disciplines that are coalescing to create an event knowledge base. Yet synthesising this knowledge is no easy task and this review examines the evolution, progress and future prospects for event research within a tourism context, focused on the notion of the planned event within tourism.

Planned events in tourism are created for a purpose, and what was once the realm of individual and community initiatives has largely become the realm of professionals and entrepreneurs. Fig. 1 provides a typology of the four main categories of planned events within an event-tourism context, including the main venues associated with each. Business events (or the MICE sector) require convention and exhibition centres, including numerous, smaller private parties and functions held in restaurants, hotels, or resorts. Sports also require special-purpose facilities including athletic parks, arenas and stadia. Festivals and other cultural celebrations are less dependent on facilities and can use parks, streets, theatres, concert halls and all other public or private venues. Entertainment events, such as concerts, are generally provided by the private sector and utilize many types of venue.

It was only a few decades ago that 'event tourism' as a phenomenon became established as a recognisable term within the tourism industry and research community, so that subsequent growth of this sector can only be described as spectacular. One indication of the progress can be gauged from searches of the scientific literature since 2007 when the initial Getz (2008) review was undertaken. SCOPUS results based on a search for the period 2008–October 2014 report over 1000 articles using the search terms 'event' + 'tourism', with an increasingly interdisciplinary focus within the literature. If one then assumes that even 10 references in each article are from events-related literature (and the bibliographies of these articles typically cite a much more comprehensive events related literature than that), then we see the field is drawing upon a burgeoning literature base of over 10,000 items as a conservative estimate, probably nearing 15,000 items. This makes producing a second synthesis of the research literature increasingly challenging and requiring certain parameters to be established to draw out the essence of growth in the field since 2007, the end date when the previous review was undertaken (Getz, 2008). For this reason, the review is necessarily selective in what it draws upon citing major studies in the leading interdisciplinary and specialist journals in cognate fields (e.g. tourism, leisure, hospitality) as well as the increasing move towards new areas (e.g. risk, travel medicine, history, planning and cultural studies) where the focus is related to event tourism. Clearly this requires a certain academic judgement on what to include and not include (i.e. excluding conference papers). Therefore, the purpose of the review is to demonstrate this expanding nucleus of knowledge that continues to assist in our understanding of event tourism, building upon elements of the previous article that remain valid in 2015. We deliberately exclude the debates on event studies (see Getz, 2012a and Getz, 2012b) and event management (Bowdin, Allen, Harris, McDonnell, & O'Toole, 2012) as sub-fields, deliberately structuring the review around germane concepts and approaches to event tourism that help in our synthesis of this large knowledge base that is expanding across the social sciences. Critics may point to missing themes, articles or studies and we accept that achieving a comprehensive coverage from the volume of material is impossible even in an article of this extended length.

1.1. Structure of the paper

To aid the reader, the paper is structured in three discrete sections: the conceptualisation of event tourism and then progress in the research literature and then a model of the event tourism system: We commence the first section by discussing the epistemology and ontology of event tourism to outline some of the key propositions around events tourism so as to highlights it significance to tourism, in much the same way that previous

1. Introduction

reviews (e.g. Ashworth & Page, 2011 on urban tourism) have done. Following on from the propositions, the conceptualisation of event tourism is discussed in terms of "the tourism perspective" is examined where event tourism is defined from both a demand and supply perspective, utilising Leiper's (1979) tourism system model. A number of event tourism career paths are identified (Fig. 3) which arises from the event tourism nexus. This is followed by a more focused discussion of the destination perspective in which an event portfolio model is examined (Fig. 4) and role of tourism organisations. This strategic approach can help shape evaluation, planning, and policy for events. An event-centric perspective on event tourism is discussed, referring to how many events are marketed to attract tourists. The conceptual section concludes with a brief review of major trends in the context of propelling and constraining forces, and this helps to explain the phenomenal growth of events and event tourism.

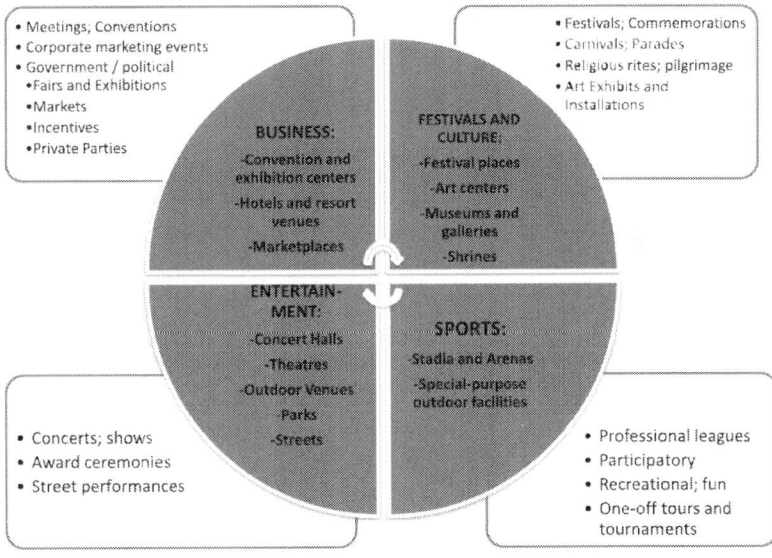

Figure 1. Typology of planned events and venues: An event-tourism perspective.

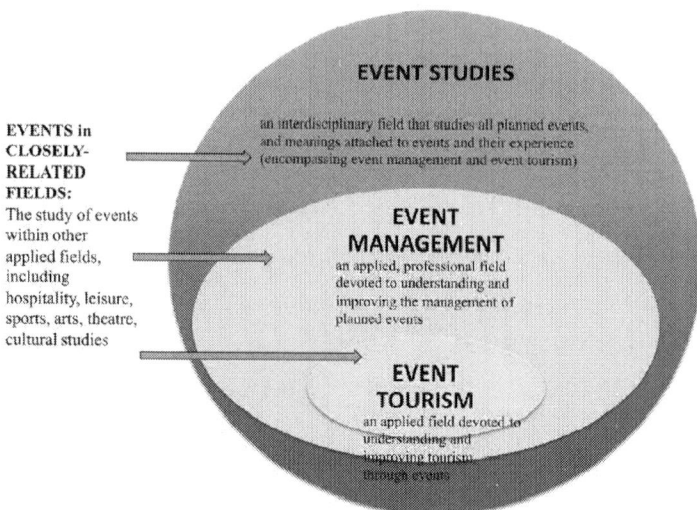

Figure 2. Event studies, event management and event tourism.

TYPICAL FUNCTIONS	MAJOR TASKS
Event Facilitator/Coordinator	-work with events in the destination to help realize their tourism potential (funding, advice, marketing) -liaison with convention/exhibition centres and other venues - liaison with sport and other organizations that produce events
Event Tourism Producer	-create and produce events specifically for their tourism value -stakeholder management (with numerous event partners)
Event Tourism Planner	-develop a strategy for the destination -integrate events with product development and image making/branding
Event Tourism Policy Analyst and Researcher	-work with policy makers to facilitate event tourism -conduct research (e.g., feasibility studies, demand forecasting, impact assessments and performance evaluations)
Event Bidding	-bid on events -develop relationships leading to winning events for the destination -conduct risk assessments and feasibility studies for each potential bid
Event Services	-provide essential and special services to events (e.g., travel and logistics; accommodation and venue bookings; supplier contacts)

Figure. 3. Event tourism career paths.

1. Introduction

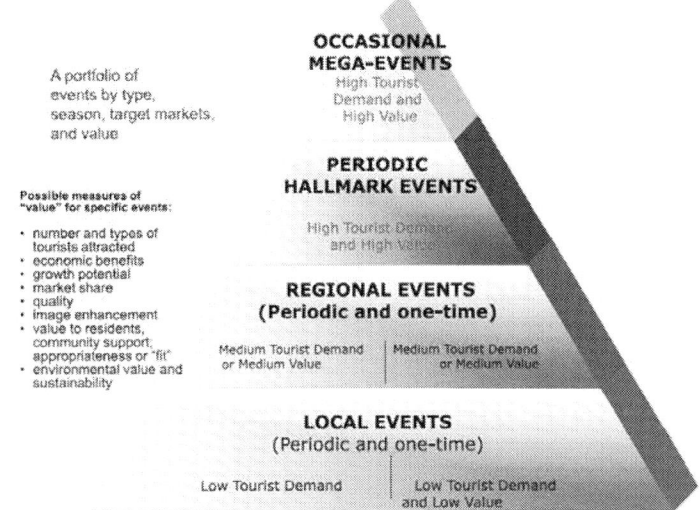

Figure 4. The Portfolio approach.

The second part of the review then reviews "Event tourism in the research literature" commencing with a review of key journals, then a chronological summary that reveals the origins and evolution of event tourism within the context of both tourism and event management. Developments subsequent to 2007 have been added, with citations, primarily from journals (although some books are referred to when they are advancing the field forward as major syntheses). A thematic approach is then taken by assessing literature specific to the four general categories of events and related venues (i.e. business, sport, festivals, and entertainment) that dominate praxis and have attracted the most attention from researchers.

The last section of the paper presents a model (Fig. 5) of the event tourism system emanating from the synthesis of the concepts and literature. The core phenomenon (event experiences and meanings) is introduced followed by the antecedents and choices (including motivation research), planning and managing event tourism, patterns and processes (including spatial, temporal, policy making and knowledge creation), outcomes and the impacted. Table 2, Table 3, Table 4, Table 5, Table 6, Table 7, Table 8, Table 9 and Table 10 provide a summary of established research themes and key concepts and terms within event tourism, categorized by reference to the main elements in the framework. Each table also considers future directions by pinpointing emerging or desired lines of research, as well as methodologies, so that these can be viewed as a research agenda. The paper concludes with a discussion of implications for the practice of event management and tourism as well as in advancing theory in event tourism.

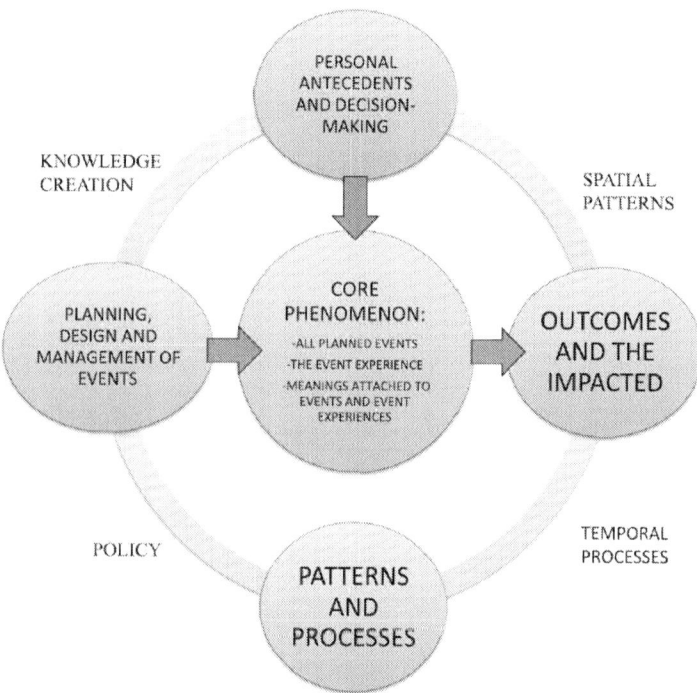

Figure 5. A framework for studying knowledge on event tourism (After Getz, 2005 and Getz, 2013a).

2. THE EPISTEMOLOGY AND ONTOLOGY OF EVENT TOURISM

From the point of view of professional praxis, or industry, we are dealing with the nexus of event and tourism management. But is there a distinct and

2. The epistemology and ontology of event tourism

substantial body of knowledge that justifies treatment of event tourism as a field of study? Ontologically, this question requires the identification of distinct claims to knowledge, core concepts and terminology, leading to the conclusion that event tourism is a sub-field of both event and tourism studies. This ontological analysis is quite different from epistemological considerations where we examine the foundation disciplines and their theories and methodologies.

A set of core propositions that started as reflections upon observable phenomena have subsequently shaped the entire event-tourism discourse. These propositions are well established in the literature, with plenty of research support, but they are not predictive or explanatory theories; they are a stage in theory development and can readily be turned into hypotheses for testing. For example, while it is obvious that events attract many tourists around the world, generating huge expenditures and helping to overcome seasonality and animate places, one cannot use this knowledge to predict the specific outcomes of any given planned event or event-tourism strategy.

The discourses associated with events and tourism and the ontological mapping of event studies have been discussed by Getz, 2012a and Getz, 2012b. In that paper consideration was given to the theoretical and methodological contributions of various foundation disciplines, along with public policy implications. This has led to the evolution of event studies that has evolved to a more social science informed subject area passing through a series of phases of growth as will be highlighted in Section 3 later. What Fig. 2 depicts is the way in which event studies now incorporates a wider perspective of event scholarship in which event management and event tourism are situated as foundation blocks for event studies. This suggests why the exploration of the nexus of event and leisure studies is of considerable relevance beyond the event tourism focus as leisure is a more all-embracing paradigm in which tourism is situated. The leisure dimensions were illustrated by Page and Connell (2010a) and Patterson and Getz (2013) where the contributing concepts and underlying focus is on entertainment, consumption and the desire to attend events in modern society.

2.1. The core propositions

The core propositions of event tourism essentially define an instrumentalist sub-field. The ontology of event tourism consists of claims to knowledge related to these propositions, together with concepts and terminology specific to them. These can also be viewed as goals to pursue or as development roles played by events. The logic is cascading, starting with the observable phenomenon that events attract tourists (See Table 1).

Table 1. Core propositions of event tourism.

a) Events can attract tourists (and others, such as sponsors and the media) who otherwise might not visit a particular place; the spending of event tourists generates economic benefits; event tourism can be leveraged for maximum value in combatting seasonality of demand, spreading tourism geographically, and assisting in other forms of urban and economic development; portfolios of events can be designed for maximum impact, especially by appealing to multiple target segments.
b) Events can create positive images for the destination and help brand or re-position cities.
c) Events contribute to place marketing by making cities more liveable and attractive.
d) Events animate cities, resorts, parks, urban spaces and venues of all kinds, making them more attractive to visit and re-visit, and utilizing them more efficiently.
e) Event tourism acts as a catalyst for other forms of desired development (including urban renewal, community capacity building, voluntarism and improved marketing), thereby generating a long-term or permanent legacy.

Embedded in the core propositions in Table 1 are many related propositions or assumptions that are rooted in event and tourism studies. Since the starting point is the observable phenomenon that events attract tourists, there has emerged research themes on motivation and what makes events needed and attractive. The benefits of event tourism are both generic to leisure and travel, and specific to special interests. Meanings attached to event tourism cover the spectrum from personal identity and development to economic, social and cultural development, but are highly dependent upon one's perspective; often it is the distribution of costs and benefits that should be of paramount concern. Distinct concepts and terms do exist in the ontology of event tourism, and these are identified throughout the paper.

2. The epistemology and ontology of event tourism

They include terms linked to the functionality of events (i.e., destination, hallmark, mega, regional and iconic event), and those related to event tourists, such as the event-tourist career trajectory.

Ontological mapping, in which claims to knowledge are identified, is quite different from the listing of things to know and skills to possess that constitute professional standards such as EMBOK (the event management body of knowledge – see www.embok.org/) and MBECS (Meeting and Business Event Competency Standards – see www.mpiweb.org/MBECS). Both these projects lend themselves to development of curriculum and certification processes, but do not establish knowledge or the means to acquire it.

2.2. The emergence of event tourism as a research paradigm

The term 'event(s) tourism' was not widely used, if at all, prior to 1987 when The New Zealand Tourist and Publicity Department (1987) reported: "Event tourism is an important and rapidly growing segment of international tourism". Getz (1989) developed a framework for planning 'events tourism'. Prior to that, conventional reference to special events, hallmark events, mega events and specific types of events dominated the discussion. Therefore, 1989 marks a watershed in the recognition of the term 'event tourism', generally recognized as being inclusive of all planned events in an integrated approach to development and marketing.

As with all forms of special-interest travel, event tourism can be categorised into demand and supply issues. 'Special' in this context refers to the experiences of tourists who are motivated by particular interests (Trauer, 2006) and not what is offered by way of products. This definition suggests that only intrinsically motivated travel qualifies, but this cannot be true as many trips to events combine intrinsic (i.e., freely-chosen leisure) and extrinsic motivations (i.e., arising from the demands or expectations of others, including work and career advancement, or in pursuit of offered rewards). A demand-side or consumer perspective (see Getz, 2013a) requires determining *who travels for events* and *why,* and also *who attends events while traveling*. Additional attributes to assess are what 'event tourists' do and spend. Included in this demand studies are the assessments of the value of events in promoting a positive destination image, place marketing in general, and co-branding with destinations. In supply terms, destinations develop, facilitate and promote events of all kinds to meet the multiple goals discussed in the core propositions. This can be called 'industry', but it is more appropriately thought of as a strategic area of tourism and place-marketing praxis (Leiper, 2008).

There is no real justification for considering event tourism as a separate field of study. The constraint is that both tourism and event studies are necessary to understand this kind of experience. There are sub areas like sport and cultural tourism (in which intrinsic motivations prevail) and business travel (mostly extrinsically motivated) that also focus on the event tourism experience (for example, see Music Events in Gibson & Connell, 2012). In a similar vein, Deery, Jago, and Fredline (2004) asked if sport tourism and event tourism are the same thing. Their conceptualization showed sport tourism as being at the nexus of event tourism and sport, with both sport tourism and event tourism being sub-sets of tourism in general. Indeed, there is almost limitless potential for sub-dividing tourism studies and management in this manner.

Event tourism is not usually recognized as a separate professional field. Mostly it is seen as an application of, or speciality within national tourism offices (NTOs) and destination marketing/management organizations (DMOs). Event development agencies (as opposed to agencies focused on protocol, arts and culture which also deal with planned events) embody event tourism completely, and there are a growing number of associated career paths or technical jobs, as illustrated in Fig. 3.

Research on event-tourism careers has been minimal. McCabe, 2008 and McCabe, 2012 has examined careers and career planning and development for the convention and exhibition industry, while Ladkin and Weber (2010) looked at the Asian context in the same sector. Baum, Deery, Hanlon, Lockstone, and Smith (2009) examined work in events and conventions. Profiles of professionals documented by Getz, 2013a and Getz,

2. The epistemology and ontology of event tourism

2013b demonstrate how direct experience in the event sector, especially organization and marketing, was essential for those occupying private and public-sector positions. A more recent review by Getz (2014) has also highlighted wider debates associated with the evolving nature of such pathways and Barron and Leask (2012) focused on the wider debates around event education. University education for event-tourism careers is similarly underdeveloped, with Indiana University – Purdue University in Indianapolis offering the only named degree in Event Tourism, being at the Masters level. Many degree programs in tourism that feature event management do combine the subjects, such as the Master's Degree in Event Management for Tourism at Universidad Europea de Madrid.

2.3. The destination perspective on event tourism: the portfolio approach

The specific role of a DMO (destination marketing organization) is generally to promote tourism – both business and leisure motivated (see Pike and Page 2014 for a detailed review). Conventions are considered business travel and participation sport events or festivals are part of leisure travel. In a study of Canadian visitor and convention bureaus (Getz, Anderson, & Sheehan, 1998), events were revealed to be one of the few areas of product development engaged in by DMOs; typically their membership (often dominated by commercial accommodation operators and attractions) want visitor demand all year round. Yet the contribution to events to inbound tourism growth through time via sport events, for example, is often one approach utilised to justify the importance of such events (see Nishio, 2013). Adopting a comprehensive portfolio approach leads to greater emphasis on creating new events and attracting them through competitive bidding although this is not without its critics as it depends upon the political economy and mandate for such bidding (see Sant, Mason and Hinch, 2013 and the Vancouver 2010 Games). Similarly, Liu, Broom and Wilson (2014) point to the legacy for Beijing 2010 and the support from residents in other cities to adopt such an approach to destination development.

The portfolio approach (see Fig. 4) is similar to how a company strategically evaluates and develops its line of products and services, or how financial assets are managed. It is goal-driven, and value-based (Ziakis 2010). Ziakas (2013:14) defined "An event portfolio ... [as] ... the strategic patterning of disparate but interrelated events taking place during the course of a year in a host community that as a whole is intended to achieve multiple outcomes through the implementation of joint event strategies." Destinations must decide what they want from events (the benefits), and how they will measure their assets' short-term and accumulating value within the prevailing political economy of public funding for such public sector interventions in a locality. Public acceptance of such strategies and localised benefits remain a contested area of study and run in parallel to the historical debates on how destinations leveraged public sector support for tourism promotion in the nineteenth century (See Pike and Page 2014) and events. Interesting studies from the history of sport have demonstrated these debates in the case of the Stockholm 1912 Olympic Games (Edvinsson, 2014) and at a local level (e.g. Noël, 2008). More recently, the debates associated with bidding for and hosting European City of Culture (Richards and Palmer, 2010) continue these debates over the use of public funds to bid for and then invest in the infrastructure in host cities to leverage long-term tourism development initially via events. Therefore, the notion of the event portfolio remains a key concept in seeking to assess "The return or benefit realised on the investment in an events portfolio is various and distributed over time" (O'Toole, 2011: 6). Portfolio management is therefore strategic and quite different from typical project management as applied to events. Asset management theory has a key role to play in event tourism, argued O'Toole, including the matter of how to dispose of non-performers.

Within the portfolio model, two terms are notable: 'Mega events', with a long history of their use to enhance tourist attractiveness and related image-making or developmental roles (e.g., Gripsrud et al., 2010 and Grix, 2012). Indeed, this was the subject of an AIEST conference in 1987. The perceived successes of mega events, including the Brisbane World's Fair and America's Cup Defence in Perth, Australia, definitely stimulated the creation of event development agencies, research, and event management programs, helping

2. The epistemology and ontology of event tourism

position Australia as a world leader. A similar consequence of staging major events has been observed in other countries, including New Zealand (Gnoth & Anwar, 2000). The other notable term is 'hallmark event' which has various meanings. Ritchie (1984, p. 2) published the first general discussion of their impacts and referred to them as "Major one-time or recurring events of limited duration, developed primarily to enhance the awareness, appeal and profitability of a tourism destination" adopting a typology that included World Fairs/Expositions, Unique Carnivals and Festivals, Major Sport Events, Significant Cultural and Religious Events, Historical Milestones, Classical Commercial and Agricultural Events and Major Political Personage Events. C.M. Hall (1989: 263) defined hallmark events this way, incorporating the key consideration of international stature: "Hallmark tourist events are major fairs, expositions, cultural and sporting events of international status which are held on either a regular or a one-off basis. A primary function of the hallmark event is to provide the host community with an opportunity to secure high prominence in the tourism market place." In his subsequent book on Hallmark Events, Hall added (1992: 1): "Hallmark events are the image builders of modern tourism …", but he also equated the term with "mega or special events".

Getz (2005, p. 16) used the term in a manner more specifically tied to image making, place marketing and destination branding where 'hallmark' describes an event that possesses such significance, in terms of tradition, attractiveness, quality, or publicity, that the event provides the host venue, community, or destination with a competitive advantage. Over time, the event and destination can become inextricably linked, such as Mardi Gras and New Orleans.

While occasional mega events are generally perceived to be a means to pump-prime or boost image, tourism, and development in general, hallmark events may, for a fraction of the cost, provide permanent benefits that are valued by the entire community (see for example, the classic sociological study by Duvignaud, 1976 on festivals). Getz and Andersson (2008) and Getz, Svensson, Pettersson, and Gunnervall (2012) argued that events that become permanent institutions have assured resources and political support, and are viewed as valued traditions, and perform essential roles within the community (see Jepson et al., 2014 and Jepson and Clarke, 2014 for recent examples of such events).

'Local' and 'regional' events, occupying the base levels of the portfolio pyramid, are problematic from a tourism perspective. Some of these events have tourism potential that can be developed, requiring investment, and some are not interested in tourism per se—perhaps even feeling threatened by it. If local events are primarily community or culturally oriented there is a good argument to be made for not exploiting them. Certainly the issue of preserving cultural authenticity (e.g. Matheson, Rimmer, & Tinsley, 2014) and local control emerges whenever tourism goals are attached to local and regional events and the Scottish Highland Games is a case in point (Brewster, Page and Connell 2009) where the tourism potential varies according to the scale of the event, its history and organising committee and its reliance upon volunteerism (e.g. Benson et al., 2014 and Wang and Yu, 2015). In addition, Brida, Disegna, and Scuderi (2014) highlight the importance of the 'local' in Christmas markets. There are also important strategic differences to be drawn between event tourism activity that has longevity versus that which is based upon a short-term fad as illustrated by Chang and Mahadevan (2014) using contingent valuation methods to distinguish between the willingness to pay for long-term performing arts compared to visual arts that were seen as faddish.

When contemplating generic event development strategies, some destinations appear to over-emphasize mega events to the detriment of a more balanced portfolio, while others pursue the promotion of one or more events as destination hallmarks to signify both quality and other brand values. Santos (2014) points to the history of Brazil with its bidding for mega events such as the 1919 South American Football Cup through to the 2014 hosting of the World Cup and 2016 Olympic Games. Santos also analysed this strategy in terms of the country's principal approach to international politics in the twenty-first century. A related strategy is to deliberately seek to elevate existing events into those with hallmark status, a process that can be said to 'institutionalize' events. This was demonstrated by Lavenda (1980)

2. The epistemology and ontology of event tourism

in the political analysis of the Carnival of Caracas, Venezuela where the old Carnival was transformed into a highly organised, European mass event. As Lavenda (1980) illustrated, the wild, rowdy and attitude of a barbaric event was pilloried in the media to justify the creation of the civilizing model which was modelled on the Carnivals in Venice, Paris and Rome (i.e. masked and with floats) to create a directed and institutionalised events in which around 30% of the population became involved. A more recent trend is for DMOs and event development agencies to create and produce their own major events as part of a sophisticated branding strategy. An important consideration is that the typology of events in the portfolio model is based on functionality; this is the degree to which certain economic, tourism or political goals can be met through hosting and marketing events. As such it represents a discourse dominated by specific developmental and political assumptions that might run counter to an events strategy based on fostering community development, culture, sport, leisure, health or other aims which may require long-term research activity to assess (e.g. see Benedict & Dobkin's, 1983 anthropology of World Fairs) for localities.

It is also possible to classify events on the basis of their place attachment, being the degree to which they are associated with, or institutionalized in a particular community or destination, epitomised in the Scottish Highland Games worldwide. Mega events are typically global in their orientation and require a competitive bid to 'win' them as a one-time event for a particular place (see Lai, 2015). By contrast, 'hallmark events' cannot exist independently of their host community, and 'local' or 'regional' events are by definition rooted in one place and appeal mostly to residents. As the number, size and significance of events and event-tourism increases, greater attention to the dynamics and management of portfolios and whole populations of events is required. Population studies have been scarce, with research in Norway leading the field (e.g. Getz et al., 2013 and Jaeger and Mykletun, 2013). The application of organizational ecology theory helps explain how the fate of individual events is dependent upon processes of competition versus co-operation, resource limits, finding a sustainable niche, or being a generalist versus a specialist in terms of resources and target markets. The event development agencies that exist in every state and major city in Australia certainly led the way in developing a collaborative model for event tourism development. For example, EventsCorp Western Australia and Queensland Events Corp (now Queensland Tourism and Events), have strategies, policies and programs for attracting, bidding, developing and assisting events primarily to foster tourism. Getz, 1997 and Getz, 2005) profiled the Queensland agency, while Getz and Fairley (2004) examined media management issues surrounding the state agency's two major 'owned' events on the Gold Coast.

A substantial part of the event tourism business of DMOs and event development agencies is bidding on events that have owners. This process has been described as a special-purpose marketplace by Getz (2004b) who studied the event bidding goals, methods and attributed success factors of Canadian DMOs. Bidding has also been studied by Emery, 2001 and Persson, 2002, and Westerbeek et al., 2002 and Berridge, 2010, and Foley, McGillivray, and McPherson (2012). Lockstone-Binney, Whitelaw, Robertson, Junek, and Michael (2014) have extended this rather neglected line of research by focussing on the roles of ambassadors in the bidding process.

Gnoth and Anwar (2000) examined New Zealand's event tourism initiatives and offered a framework for developing an effective strategy. Although it is obvious that resources have to be committed, perhaps a more important issue was determining how to measure the country's return on investment and to coordinate the various stakeholders necessary to become competitive. Getz (2003b) provided specific advice on planning and developing sport event tourism, including a case study from Seminole County, Florida, to illustrate supply, demand and process issues. Higham (2005) is an excellent source of practical planning and marketing advice on that sector, while O'Toole (2011) features festival tourism. Additional examples include the study by Getz (2013) which derived examples from EventScotland, representing a national event tourism strategy and program, Visit Denver as a city that produces its own events, and also Calgary Sports Tourism Authority and Northern Ireland Tourist Board. A systematic comparison of organizational and strategic approaches to event tourism should be a

2. The epistemology and ontology of event tourism

research priority to begin to appreciate the various approaches possible for organisations to aid organisational learning for destinations. For example, in addressing peripherality and to achieve place branding outcomes, Jutland's food and lobster festival created both local and external audiences (Blichfeldt & Halkier, 2014).

To be most effective, the DMO or event agency has to establish relationships with the event sector and individual events, hence a network approach is useful as Frew and Williams (2014) study of the tourism relationships with sport, dance and events. This reiterates many of the findings of other studies such as Whitford, 2004a and Whitford, 2004b research in Australia which documented the development of event tourism policies and programs, particularly as a tool in regional development, a classic theme that can be dated to Coppock's (1977) seminal study of tourism. In one region, Whitford (2004a) found that policies did not give adequate recognition to the roles of events in fostering regional growth, being largely socio-cultural in nature. This revealed a gap and disconnect between local authority policies and those of states and the nation that aggressively pursued event tourism for its economic benefits.

Similarly, Stokes (2004) studied the Australian event development agencies from the perspective of stakeholder networks, collaboration, and strategy making, and specifically looked at the relevance of the concept of knowledge networks. Stokes' (2008) analysis revealed the dominance of a corporate orientation in which event-related strategy and decisions were made at the state level. A 'soft' or informal network of stakeholders existed, dominated by a core of influential governmental agencies which varied depending on whether the agency was engaged in event bidding, development or marketing. This approach contrasted with the community orientation found at the regional-authority level, where more formal networking occurred between public agencies and private organizations for the purpose of actually producing events. Phi, Dredge, and Whitford (2014) also examined the role of Q method in resolving problems in the planning and management of events amongst different stakeholders (also see Andersson & Getz, 2009 for a fuller discussion of the scope of organizations involved in event tourism).

Most strategy, development and marketing research in the event-tourism sector has been developed from a supply perspective, that is focused on selling space in venues, marketing existing events, and bidding on one-time events. Achieving competitive advantage in this global marketplace requires increasing investment in infrastructure and bidding capabilities, with the result of increasing power for the owners of popular events and the division of destinations into 'leagues' defined by the size and quality of their venues and other resources. Global cities are now the only ones able to compete for the most desired mega events. A demand perspective offers greater scope for innovation and growth, potentially at much lower costs and risks. This strategy is based on market intelligence to gain greater understanding of the motivations and social worlds of special-interest groups. Leisure and sports in particular present almost unlimited potential for growth, as involvement in any sport, hobby, artistic expression or lifestyle pursuit leads to event-related travel (Getz & Patterson, 2013), often mediated by formal organizations such as DMOs, travel companies and clubs. For example, DMOs such as Visit Denver produce their own events to fill gaps in its event portfolio.

2.4. An event-centric perspective on event tourism

Many planned events are produced with little or no thought given to their tourism appeal or potential since that is not always the intended outcome. Sometimes this is due to the organizers' specific aims, and sometimes there is simply no relationship established between events and tourism. In Calgary, case study research (Getz, Andersson, & Larson, 2007) found that seven festivals were overlooked by the DMO that had limited or no interest in their tourism potential. This situation had evidently arisen because of the absence of both a tourism plan and a comprehensive events policy. It appeared that the long-standing promotion of one hallmark event, the annual Calgary Exhibition and Stampede, results in small festivals being perceived as insignificant, overshadowed in the media, and somewhat deprived of sponsorship—according to the festival managers.

Festivals and events desiring the support or cooperation of tourism agencies, or simply looking for increased respect, tend to conduct tourism and

2. The epistemology and ontology of event tourism

economic impact studies to 'prove' their value in economic terms. A significant number of such studies were also undertaken in the early new millennium by Scottish Enterprise with its involvement in using events as a means to address market failure in visitor markets and to intervene to stimulate off-peak growth (Page and Connell 2012). This saw many events reach the first stage of growth, which can broadly be labelled as becoming a tourist attraction, then the next stage would be to use that positioning to gain legitimacy or foster growth. In the context of stakeholder and resource dependency theory, events must secure tangible resources and political support to become sustainable, giving up a degree of independence in the process and creating long-term value in the event transaction and offer.

Creating and marketing events as tourist attractions, or as image makers and catalysts, requires a marketing orientation and commitment to customer service. Destination events, those that are intended to attract tourists, have to be positioned and branded in such a way as to be attractive both to those seeking generic benefits such as entertainment, socializing and escapism – often these will be residents – and those with special interests who seek very specific benefits. For example, Tkaczynski and Toh's (2014) analysis of visitor motivation at a multicultural event highlighted many of these broad benefits in their factor analysis identifying people, escape, culture and enjoyment as key elements within their segmentation study. In this context hallmark events satisfy both benefit categories, being established traditions for residents, while iconic events hold symbolic value for special-interest travellers.

Models consisting of goals and processes for developing hallmark and iconic destination events have been published in Getz et al. (2012) and Getz, 2013a and Getz, 2013b. When events are purposefully established to be tourist attractions, market intelligence – especially to understand special interest groups within their social worlds – is the key. But when they are also intended to become permanent institutions, a much broader consideration of stakeholders and community benefits is necessary as Pappas (2014) observed with the London Olympics in 2012. Yet such planned growth also has to be set against constraining forces or inhibitors that will certainly slow or halt growth. The most problematic are climate change (see Scott, Steiger, Rutty, & Johnson, 2014) and rising fuel costs which can impact leisure and tourism simultaneously. Jones (2012) argued that a combination of public-sector cuts owing to recession, higher energy costs and regulations stemming from worries about climate change make the events sector vulnerable; retrenchment might be imminent. Hall (2012), amongst others, questioned the sustainability of mega events and noted that it is unlikely that the convergence of political and corporate interests that favour mega events will adapt steady-state sustainability principles. New research perspectives on constraints or inhibiting factors are being better understood from the interdisciplinary growth of work from tourist health and safety, especially travel medicine with its focus on risk and disease. For example, van Panhuis et al. (2014) examined the risk of dengue fever among visitors to the World Cup 2014 estimating risk rates of getting the disease given the local immunity to the disease.

3. EVENT TOURISM IN THE RESEARCH LITERATURE

The ensuing discussion aims to be systematically comprehensive, leading to identification of theoretical and research themes and gaps. First, a chronological review is provided, showing how this sub-field originated and evolved. Then a thematic review is undertaken, looking specifically at types of events and 'mega-events'. Earlier literature reviews have been consulted. These include reviews of research in the event management field by Formica, 1998 and Getz, 2000b, and Harris, Jago, Allen, and Huyskens (2001). Hede, Jago, and Deery (2003) reviewed special events research for the period of 1990–2002. Sherwood's doctoral dissertation (2007) also entailed a large-scale review of pertinent literature and specifically examined 85 event economic impact studies prepared in Australia. A review by Mair and Whitford (2013), employing the Q-sort technique to elicit expert opinion, concluded that the majority of outputs had focused on economic impacts. Yet Mair and Whitford (2013) acknowledged that the most important topics for future research were socio-cultural and community impacts, then environmental impacts and sustainability issues as well as the policy dimensions of event tourism.

3.1. Review of key journals

Events are a very well established theme within tourism, with the first event-related articles appearing in the 1970s. Most tourism journals contain articles of relevance to event-tourism, but a complete review of all of them has not been attempted. Geographic coverage is expanding, in part owing to the spread of event tourism around the globe, and in a part a reflection of the growing number of scholars interested in event studies. There have been surges in research tied to mega-events in particular, and this has had the effect of greatly increasing event scholarship within, and external interest in host countries.

The most consulted tourism journals are *Annals of Tourism Research (Annals), Journal of Travel Research (JTR) and Tourism Management (TM)*. Kim, Boo, and Kim (2013) conducted a review of event-related articles published in these three, consisting of 178 research papers from 1980 to 2010. They found 25% were published in Annals, 30% in JTR and 45% in TM. Of these, the most frequently covered were sporting events (22.5%). The authors were surprised to learn that only 41% of the papers examined the event-tourism perspective (as opposed to organisational/management topics) with impact assessments dominating as well as a focus on international-scale events, such as the Olympics which have created their own genre of books and journal outputs across the social sciences. Interest in participants, residents and tourists had also increased over the 30 years, JTR having a greater focus on behaviour and economic impacts, while Annals addressed a wider range of topics, and TM published papers focused on the planning and management of event tourism. Kim et al. (2013) argued that more research was needed on the individual behavioural and psychological factors of event tourists and on non-economic impacts.

The event-focused journals are primarily management oriented, but *Event Management* (called Festival Management and Event Tourism from its inception in 1973 through the first five volumes until 2000) has contained tourism-specific articles from the very first issue. North American content has dominated, but it has become increasingly international in scope. Tourism-related content from 2012 through 2014 (vols. 16–18) covered a wide range of topics including papers on: golf, music and festival tourism, social, image and economic impacts, visitor motivation, experience and spending, event quality and tourist satisfaction, capacity building, event-tourism development and legacies, support for events, mega-event sustainability, and destination and venue selection for conventions. As examples of geographic coverage, Pechlaner, Dal Bò, and Pichler (2013) examined destination images and cultural events in Italy, Blešić, Pivac, Stamenković, and Besermenji (2013) researched music festival motives in Serbia, and Avgousti (2012) reviewed event tourism in Cyprus. Oh and Lee (2012) studied festival tourism in Korea, Trost and Milohnic (2012) looked at management attitudes towards event impacts in Croatia, Mohan and Thomas (2012) focused on team identification and sport-event travel in the USA. Kruger and Saayman (2013) undertook research on music festivals in South Africa, while Whitford and Dunn (2014) reported on research concerning indigenous festivals in Papua New Guinea and Tikkanen (2008) examined the internationalisation of such festivals. There were other articles focused on Singapore, Canada, Australia, Germany, Switzerland, Sweden, Finland, Norway, Italy, Serbia, England and Scotland. There are now many Chinese scholars studying events and tourism but China-specific publications in English-language journals have so far been few.

Convention and Event Tourism has a mandate to cover both event operations and event tourism. Given its previous name, Convention and Exhibition Management, this journal features more content on MICE or business events, and somewhat less on sports and cultural productions. The *International Journal of Event and Festival Management* does not focus on tourism, yet many articles are pertinent. Event-tourism content is also found in the *International Journal of Event Management Research* (a free, online journal) and the *Journal of Policy Research in Tourism, Leisure and Events*. Given the magnitude of sport-event tourism it is no surprise that various

3. Event tourism in the research literature

sport journals extensively cover event-tourism topics, with the *Journal of Sport and Tourism* being the most prominent.

3.2. A chronological review of the event tourism literature

Important concepts and terminology are highlighted, with particular reference to seminal contributions. The review ends with articles and books published by early 2015, including volume 18(3) in Event Management. Given the ever-increasing number of pertinent articles and books complete coverage has become nearly impossible, but more important is the identification of themes, issues and trends.

3.2.1. The formative years

Formica (1998) found a limited number of papers on events management or tourism published in the 1970s—with a total of four in *Annals of Tourism Research* and *Journal of Travel Research*. Events were not yet 'attractions' within the tourism system of Gunn's (1979) landmark book, *Tourism Planning*, although in passing he did mention 'places for festivals and conventions'. In the 1960s and 1970s the events sector was not recognized as an area of separate study within leisure (Page and Connell 2010b), tourism or recreation, all of which were rapidly growing in the academic community and in professional practice. Boorstin (1961), a historian, first drew attention to the phenomenon of 'pseudo events' created for publicity and political purposes. Attention was paid to festivals as anthropology, sociology and art. For example, Greenwood's (1972) study of a Basque festival from an anthropological perspective lamented the negative influence of commodification for tourism purposes on authentic cultural celebrations. The authenticity of events, their social–cultural impacts, and effects of tourism on events remain enduring themes as the antecedents were outlined earlier in this paper with classic studies by Benedict and Dobkin, 1983 and Duvignaud, 1976 and Lavenda, 1980). J.R.B. Ritchie and Beliveau (1974) was the first to specifically focus on event tourism illustrating how 'hallmark events' could combat seasonality of tourism demand, with a case study of the Quebec Winter Carnival. The paper included citation of an unpublished study of the economic impacts of the Quebec Winter Carnival dated 1962, which is perhaps the earliest such study recorded in the research literature. Most of the pioneering published studies were single-event economic impact assessments, notably Vaughan's (1979) study of the 1976 Edinburgh Festival and Della Bitta, Loudon, Booth, and Weeks (1978) which has now developed as the Edinburgh Festival Impact Study (Edinburgh Impact Study).

3.2.2. The 1980s

Event Tourism expanded dramatically as a research topic in the 1980s. A number of extension studies at Texas A&M focused on events and tourism including the Gunn and Wicks (1982) report on visitors to a festival in Galveston. Two notable research articles from early in this decade include those by Gartner and Holocek (1983) on the economic impact of an annual tourism industry exposition, and J.R.B. Ritchie's (1984) treatise on the nature of impacts from hallmark events, which remains a classic in terms of citations and influence on the subject. A major study of festival visitors and the economic impacts of multiple festivals in Canada's National Capital region was conducted in the latter part of this decade (Coopers and Lybrand Consulting Group, 1989), followed by a similar major study in Edinburgh (Scotinform Ltd., 1991). These remain landmarks in terms of their scope and cross-event comparisons.

By the mid 1980s, Mill and Morrison's (1985) USA-based text *The Tourism System* explicitly recognized the power of events. The 1985 TTRA Canada Chapter conference was themed 'International Events: The Real Tourism Impact' (Travel and Tourism Research Association & Canada Chapter (TTRA), 1986), with the impetus coming from the planned 1986 Vancouver World's Fair and the 1988 Calgary Winter Olympics. Internationally, the AIEST (1987) conference produced a notable collection of material on the general subject of mega events. Hall (1989) also reviewed the growing number of student dissertations in the field in the 1980s.

Australian scholars have always been prominent in event-tourism research. Prior to the America's Cup Defence in Perth in 1988, the People and Physical

Environment Research Conference, 1987, was held under the theme of the Effects of Hallmark Events on Cities and other papers (e.g. Cowie, 1985 and Shaw, 1985) examined impacts. Soutar and McLeod (1993) later published research on residents' perceptions of that major event. One of the most influential research projects of that period was the comprehensive assessment of impacts from the first Adelaide Grand Prix (Burns et al., 1986 and Burns and Mules, 1989). At the end of the 1980s, Syme, Shaw, Fenton, and Mueller (1989)*The Planning and Evaluation of Hallmark Events* was a landmark review of the research field that has seen a substantial growth. Another seminal study was also published by C.M. Hall (1989) which noted the need for greater attention to social and cultural impacts. Other key studies were by Ley and Olds (1988) on world fairs and Jackson's (1988) analysis of carnivals, a theme expanded substantially by Ferdinand and Williams (2013).

3.2.3. The 1990s

1990 was a pivotal year in the event management literature. Goldblatt's (1990)*Special Events: The Art and Science of Celebration* was published, followed by *Festivals, Special Events and Tourism* (Getz, 1991) and a year later Hall's (1992)*Hallmark Tourist Events*. In the early 1990s academics were clearly leading the way, as at that time there were few if any degree programs, and few courses available anywhere, that featured event management or tourism. In the USA George Washington University pioneered event management education, leading Hawkins and Goldblatt (1995) to address the need for event management education. They also asked how events should be treated within a tourism curriculum. The mid-to-late-1990s were the 'take-off' years for academic institutionalization of event management, and with it a more legitimized advancement of scholarship on event tourism and event studies. This process has been roughly 25–30-years behind the equivalent for tourism, hospitality and leisure. There is also no doubt that leisure, tourism and hospitality provided a large part of the foundation, by having adapted discipline-based theory and methodology, supporting event-specific courses, and by spinning off event management degree programs.

Festival Management and Event Tourism (later renamed Event Management) started publishing in 1993, and many of its articles have advanced event tourism research and theory. Uysal, Gahan, and Martin (1993) in the very first issue began an enduring discourse on why people attend and travel to festivals and events. Two other vital event tourism research themes were established early in this journal, including Bos (1994) who examined the importance of mega-events in generating tourism demand, and Crompton and McKay's (1994) on measuring the economic impacts of events. Crompton's (1999) related contributions Measuring the Economic Impact of Visitors to Sport Tournaments and Special Events continued this theme. A very large number of research projects were commenced in Australia in preparation for the Sydney 2000 Summer Olympic Games. Faulkner et al. (2000) reported on this impressive initiative; Australian research on tourism and events has remained substantial, in large part funded by the Sustainable Tourism Cooperative Research Centre (STCRC) which operated from 1997 through 2010. The permanent site www.sustainabletourismonline.com provides a repository of relevant research reports including the pioneering ENCORE Event Evaluation Toolkit.

3.2.4. After 2000

As the 20th century closed, the world celebrated with numerous special events. This undoubtedly boosted the events sector and its tourism value. Since then we have witnessed unprecedented growth in event tourism accompanied by immense investments in bidding, especially for mega events, and destinations relying on impressive new infrastructure for events. Global media coverage of events (see Ritchie, Shipway, & Chien 2010) has been facilitated by both mass and social media (e.g. see Potwarka, Nunkoo and McCarville's 2014 use of planned theory to understand television of the 2010 Winter Olympics), and the rise of a vibrant private sector in event production and marketing is noteworthy. Events and event-tourism have become mainstream, accepted as ordinary parts of contemporary lifestyle, and as legitimate research topics in many disciplines and applied fields

3. Event tourism in the research literature

reflected in the growing nucleus of articles in wider social science journals such as *Urban Studies*.

Signalling this acceptance, the peer reviewed *Routledge Handbook of Events* (Page and Connell, 2012) is a compendium largely concerned with event management topics. The Getz article on event studies discusses the event-tourism "discourse", and Weed's article on an interdisciplinary research agenda for sport, tourism, leisure and health stresses the outcomes expected of events and event tourism. Frost's article emphasises events within a destination planning and development context, with mega and hallmark events being featured, plus business events and sports. Other tourism-related topics discussed in the Handbook include events as attractions, safety and security, and tourist development and urban regeneration.

Other recent advances in the field include *Event Tourism* (Getz, 2013), designed as a comprehensive textbook on the subject, and *Eventful Cities* (Richards & Palmer, 2010), which assesses the multiple roles and planning of events in city cultures and tourism. *Tourism Economics and Policy* (Dwyer, Forsyth, & Dwyer, 2010) essentially codifies all knowledge of tourism and event-tourism economics. Books devoted to meetings, conventions and exhibitions contribute significantly, although they are seldom framed as event tourism. General and topic-specific textbooks on event management, as updated periodically, also help advance the study of events and tourism. Perhaps surprisingly, *The Routledge Handbook of Tourism Research* (Hsu & Gartner, 2012) has no content specific to events, but the six volume collection of re-printed articles called *Tourism* (Page and Connell, 2010b) includes a six-article section on Event Tourism in Vol. 6. In the editors' introduction entitled The Evolution of the Subject of Study, Page and Connell (2010b,p. xiv) argue that "The period since the 1970s saw the gradual evolution of an enduring area of study – event tourism" marking its legitimacy within the comparative recent history of tourism research. Given the vastness of relevant literature, a thematic review now follows, starting with event types and continuing through a systematic framework for understanding events and tourism, emphasising ontological development, that is key concepts and terminology.

3.3. Literature specific to event types

Although all types of planned events have tourism potential, including even the smallest wedding or reunion (see Kruger, Saayman, & Ellis, 2014), larger events dominate in the literature and in event tourism development. In this section attention is given to the four general categories illustrated in Fig. 1.

3.3.1. Business events and tourism

Interest in the tourism value of business events, including meetings, conventions, and exhibitions (both trade and consumer shows) has a long pedigree given that almost all major cities now possess impressive convention and exhibition facilities (see Boo, Koh, & Jones 2008), along with agencies devoted to selling the space and bidding on events (see Kim, Yoon, & Kim, 2011 on the competitive positioning strategies of event convention centres in East Asia). The first convention bureau in the USA was established as far back as 1896 (Spiller, 2002) and the Destination Marketing Association International traces its origins to 1915 (www.destinationmarketing.org/) as highlighted in the *Introduction*. Often referred to as the MICE industry, that is, meetings, incentives, conventions and events/exhibitions (Schlentrich, 2008), there is some doubt about the validity of including incentive tours; Fenich (2005) prefers MEEC for Meetings, Expositions, Events and Conventions. Weber and Chon (2002) assessed this sector in *Convention Tourism: International Research and Industry Perspectives*. Other books on the subject of business-event tourism include those by Davidson and Cope, 2003, Davidson and Rogers, 2006 and Davidson, 2009, and Mair (2013).

Weber and Ladkin, 2004 and Weber and Ladkin, 2009 explored trends in the convention industry including government's increasing awareness of its economic benefits. Review articles have covered convention tourism research (Yoo and Weber, 2005 and Mair, 2012) and convention and meeting management research (Lee & Back, 2005) including the tourism dimension. Lee and Lee (2014) reviewed research articles on exhibitions and discussed themes related to exhibitors in particular. Lee and Palakurthi (2013) conducted research on how constraints influenced exhibition attendance. Mair (2012) reviewed 144 articles from the business-event

3. Event tourism in the research literature

literature for the period 2000 through 2009; the vast majority of which were published in the *Journal of Convention and Event Tourism*. Mair concluded that it remained difficult to obtain adequate statistics on this sector. Major themes in this literature include the meeting planner, technology, economic impact assessments, venue selection, evaluation of satisfaction, the role of destination image on attendance, and the decision-making processes of attendees (also see Jin and Weber, 2013 and Whitfield et al., 2014). Research needs were identified by Mair (2012) focused on social and environmental impacts; climate change effects; incentive travel, and qualitative research such as the experience and meanings attached to events.

The economic value of business events has been the subject of many studies (e.g. Deng & Li, 2014; Wang, Li and Peng 2014), including at the city level (e.g., Melbourne Convention and Exhibition Centre, 2010), for rural areas (Grado, Strauss, & Lord, 1998) and for entire countries (e.g., Convention Industry Council, 2012 for the USA, and Hanly (2012) for Ireland. Dwyer (2002) discussed six facts about MICE impacts, all of which support and elaborate upon the core propositions of event tourism. The fundamental "facts" are that international convention-goers spend more and are often accompanied by others. The "yield" of event tourists is generally found to be higher than visitors with more general travel motives. In 2011 the Convention Industry Council released a study entitled *The Economic Significance of Meetings to the U.S. Economy*, revealed that "the U.S. meetings industry directly supports 1.7 million jobs, a $106 billion contribution to GDP, $263 billion in spending, $60 billion in revenue, $14.3 billion in federal tax revenue and $11.3 billion in state and local tax revenue". Interest has recently expanded to include a broader social and economic legacy, with Foley, Schlenker, Edwards, and Lewis-Smith (2013) focussing on how business events promote knowledge diffusion, networking, new collaborations leading to innovation, and educational outcomes. There can also be effects on raising awareness and profiling, showcasing and destination reputation building, and providing a platform for intercultural understanding.

It might be assumed that extrinsic motivators explain most business-event travel, that it is necessitated by doing business or to advance one's career. However, business events and pleasure travel do mix, and the connection has been examined by Davidson (2003). Research in the UK suggested that 40% of business travellers and their families or colleagues return to the hosting destination as leisure visitors in the future (Business Tourism Partnership, 2004). Furthermore, the mix of business and pleasure, particularly the lure of specific destinations, can be a crucial factor in decisions. The motives and decision-making of business-event travellers have been frequently studied, such as for trade or consumer shows (Lee et al., 2010a, Lee et al., 2010b, Park, 2009 and Rittichainuwat and Mair, 2012) and convention attendance (Jago and Deery, 2005 and Mair and Thompson, 2009). Attendee decision-making, satisfaction and loyalty are also major topics of researchers (e.g., Bauer et al., 2008, Breiter and Milman, 2006, Lu and Cai, 2011, Mair and Thompson, 2009, Severt et al., 2009, Severt et al., 2007, Tanford et al., 2012 and Yoo and Chon, 2010). Rittichainuwat, Beck, and LaLopa (2001) studied motivations, inhibitors and facilitators for attending international conferences (also see Ramirez, Laing, & Mair, 2013 on inentions to visit). Oppermann and Chon (1997) examined convention tourism from the perspectives of both association and attendees' decision making including locational factors, intervening opportunities, personal and business factors, association and conference factors, experiences and their evaluation. Ngamsom and Beck (2000) researched motivations and inhibitors affecting event-travel decisions by association members. Recent research includes a study of the Millenial Generation and what they want from meetings and events (Fenich, Halsell, Ogbeide, & Hashimoto, 2014).

MICE tourism cannot exist without special-purpose venues, but research on this critical input has been limited. Nelson (2006) and Nelson, Baltin, and Feighner (2012) examined issues surrounding public financing of convention hotels and the benefits cities can realize, while Clark (2006) looked at the additional requirements imposed on cities once a convention centre has been built. Krugman and Wright (2006) discussed special considerations for international business events, and Wan (2011) assessed Macao's competitiveness as a MICE destination. The competitiveness of the

3. Event tourism in the research literature

Italian convention industry was framed in the context of clusters by Bernini (2009). How meeting planners and associations or corporations make locational and venue decisions has been the subject of repeated studies, including those by Oppermann and Chon, 1997, Crouch and Ritchie, 1997, Jago and Deery, 2005 and DiPietro et al., 2008 and Elston and Draper (2012) (also see Meeting Planners International, n.d.).

Jin and Weber (cited in Getz, 2013a and Getz, 2013b) documented how China has become a dominant player in the Asia Pacific exhibition market, with indoor exhibition space in 2011 totalling 4.7 million square meters, or 15% of the world total. China ranks 2nd among the countries with the highest venue capacity, following the U.S.A and preceding Germany (UFI, 2011 and UFI, 2012), but actual utilization is much less than global averages. Park, Wu, Ye, Morrison, and Kong (2014) specifically studied meeting-planner perceptions of Beijing as a destination. Major international congresses attract large enough attendance for them to be considered mega events, at least from the tourism perspective; media coverage is often minimal, however. World's Fairs, or expos, certainly qualify as mega events, particularly as they often last six months and can therefore draw huge numbers of tourists as well as media.

World's fairs and their tourism connections have been examined by Mendell et al., 1983, Dungan, 1984, Lee, 1987, Hatten, 1987 and Dimanche, 1996, and de Groote (2005), yet tourism remains a minor topic in this context. Most literature on world's fairs deals with historical, sociological and anthropological themes as already discussed with reference to classic studies, urban development or advances in technology. Tourism is, however, discussed in the *Encyclopaedia of World's Fairs and Expositions* (Findling & Pelle, 2008). In a recent contribution to this sub-field, Lee, Mjelde, Kim, and Lee (2014) examined the intention-behaviour gap for attendance at a major exposition in Korea.

3.3.2. Sport events and tourism

In a recent review article, Alexandris and Kaplanidou (2014) intimated that "Sport tourism is one of the fastest growing forms of special tourism internationally", and certainly the pertinent literature on sport event tourism has been expanding at a fast pace. Sport tourism is globally well established (e.g. Presenza & Sheehan, 2013), and sports as 'big business' is an enduring theme. A growing number of books now exist on sport tourism, both theoretical and applied in nature, and sport events figure prominently (see: Gammon and Kurtzman, 2002, Gibson, 2006 and Higham, 2005; Hinch and Higham, 2011, Hudson, 2002, Ritchie and Adair, 2004, Standeven and De Knop, 1999, Turco et al., 2002 and Weed and Bull, 2009). Weed, 2012a and Weed, 2012b*Olympic Tourism* and *Sport and Tourism: A Reader* (2009) provide extensive coverage of sport-event tourism. Throughout North America, almost every city has a sport tourism initiative, often with dedicated personnel and agencies, and global competition to bid on events and attract the sport event tourist is fierce. In 1992 the US National Association of Sport Commissions was established, with the well-publicized experience of Indianapolis leading the way. The Travel Industry Association of America (TIAA) in 1997 conducted a survey that examined sport-related travel, providing vastly improved understanding of this market (TIAA, 1999). An early published contribution by Rooney (1988) remains a classic geographical analysis of sport (Page, 1990), specifically in the form of a paper on mega sport events as tourist attractions at the 1988 TTRA Montreal conference. The journal of *Sport and Tourism*, founded in 1993 (after 7 years in electronic format) as the Journal of Sport Tourism and edited by Joseph Kurtzman as an initiative of the new Sports Tourism International Council. Gibson (1998) provided the first assessment of sport tourism research and Weed (2006) reviewed the literature from 2000 to 2004, examining what exactly sport tourism is and its place in academia.

The nature of the sport tourism motivation has received considerable attention. Active and passive sport tourists were identified by Gibson, 1998 and Gibson, 2006, while Robinson and Gammon (2004) examined primary and secondary motives for sports-related travel. Nostalgia as a motivator has been examined by Fairley and Gammon (2006), and this ties in with the notion of community of interests or sub-cultures (Green, 2008, Green and Chalip, 1998 and Pitts, 1999). Petrick, Bennett, and Yosuke (2013) examined satisfaction and loyalty. Runners and triathletes have been

3. Event tourism in the research literature

studied repeatedly in the event-tourism context, including papers by Miller, 2012, Shipway and Jones, 2007 and Shipway and Jones, 2008. Motives, satisfaction and behaviour intentions of youth sport-event tourists were examined by Prayag and Grivel (2014).

According to Shipway and Jones (2007) amateur distance running is often serious leisure. Applying social identity theory and employing a quasi-ethnographic methodology, greater explanation of the running careers of participants was achieved. Berridge (2014) also employed ethnographic research to examine the event experiences of cyclists in a Gran Fondo event. Chen (2006) provided a review of the literature on sport fans behaviour, experiences and values, concluding that most studies suggest that personally relevant values (from needs and the benefits sought), and "identifications" (such as social identity) explain why fans become highly involved and committed to teams. Theoretical perspectives on the sport fan include the following: Wann (1995) and Wann, Schrader, and Wilson (1999) who developed the most frequently-cited Sports Fan Motivation Scale; Wann, Grieve, Zapalac, and Pease (2008) compared motivations across different sports. Related work has been completed by Milne and McDonald (1999) on the motivation of sport consumers, Trail and James (2001) who developed a Motivation Scale for Sport Consumption, Funk et al., 2001 and Funk et al., 2003 and their Sport Interest Inventory, a motivational comparison across sports by James and Ross, 2004 and Mehus, 2005 with an Entertainment Sport Motivation Scale, and Koo and Hardin (2008) on emotional attachment.

Funk (2008; 2012) and Funk and James, 2001 and Funk and James, 2006 employ self-concept in their approach to studying consumer behaviour for sports and events, namely as an integral part of the "psychological continuum model". Awareness, attraction, attachment or allegiances are the progressive steps open to consumers and participants. Self-concept connects specifically to identifying with a sport, event or team. Funk, Toohey, and Bruun (2007) employed structural equation modelling to demonstrate that international participation was motivated by a combination of factors including prior running involvement and a favourable image of the host destination. Weed and Bull (2009) elaborated upon and revised a Sports Tourism Participation Model (its origins being with the English Sports Council) for different products, including both spectator and participation sports tourism. A number of studies of sport events suggest that this market is quite different from cultural tourism. Leibold and van Zyl (1994) noted that sport enthusiasts attending the Los Angeles Olympics in 1984 came primarily to see the Games but generated very little revenue in dining and sightseeing. They concluded that sport tourists might be less affluent and spend less on entertainment than average travellers.

The connections between sport, events, venues and urban development or renewal have become a major theme. For example, Rozin (2000) described Indianapolis as a 'classic case' of how sports can generate a civic turnaround. Sports Business Market Research Inc. (2000, p. 167) observed that in the 1980s and 1990s American cities "put heavy emphasis on sports, entertainment and tourism as a source of revenue for the cities." Gratton and Kokolakakis (1997) believed that UK sports events had become the main platform for economic regeneration in many cities, with a wide ranging reviews apparent in Smith (2012). Carlsen and Taylor (2003) examined how Manchester used the Commonwealth Games to heighten the city's profile, giving impetus to urban renewal through sport and commercial developments, creating a social legacy through cultural and educational programming. Legacy is a major theme in Shipway and Fyall (2012).

Venues for sport events are often a controversial topic. While all communities want facilities for their own residents, there are substantial additional costs involved to be able to compete for major sport events, to subsidize professional teams, and to employ sport tourism as a tool in urban re-development. Rosentraub (2009) and Rosentraub and Joo (2009) have argued for sports as a tool is inner-city redevelopment (as opposed to locating arenas and stadia in the suburbs) and for public-private partnerships to finance them. Cities can be ranked according to their sport venues (i.e., their variety, size and quality) and their success in hosting major events, leading to competitions for titles such as Ultimate Sport City. Dolles, quoted in Getz (2013a: pp.239) observed that: "Today's modern stadia and

3. Event tourism in the research literature

large-scale events usually have an extraordinary positive impact on the host region in terms of one or more of the following dimensions: tourist volumes, visitor expenditures, publicity, and related infrastructural and administrative developments which substantially increase the destination's capacity and attractiveness." Cities might want to be renowned as a sport capital or win an accolade like "SportBusiness Ultimate Sport Cities" (www. sportbusiness. com). SportBusiness provides internationally recognized rankings of the world's top sports hosts, as initiated by independent industry consultant Rachael Church-Sanders in 2006, and such rankings reward cities possessing the best infrastructure and track records in bidding and hosting major events.

Economic impacts of sport events is a major research theme and the subject of a growing discourse on costs and benefits as illustrated in Table 2.

Researchers have been attracted to the subject of resident perceptions of impacts and support for, or opposition to sport events (see Backman, Hsu, & Backman, 2011). Lee and Krohn (2013) studied Indianapolis residents prior to hosting the Super Bowl, and this and other studies point out how perceptions of positive impacts dominate in the pre-event stage. A considerable and disproportionate amount of research has been directed at mega sport events, given their global prominence and infrastructure costs, but is somewhat irrelevant when it comes to providing useful lessons for developing an event-tourism portfolio in most cities and destinations. Historically, a great deal of attention has been paid by researchers and theorists to the Olympics. Their magnitude, political and economic importance, prominence in the media and frequent controversies make them popular subjects. Although mega-sporting events have never been more sought after, the discourse has increasingly featured scepticism over claimed benefits and legacies (see for example, reviews by Coates and Humphreys (2008) and Mills and Rosentraub (2013)). Exaggeration of expected benefits seems to be the norm, and there are few legitimate attempts made to demonstrate achievement of forecasts; costs are often hidden, and distributional effects (i.e., who gains and who pays) and externalities are typically ignored.

Table 2. Key studies on the economic impacts of event tourism.

Reviews on sport-event impacts include:
- Andersson, T., Armbrecht, J., & Lundberg, E. (2012). Estimating use and non-use values of a music festival. Scandinavian Journal of Hospitality and Tourism. 12 (3), 215–231.
- Crompton, J. (1999). Measuring the economic impact of visitors to sports tournaments and special events. Ashburn, Virginia: Division of Professional Services, National Recreation and Park Association.
- Turco, D., Riley, R., & Swart, K. (2002). Sport tourism. Morgantown, WV: Fitness Information Technology Inc.
- Gratton, C., Dobson, N., & Shibli, S. (2000). The economic importance of major sports events: A case study of six events. Managing Leisure, 5, 14–28.
- Daniels, M. & Norman, W. (2003). Estimating the economic impacts of seven regular sport tourism events. Journal of Sport Tourism, 8(4), 214–222.
- Preuss, H. (2004). The economics of staging the Olympics: a comparison of the Games 1972–2008. Edward Elgar Publishing.
- Preuss, H. (2005). The economic impact of visitors at major multi-sport events. European Sport Management Quarterly, 5(3), 281–301.
- Preuss, H. (ed.) (2007). The Impact and Evaluation of Major Sporting Events London: Routledge.
- Preuss, H. (2009). Opportunity costs and efficiency of investments in mega sport events. Journal of Policy Research in Tourism, Leisure and Events, 1 (2), 131–140.
- Higham, J. (1999). Commentary: Sport as an avenue of tourism development: An analysis of the positive and negative impacts of sport tourism. Current Issues in Tourism, 2(1), 82–90.
- Gibson, H. (Ed.). (2006). Sport tourism: Concepts and theories. London: Routledge.
- Weed, M. (2006). Sports tourism research 2000–2004: A systematic review of knowledge and a meta-evaluation of methods. Journal of Sport and Tourism, 11(1), 5–30.
- Weed, M. (2009). Sport and Tourism: A Reader. London: Routledge.
- Dwyer, L. (2002). Economic contribution of convention tourism: Conceptual and empirical issues. In K. Weber, & K. Chon (Eds.), Convention tourism: International research and industry perspectives (pp. 21–35). New York: Haworth.
- Dwyer, L., Forsyth, P., & Spurr, R. (2005). Estimating the impacts of special events on an economy. Journal of Travel Research, 43(4), 351–359.
- Dwyer, L., Forsyth, P., & Spurr, R. (2006). Assessing the economic impacts of events: A computable general equilibrium approach. Journal of Travel Research, 45(1), 59–66.
- Dwyer, L., & Forsyth, P. (2009). Public sector support for special events. Eastern Economic Journal, 35(4), 481–499.
- Dwyer, L., Mellor, R., Mistillis, N., & Mules, T. (2000a). A framework for assessing 'tangible' and 'intangible' impacts of events and conventions. Event Management, 6(3), 175–189.
- Dwyer, L., Mellor, R., Mistillis, N., & Mules, T. (2000b). Forecasting the economic impacts of events and conventions. Event Management, 6(3), 191–204.
- Masterman, G. (2009). Strategic sports event management. Abingdon: RoutledgeAbingdon: Routledge.
- Hinch, T., & Higham, J. (2011, 2nd. ed.). Sport tourism development. Bristol: Channel View.
- More methodological chapters are contained in Page and Connell (2012).

Table 2. (Continued)

- Kennelly and Toohey (2014) focused on how strategic alliances between events, sports and tour operators could enhance financial outcomes and benefits to sport tourists.
- Lee et al. (2015) have examined how repeat attendance and travel distance link to event-related expenditure.

Legacy planned as non-economic benefits is a relatively new emphasis. Harris (2014) was able to demonstrate how the Olympics in Sydney was leveraged for education on sustainability. Gibson et al. (2014) examined the legacy of the World Cup on South Africans in terms of psychic income finding significant increases in pride and euphoria post-event, while there were mixed results regarding social capital formation (also see Jamieson 2014). Less optimistically, the much-claimed benefit of increasing sport participation through the demonstration effects of sport events has yet to be proven.

The Olympics-related literature is huge, fuelled in part by Olympic research centres around the world. Numerous themes are covered as documented in Table 3.

Hosting the Olympics and other mega events has become controversial. Gursoy and Kendall (2006) found that resident support depends most on perceived benefits, not negative impacts. Prayag, Hosany, Nunkoo, and Alders (2013) addressed the issue of resident attitudes towards the London Olympics, determining that perceived economic and socio-cultural impacts (positive and negative) influence overall attitude, but perceived environmental impacts were not. Without doubt the Olympics are a fertile ground for research, but this has tended to overshadow other mega events like world's fairs and international sport championships. Mega-events, as discussed earlier, were the subject of an AIEST conference in 1987, and a conference with subsequent book edited by Andersson, Persson, Sahlberg, and Strom (1999). Roche, 2000 and Roche, 2006 widely cited papers on both the Olympics and mega events in general have been situated in an analysis of globalization. Hiller (2000b) adopted an urban sociological perspective on mega events, while Carlsen and Taylor (2003) focused on mega events and urban renewal. A special issue of *Asian Business and Management*, edited by Dolles and Söderman (2008), summarized the history of mega-sporting events in Asia. Maennig and Zimbalist's (2012) edited handbook on the economics of mega sporting events, featured articles on bidding, costs, benefits and methods.

Table 3. Key studies on the Olympics and Event Tourism.

- Economic costs and impacts (e.g., Cicarelli & Kowarsky, 1973; Glos, 2005; Kasimati, 2003; Taylor & Gratton, 1988).
- Tourism markets for Olympics have been explored by Pyo et al. (1988)
- Tourism and urban regeneration issues (Hughes, 1993)
- Tourism impacts of the Olympics (Faulkner et al. 2000; Kang and Perdue 1994; Teigland, 1996)
- Sociological perspectives (e.g. Hiller, 2012 adopted an interactionist approach to the study of Olympic host cities); Poynter and MacRury (2009) examined London as an Olympic host.
- Gold and Gold (2010) adopted an historical approach to Olympic cities, covering the period 1896 through 2016.
- Tourism marketing and Olympics was reviewed by Leibold and van Zyl (1994).
- Other topics include Olympic bids, politics, and urban boosterism (Hiller, 2000a); the Olympic legacy (Ritchie, 2000); host population perceptions of Olympics (Hiller & Wanner, 2011; Mihalik, 2001; Ritchie & Smith, 1991); sponsorship and Olympic impacts (Brown, 2002) and business leveraging surrounding the Olympics (O'Brien, 2006). Toohey and Veal (2007) took a general social science perspective to Olympic studies, and a critical evaluation of the Olympics has been provided by Waitt (2003; 2004). Dansero and Puttilli (2010) considered the Turin Olympic legacy, while Ziakas and Boukas (2012) similarly focused on Athens. Sant, Mason, and Hinch (2013) considered leveraging and the legacy of the Vancouver winter games. A special issue of the *Journal of Tourism and Cultural Change* has been devoted to the Olympics and tourism (see Ploner & Robinson, 2012).

3.3.3. Festivals and other cultural celebrations

Festivals in society and culture, pertaining to their roles, meanings and impacts, is the best developed discourse, rooted firmly in sociology and anthropology. Festivals and tourism has been reviewed in depth by Getz (2011) in a study that identified the following classical themes pertaining to festivity: myth, ritual and symbolism; ceremony and celebration; spectacle; communitas; host-guest interactions (and the role of the stranger); liminality; the carnivalesque; authenticity and commodification; pilgrimage;

3. Event tourism in the research literature

and a considerable amount of political debate over impacts and meanings. Festival tourism is a mainstream subject of research (e.g., Anwar and Sohail, 2004, Donovan and Debres, 2006, Formica and Uysal, 1998, McKercher et al., 2006, Nurse, 2004, Robinson et al., 2004 and Saleh and Ryan, 1993). Occasionally art exhibitions and tourism have been examined (e.g., Mihalik & Wing-Vogelbacher, 1992). Much of the discourse has been subsumed in the literature on cultural tourism (e.g., McKercher and du Cros, 2002, Richards, 1996 and Richards, 2007). Festivals have been examined in the context of place marketing, urban development, tourism and more recently social change (e.g., Picard & Robinson, 2006a). Some of the growth in festival numbers and variety has been attributed to diaspora (Basu, 2005 and Laing and Frost, 2013), that is the mass-migration of people who carry traditions with them.

A special issue of *Tourist Studies* on music and tourism (see Lashua, Spracklen, & Long, 2014) points to the rapid growth of music festivals and concerts as cultural and touristic phenomena. Topics covered by contributors include the festival experience, fandom, authenticity versus commercialisation, self-identity and place marketing. Gibson and Connell's (2012:16) *Music Festivals and Regional Development in Australia* observed that "Since the 1980s, in Australia and elsewhere, some music festivals have been linked to local tourism strategies, their growth nothing short of dramatic and their economic potential considerable." Pegg and Patterson (2010) conducted visitor research on one such festival that has become the hallmark event of a country town. Gelder and Robinson (2009) compared motivations at two popular UK music festivals. The vital importance of festivals in other forms of special-purpose tourism has been examined in the context of fashion (Williams, Laing, & Frost, 2013), foodies and food tourism (Getz, Robinson, Andersson, & Vujicic, 2014), and food and wine (Cavicchi & Santini, 2014).

Critical discourse is expanding illustrated by the chapters in Merkel (2013). Themes discussed by contributors include contested mega events, power and politics, events as propaganda, discourses, ideology, protests, and festival impacts. Festival tourism and 'festivalization' has become an issue in cultural studies (Quinn, 2006 and Richards, 2007). Generally framed in a negative sense, festivalization has no precise meaning, but generally authors refer to the festivalization of urban policy and spaces and resultant authenticity loss, and an overemphasis on tourist demand versus resident needs and benefits. Concern over commodification and loss of authenticity through festival tourism was an early theme, and often-cited is Greenwood's (1972) study of a Basque festival from an anthropological perspective which lamented the negative influence of tourism on authentic cultural celebrations. MacCannell (1976) is almost always cited in discussions of tourism authenticity. In another early study, Buck (1977) advocated staged tourist attractions, such as festivals, for protecting vulnerable cultural groups.

Cohen (1988) addressed commodification and staged authenticity in the context of tourism, and whether tourists could have authentic experiences, arguing that authenticity is negotiable and depends on the visitor's desires. Emergent authenticity occurs when new cultural developments (like festivals) acquire the "patina of authenticity over time". Although Cohen (1988) is not explicitly about festivals it is highly relevant, as is Cohen (2007) as it addressed the authenticity of a mythical event in Thailand. A limited number of authors have examined the authenticity of ethnic festivals, including Hinch and Delamere (1993) on Canadian native festivals that served as tourist attractions. Xie (2003) studied traditional ethnic performances in Hainan, China in terms of the relationship between commodification and authenticity. Chhabra, Healy, and Sills (2003) and Chhabra (2005) addressed authenticity issues by reference to goods sold at a festival and the perceptions of visitors. Muller and Pettersson (2006) focused on a Sami festival in Sweden, while Neuenfeldt (1995) adopted a sociological approach to the study of an aboriginal festival in Australia, viewing the performance as social text. Aboriginal Corroborees were also examined by Cahir and Clark (2010).

On the positive side are links to the preservation of traditions, group identity building and legitimation, communitas among sub-cultural groups, generation of social and cultural capital, and the value of celebrations in establishing place identity and civic pride. De Bres and Davis (2001) determined that events held as part of the Rollin' Down the River festival led to positive self-identification for local communities. Derrett (2003) argued that community-based festivals in New South Wales, Australia, demonstrate a community's sense of community and place. Costa (2002) described "festive sociability" at the Fire Festival in Valencia, Spain, as being central to the transmission of tradition. Matheson (2005) discussed festivals and sociability in the context of a Celtic music festival; the backstage space is the realm of authentic experiences and communitas. Hannam and Halewood (2006) determined that Viking themed festivals gave participants as sense of identity and reflected an authentic way of life. Other studies (e.g. the Travel Industry Association of America and Smithsonian Magazine (2003) and The National Endowment for the Arts (USA) profile events as part of traveller experience). Some festivals attract pilgrims, and others are an essential part of pilgrimage (in a religious or spiritual sense) to holy places. Ahmed (1992) studied the Hajj in terms of its tourism importance and organizational challenges. Díaz-Barriga (2003) studied a pilgrimage festival in Bolivia which has become a point for political controversy and contested meaning. Nolan and Nolan (1992) studied religious sites in Europe that act both as festival-pilgrimage and secular tourist attractions, stressing management implications. Ruback, Pandey, and Kohli (2008) compared the differences between religious pilgrims to a festival in India and non-religious visitors on their perception of the Mela. Hindu festivals at sacred sites were the subject of Shinde (2010), and Buzinde, Kalavar, Kohli, and Manuel-Navarrete (2014) interrogated pilgrims' motivations, activities and experiences of the 2013 Kumbh Mela. Matheson et al. (2014) sought to determine if spirituality motivated visitors to a Celtic-themed fire festival in Edinburgh.

Several researchers have sought to determine the marketing orientation of festivals (Mayfield and Crompton, 1995, Mehmetoglu and Ellingsen, 2005 and Tomljenovic and Weber, 2004). Carlsen and Getz (2006) provided a strategic planning approach for enhancing the tourism orientation of a regional wine festival. Why people attend festivals was quickly established as an enduring research theme (e.g., Backman et al., 1995, Baker and Draper, 2013, Brown, 2010, Chang and Yuan, 2011, Crompton and McKay, 1997, Formica and Murrmann, 1998, Getz and Cheyne, 2002, Lee et al., 2012a, Lee et al., 2012b, Lee et al., 2012d, Lee et al., 2012c, Mohr et al., 1993, Scott, 1996 and Thrane, 2002). Review articles on festival motivation have been published by Lee et al., 2004 and Li and Petrick, 2006, and Wooten and Norman (2008). Quinn (2010) examined festival tourism growth in the context of cultural policy and urban studies. Savinovic, Kim, and Long (2012) reviewed at audience motivation, satisfaction, and intention to revisit an ethnic minority cultural festival. The links between motivation, satisfaction and behaviour remain a popular topic, including a segmentation of festival-goers on the basis of psychological commitment by Lee and Kyle (2014).

Cross-cultural festival motivation research has been conducted by Kay, 2004 and Schneider and Backman, 1996, and Dewar, Meyer, and Li (2001) who collectively found that there are generic motivations. Seeking and escaping theory (Iso-Ahola, 1980 and Iso-Ahola, 1983) has largely been confirmed as explaining festival tourism motivation in many studies. Researchers have demonstrated that escapism leads people to events for the generic benefits of entertainment and diversion, socializing, learning and doing something new (i.e., novelty seeking). Nicholson and Pearce (2001) studied motivations to attend four quite different events in New Zealand: an air show, award ceremony, wild food festival, and a wine, food and music festival and concluded that multiple motivations were the norm. While socialization was common to them all, it varied in its nature. Event-specific reasons (or targeted benefits) were tied to the novelty or uniqueness of each event.

Serious leisure (as used by Mackellar, 2009 and Mackellar, 2013 based on the theories of Stebbins, 1982 and Stebbins, 2006 and involvement theory (e.g., as employed by Kim, Scott, & Crompton, 1997) offer great potential for exploring event-specific motives, connecting directly to the event-tourist career trajectory. A number of specific motivational issues have been researched. Junge (2008) looked the motivations explaining heterosexual attendance at Gay events, while Kim, Borges, and Chon (2006) employed the New Environmental Paradigm Scale to examined motivations of people attending a film festival in Brazil that was created to foster awareness of environmental issues. Yuan et al., 2005 and Yuan et al., 2008 and Dodd, Yuan, Adams, and Kolyesnikova (2006) studied wine festival attendees on their motivations, while Yuan and Jang (2008) explored the wine festival attendee's satisfaction and behavioural intentions. With regard to host-guest encounters during a festival, Giovanardi, Lucarelli, and Decosta, (2014) studied an Italian event in a mass tourism destination, and this is a neglected line of research.

Despite their enduring popularity and institutional status in many cities across North America (they are called shows in Australia and New Zealand), the tourism appeal and roles of state fairs and similar exhibitions has been an under-researched topic. Mihalik and Ferguson (1995) indicated that hundreds of millions of visitors have attended a state fair in the past 12 months making it one of the largest leisure spectator activities in the United States. State Fairs seem to capture the essence of agricultural shows, festivals, and exhibitions rolled into one. In the case of the Calgary Exhibition and Stampede (see Getz, 1993 and Getz, 2005 for cases on its marketing) it also includes a rodeo, parade and community events. In a 2013 article Lillywhite et al. provided a "portrait" of the US fair sector.

3.3.4. Entertainment

Stein and Evans (2009) defined the entertainment industry as including media (TV, radio), recorded music, video games, film, publishing, theatre, sports, theme parks, casinos and gambling, travel and tourism, museums, shopping, and special events. This is so inclusive that it defies measurement, but there is no doubt that theatre, concerts, shows and spectacles are a big part of event tourism. Certainly any aspect of sport or celebration has entertainment value, as does the spectacle associated with fireworks or parades. Anything can be entertaining that is found to be pleasurable, diverting or fun, although Hughes (2000), made the case that the "arts" are usually associated with refinement and "high culture" while entertainment performances are more mainstream, or popular. Entertainment is usually provided by the private sector, for profit, but distinctions between arts and entertainment are mostly a matter of judgement (Du Cros & Jolliffe, 2014). Research on entertainment events has lagged, perhaps because of the fact that so many events contain or feature entertainment, and many so-called festivals are really packages of concerts.

The classic study by Easto and Truzzi (1973) examined the nature of carnivals in the USA 'as an entertainment with side shows, rides, games and refreshments, usually operated by a commercial enterprise' that were travelling shows. Easto and Marcello tracked the historiography of research to reviews dating to 1881 and 1932, which were differentiated from circuses. These carnivals ranged in scale from small events to those employing up to 800 staff, with 45 railroad cars to transport the event around the USA. Easto and Marcello estimated that in the 1950s these events attracted 85 million visitors a year each time they visited towns and cities. In contrast, Reid (2006) studied the politics of city imaging in the context of the MTV Europe Music Awards in Edinburgh and Kruger and Saayman, 2012a and Kruger and Saayman, 2012b researched motivation and segmentation for music tourism in South Africa. Conversely, Che (2008) examined the case of Detroit to illustrate how entertainment districts are being employed to re-position cities.

The term "destination music" was used in a UK study of music festivals and concerts – *The Contribution of Music Festivals and Major Concerts to Tourism in the UK* by UK Music (available online at: www.ukmusic.org). From analysis of 2.5 million ticket purchases, the study conservatively estimated that 7.7 million visits were made to over 5000 festivals and concerts; music tourists made up 41% of audiences at large concerts and 48% at music festivals. Research has also been conducted in North America in the form of Attending Rock Concerts and Recreational Dancing, from the Travel Attitudes and Motivation Study (TAMS) by Tourism Canada (Lang Research, 2007). That large-scale study concluded that over the previous two years 11.8% (26,005,373) of adult Americans went to rock or pop music concerts or went recreational dancing while on an out-of-town, overnight trip of one or more nights. In some cities, notably London, Las Vegas, Toronto and New York, theatrical performances and a permanent theatre district are huge tourism attractions. Special events are a part of the attraction, ranging from touring shows to film festivals (Gilbert & Lizotte,

1998). Touring art exhibitions are associated with museums and galleries and have been studied by Mihalik and Wing-Vogelbacher, 1992 and Carmichael, 2002, and Bracalente et al. (2011). Small towns that feature theatre tourism have also been researched (Mitchell and Wall, 1986 and Mitchell and Wall, 1989). Historical re-enactments straddle the boundaries of cultural celebrations and entertainment. Carnegie and McCabe (2008) focused on how re-enactments as presentation of cultural heritage create interactions between landscapes, local communities, tourists and heritage organisations. Other studies have been undertaken by Ray, McCain, Davis, and Melin (2006) on re-enactment tourists and Wallace (2007) on war weekends.

4. A FRAMEWORK FOR KNOWLEDGE CREATION AND THEORY DEVELOPMENT IN EVENT TOURISM

Fig. 5 provides a framework for systematically studying and creating knowledge about event tourism. The discussion begins with the core phenomenon which defines event tourism as a sub field. Figure 6 through 10 summarize the discussion by listing major themes, concepts and terms associated with each element in the framework, including future research directions at the intersection of each section of the review.

4.1. The core phenomenon: event tourism experiences and meanings

Pine and Gilmore's (199)*The Experience Economy* ushered in an era of research on tourism and event experience dimensions, epitomised in marketing with the move to service dominant logic approaches (Vargo and Lusch, 2004 and Vargo and Lusch, 2008. Both the event and the travel experience have to be understood in parallel. Attending an event in one's own home community is experientially different from travelling to an event, both where travel is a necessary condition (i.e., the event motivates travel, and the costs/risks of travel might deter attendance) and where travel to an event is an integral part of a pleasurable experience. Application of theory and methods from psychology and anthropology are particularly helpful in this regard. Theorists, relying heavily on social psychology, have provided many of the insights we need, at least with regard to intrinsically motivated event tourism behaviour. Much less is known about extrinsically motivated event and travel experiences. The range of potential event experiences is quite broad, from the fun and revelry of entertainment, carnival and party, to the solemn spirituality of religious pilgrimage and celebratory rituals. Many events are about learning or aesthetic appreciation, while others foster competitiveness and commerce. For example, sport participation is about challenge, yet sport events encompass sub-cultural identity as well as nostalgia on the part of fans. Pilgrimage is a journey by definition, and generally entails a visit to a sacred site plus a special event. Other forms of event tourism can take on the form of secular pilgrimages, with events or places of high symbolic value and personal meaning becoming destinations. For example, cities that host mega events have, like Barcelona and the Olympics, turned event venues into places of pilgrimage. In the discourse pertaining to pilgrimage and event tourism, so-called secular pilgrimages (e.g., Gammon, 2004) are sometimes contrasted with religious and spiritual pilgrimages (e.g., Singh, 2006 and Timothy and Olsen, 2006), raising the issue of authenticity.

Experiences should be conceptualized and studied in terms of three inter-related dimensions (Mannell & Iso-Ahola, 1987): what people are doing, or behaviour (the 'conative' dimension), their emotions, moods, or attitudes (the 'affective' dimension), and cognition (awareness, perception, understanding). And we want to understand the event tourism experience holistically, from the needs, motivations, attitudes and expectations brought to the event, through the actual living experience (the 'doing', or 'being there') all the way to reflections on the event, including meanings attached to it and influences on future behaviour. A starting point can be the classical work of anthropologists Van Gennep (1909) and Turner, 1969, Turner, 1974 and Turner, 1982 who advanced the concept of 'liminality'. This has been found to be relevant to both travel and event experiences (Ryan, 2002).

4. A framework for knowledge creation and theory development in event tourism

In terms of one's involvement in rituals this state is characterized by humility, seclusion, tests, sexual ambiguity and 'communitas' (everyone becoming the same). 'Liminoid' described the same state but in profane rather than sacred terms, so that it might apply to carnivals and festivals, emphasizing the notion of separation, loss of identity and social status, and role reversals. In this state people are more relaxed, uninhibited, and open to new ideas. Jafari's (1987) 'tourist culture' is based on socio-anthropological theory concerning liminality and Falassi's (1987) notion of festivity as a time that is 'out of ordinary time'. Essentially, people willingly travel to, or enter into an event-specific place for defined periods of time, to engage in activities that are out of the ordinary and to have experiences that transcend the ordinary—experiences only available to the traveller or the event-goer. Csikszentmihalyi (1990) and Csikszentmihalyi and Csikszentmihalyi's (1988) concept of 'flow' or peak experiences, from leisure studies fits into this model. Facilitating 'flow' might be something the event designer wants to achieve, for maximum engagement, and something the highly involved might be more inclined to experience because of their predispositions.

Research supports the existence and importance of communitas at events. Pitts (1999) studied lesbian and gay sports tourism niche markets, and Fairley and Gammon (2006) identified the importance of sport fan communities, while Hannam and Halewood (2006) in a study of participants in Viking festivals, concluded that group identity was fostered, even to the point of establishing a 'neo-tribal' community. Green and Chalip (1998) study of women athletes determined that the event was a celebration of sub-cultural values. In a continuation of that line of research, Xing, Chalip, and Green (2014) validated a model in which social motivation is influenced by identification with the women's football subculture, which predicted the sense of community; sense of community significantly predicted spending in the destination.

The meanings attached to planned events and event tourism experiences are both an integral part of the experience and are antecedents to future event tourism behaviour. To the extent that event tourism experiences are transforming, that is they change beliefs, values or attitudes, then individuals will likely adopt new behaviours in the future. It may be that multiple event experiences are required for transformation, or it might occur as part of a social bonding. Meanings are given to events by social groups, communities and society as a whole, and are often contested. Individuals are affected by these meanings, but are also able to make their own interpretations of events. Event types are to a large extent social constructs, with collectively assigned and generally recognized meanings. Roche (2000, p. 7, see also 2006) saw events, like the global Millennium celebrations, acting as "important elements in the orientation of national societies to international or global society." Indeed, many countries have used mega events to gain legitimacy and prestige, draw attention to their accomplishments, foster trade and tourism, or to help open their countries to global influences. This is much more than place marketing—it is more like national identity building. Whitson and Macintosh (1996, p. 279) argued that countries and cities compete for mega sport events to demonstrate their 'modernity and economic dynamism. The on-going discourse on the cultural 'authenticity' of events, started with the particular concern that tourism commodifies events and corrupts their authenticity, but now also reflects the fact that many events are created for commercial and exploitive reasons (see, for example, Boorstin, 1961, Getz, 2000a, Getz, 2000b, Greenwood, 1972, Picard and Robinson, 2006a, Picard and Robinson, 2006b, Ray et al., 2006, Sofield, 1991, Xie, 2003 and Xie, 2004). However, an alternative view is that tourism helps preserve traditions and meanings, with festivals and other cultural celebrations being prime examples. In the context of foodies seeking authentic experiences, research by Getz et al. (2014) determined that festivals and other food-related events must employ interpretation to ensure that tourists understand their experiences from a local, cultural perspective.

While the event experience is a well-established theme, new theoretical perspectives and methods are being employed. Traditional consumer research is still relevant, but there is clearly a need to look deeper into the experiential realm through anthropological methods like participant observation (as employed by Getz, O'Neill, & Carlsen, 2001, at a surfing event), phenomenology (Chen, 2006) and to use experiential sampling as employed in leisure studies (Hektner and Csikszentmihalyi, 2002 and Larson

4. A framework for knowledge creation and theory development in event tourism

and Csikszentmihalyi, 1983). Coughlan and Filo (2013) employed ethnological and autoethnological research to understand participants' experiences at tourism, sport and charity events. Berridge (2014) has employed observational techniques to apply experience theory to event design and management, including an examination of the cultural experiences of cyclists. Emerging themes include the emotional aspects of event experiences (Lee and Kyle, 2012 and Robinson and Clifford, 2012), events designed to facilitate social experiences and outcomes (Nordvall, Pettersson, Svensson, & Brown, 2014), the roles of social media in marketing and shaping the event experience (Bolan, 2014 and Hudson et al., 2015), and the potential influences of technology (Sadd, 2014). Scholars have been looking at how engagement affects experience, and this is fruitful ground for theory development. For example, how do volunteering, officiating, organizing or performing in an event differ in shaping the experience? Designers are increasingly interested in how they can deliberately shape program, setting and management to heighten or lessen emotional responses and thereby affect both satisfaction and behaviour. Biometrics can be employed to test design and monitor experiences at events. These and other research challenges are outlined in Table 4.

4.2. Antecedents to event tourism

Planned events have been integral to, and endured in all societies, and it is therefore reasonable to conclude the people need events. This position is supported by the continued growth in numbers, diversity, size and ascribed significance of events and event tourism around the world. Theoretical support comes first from economic exchange theory, in which events facilitate direct exchange of goods (i.e., markets, exhibitions and fairs) plus networking, marketing, and professional-development (i.e., meetings and conventions). Hedonism and personal development occurs through different modes of participation and engagement in events, including event-tourism careers. Anthropological or symbolic exchange theory (Marshall, 1998) embodies the symbolic meanings held by events in different cultures, sub cultures, and social worlds. Social exchange theory is often used to explain attitude towards events, and perceptions of impacts, but also includes the need for socializing and group experiences. In Maslow's (1954) well-known hierarchy of needs, planned events can be viewed as mechanisms for need fulfilment at all levels: physiological (health; earning a living); safety and security (stability, order in society) love and belonging (socializing, family time, communitas, group identity building); esteem (gaining recognition and advancing one's status), and self-actualisation (through realizing one's potential, meeting challenges and gaining mastery, aesthetic appreciation, learning). Although the ability to meet needs does not in itself justify the claim that events of all types are fundamental human needs, it is obvious that events are so well embedded that doing without them is simply impossible. While it is normal to talk about push and pull factors in tourism (e.g., Crompton, 1979, Dann, 1977 and Dann, 1981), seeking and escaping theory (Iso-Ahola, 1980, Iso-Ahola, 1983 and Mannell and Iso-Ahola, 1987) offers greater explanatory power. The main proposition is that people are both seeking to find personal and interpersonal rewards and hoping to escape aspects of personal and interpersonal environments that bother or bore them.

Many personal, social and cultural factors will affect event tourism behaviour, and although there is a substantial body of literature on leisure and travel in general, the various factors specifically affecting event tourism have not been well explored. Both leisure and work-related factors have to be examined. Benckendorff and Pearce (2012) discussed the psychology of events and developed a very useful framework for applying theory to event participation, applicable to both fans and athletes. They consider pre, on-site and post-event experiences, for spectators, attendees, performers and elite participants. Psychological theory on personality, motivation and involvement are important when looking at antecedents, while role theory, identity, liminality, flow, mindfulness, emotional and performative labour, and experience analysis can be applied to the event experience itself. Satisfaction, loyalty, self-actualization and personal development apply after an event experience. Personal values have been viewed as antecedents to event tourism (Hede, Jago, & Deery, 2004), and this line of theory building connects to ego-involvement and social worlds or sub-cultures. Researchers have only recently turned their attention from general motivational studies

4. A framework for knowledge creation and theory development in event tourism

Table 4. Future research directions on the experience and meaning of event tourism.

Major themes, concepts, terms: experiences and meanings:	Future directions
1: Events + travel creates a unique experience; the destination plays an important role; travel in groups is often part of the attraction (i.e., socializing, identity building, and nostalgia)	- Limited research has been undertaken on the actual travel component of event tourism (with an event as the goal, is the trip different? consider arousal and flow within different event-travel situations)
2: Events attract tourists both for generic benefits and those appealing to special interests; destination events attract people to places they would not otherwise travel	- Continued theoretical work is needed on the nature of experience (e.g., exploring emotions through phenomenology: what makes event-tourism experiences memorable and transforming? consider each dimension of experience: conative (behaviour); affective (emotional) and cognitive)
3: All events are unique owing to combinations of setting, people and management; 'being there' for a special, time-limited experience is part of the allure	- Examine types of engagement (e.g., volunteering vs. organising) and ego-involvement as factors shaping, and influenced by event-tourism experiences; use ethnological and auto-ethnological research methods; use netnography to learn how people describe, explain and assign meaning to event-tourism experiences
4: Iconic events hold symbolic value in attracting special-interest tourists (from subcultures and social worlds); both religious and secular pilgrimage are dependent upon the symbolic meaning of places and events	- Examine how design can influence experience and behaviour and to attract tourists
5: Events can offer authentic cultural experiences, especially when hosts and guests share experiences on common ground; the widespread creation of "pseudo-events" and "staged authenticity" can confound people seeking authenticity	- Can we describe and explain the formation of personal and social constructs regarding event tourism experiences?
6: Tourism poses a threat to authenticity by commodifying culture, but can also be the mechanism for preserving traditions; emergent authenticity occurs as events become traditions	- How does communitas form and evolve at and after events? can it be facilitated?
7: The ritualistic, symbolic and celebratory meanings of events are important; individuals, groups, and whole societies value events and enjoy sharing them with visitors; events offer potential for legitimation and pride in one's place or identity	- Systematically compare different event experiences (for all stakeholders, from paying customers and guests to the general public, and between types of event, from sport to carnival)
8: Stakeholders perceive events differently in terms of meanings and value	- other under-utilized methods: hermeneutics (analysis of texts; self-reporting); direct and participant observation; experiential sampling (diary or time-sampling with standard questions)
9: One can view events as texts revealing much about host society and culture	- Biometrics offer great scope for testing reactions to design and understanding behaviour at events (e.g., monitoring the pulse, temperature, sweating)
10: Communitas, the belonging and sharing among participants or attendees, is a powerful experience and a motive for event travel	
11: Types and levels of engagement shape the event-tourism experience; events have separate appeal for fans, active participants, volunteers, officials, media, sponsors	
12: Anything can be entertaining; entertainment and spectacle can threaten the cultural significance of events	

to the issue of special-interest benefits. Mackellar, 2006a and Mackellar, 2006b research specifically addressed the differences between special-interest and generic motivations in attracting people to travel to events. Progress will follow from established lines of leisure and lifestyle research, and of necessity will utilize and adapt their theoretical constructs and methodologies.

Demand for events is notoriously difficult to predict (Mules and McDonald, 1994, Pyo et al., 1988, Spilling, 1998 and Teigland, 1996). Major events use long-term tracking studies and market penetration estimates to forecast attendance, but there have been notable failures including the New Orleans World Fair (Dimanche, 1996). An interesting study by Lee et al. (2014) examining intention-behaviour found that only 50% of those intending to visit actually did, which highlights the importance of in-depth understanding of this issue. Lee and Kim (1998) examined event forecasting, and Xiao and Smith's (2004b) study of world's fair attendance forecasting concluded with an improved approach. Boo and Busser (2006) looked at how image enhancement from events can induce tourist demand to destinations.

Motivational research in the events sector is very well established. Li and Petrick (2006) reviewed the literature pertaining to festival and event motivations and concluded from many studies that the seeking and escaping theory (Iso-Ahola, 1980 and Iso-Ahola, 1983) is largely confirmed. These are 'intrinsic motivators', with the event being a desired leisure pursuit. Researchers have demonstrated that escapism leads people to events for the generic benefits of entertainment and diversion, socializing, learning and doing something new, and just plain getting away from it all. Motivational studies are often combined with, or lead to event-consumer segmentation. The event segmentation literature has been reviewed by Tkaczynski and Rundle-Thiele (2011), with an unusual approach being taken by Tkaczynski (2013) who took into account stakeholder perceptions of their clients. The pull or seeking factors apply more to those with special interests who want a specific set of benefits offered by the event. For example, highly involved runners need events to compete in (McGehee, Yoon, & Cardenas, 2003) and professionals have to attend certain conferences because of their educational

4. A framework for knowledge creation and theory development in event tourism

content or the unique networking possibilities (Severt et al., 2007). The exact balance between generic and specific desired benefits obtained at any given event will depend on many personal factors with special considerations including family decision-making (Foster & Robinson, 2010).

A combination of an event's image and the destinations can influence the decision to travel (Kim et al., 2014 and Lai and Li, 2014). Pechlaner et al. (2013) analysed differences in perceived destination image and event satisfaction, concluding that destination image, quality of event, and customer satisfaction are highly related. Interrelationships between resident participation in events, event images and destination image in Portugal were explored by do Valle, Mendes, and Guerreiro (2012). When examining why people attend particular events for targeted benefits, it is essential to consult several underlying theoretical constructs. Recreation specialization (Bryan, 1977 and Bryan, 2000) has been examined in the context of birders attending festivals (Burr & Scott, 2004), while Lamont and Jenkins (2013) employed this construct in segmenting cyclists in a participation event. Serious leisure (Stebbins, 1982 and Stebbins, 2006) helps to explain the nature of involvement and commitment to various leisure pursuits, and it has been increasingly employed in an event-tourism context (e.g., Jones and Green, 2006, Mackellar, 2009 and Mackellar, 2013). Closely linked is social-world theory (Unruh, 1980) which has applications to event-tourist careers (Getz & Patterson, 2013). Ego-involvement theory, has been well-examined in leisure studies (see: Havitz and Dimanche, 1999, Kyle and Chick, 2002 and Kyle et al., 2007) and has been applied to event-tourism (e.g., Kim et al., 1997 and Ryan and Lockyer, 2002). Involvement is essential to the development of event-travel career theory but is also employed when examining levels of involvement with particular products, events or destinations (Filo, Chen, & Funk, 2013).

One additional area of research of significance is leisure constraints theory (Hinch et al., 2006, Jackson, 2005 and Jackson et al., 1992) which examines generic categories of constraints including the intrapersonal (one's perceptions and attitudes), interpersonal (such as a lack of leisure partners), and structural (time, money, supply and accessibility). Constraints are increasingly being addressed in the context of event tourism, with contributions being made by Van Zyl and Botha (2004) who considered the needs and motivational factors influencing decisions of residents to attend an arts festival, including "situational inhibitors". Milner, Jago, and Deery (2004) conducted a study of why people did not attend festivals and events. Perceived constraints on attending the Olympics were researched by Funk, Alexandris, and Ping (2009). Lamont and Kennelly (2011) and Lamont, Kennelly, and Wilson (2012) explicitly examined constraints within triathlon event careers, and Santos-Lewis and Moital (2013) focused on constraints to attend events across specialization levels.

Pearce's (2005) "Travel Career Trajectory" is the starting point for the hypothetical 'event travel career'. For example, there is reason to believe that business and professional practice leads to a career of necessary and/or desirable meetings and conventions, eventually resulting in a community of interest shared with others following similar career paths. The concepts of serious leisure, recreation specialization and ego-involvement suggest that many people will find intrinsic motivation to travel to events, such as amateur athletes and competitive events, or art lovers pursuing a career of volunteer experiences at music festivals. An event travel career should be evident first in terms of motivations (i.e., the underlying drive to attend events), and precise motives (for specific event experiences and events). There should be a progression through time such as participation in more and different events, looking for higher-order benefits. Geographic preferences and patterns should emerge, and this is where destinations can directly influence the process, through bidding and developing iconic and hallmark events. Hypothetically the event travel career will also be manifested in a progression from local to national and ultimately an international scale of travel. Evolving preferences for event characteristics and travel arrangements, and ultimately modified behaviour are to be expected from the dedicated and experienced event tourist (e.g., higher-level competition; travel with family and friends versus alone; combining holidays with events; behaving differently during events). Evidence supporting the concept and elements of the event travel career has been accruing, stemming from research on runners (Getz & Andersson, 2010), mountain bikers (Getz and McConnell, 2011 and Getz and McConnell, 2014) and triathletes (Lamont and Kennelly, 2012 and Lamont et al., 2012). Future research direction are outlined in Table 5.

4. A framework for knowledge creation and theory development in event tourism

Table 5. Research themes associated with the antecedents on event tourism.

Major themes, concepts and terms: antecedents	Future directions
1: There are major propelling forces shaping continued growth in the events sector, both from supply and demand perspectives: globalisation; diaspora; mass and social media; rising disposable incomes; the experience economy; destination competitiveness; the legitimation and mainstreaming of all forms of entertainment and celebration events	– Needs: do people believe they 'need' events? to travel to events? – Culture: more is needed on cross-cultural comparisons of the antecedents to event tourism as rooted in culture, sub-cultures, or social worlds and the roles of events in different lifestyle pursuits and hobbies; what is considered entertaining or socially acceptable is in part culturally determined and therefore highly variable
2: Planned events meet fundamental human needs for social, symbolic and economic exchange, plus personal development	– Economic demand for event tourism: how is it shaped by price, competition, substitution, policy and other factors?
3: Cultural differences and personal values affect perceived benefits and desired types of event experiences	– Constraints on attendance or participation in various events is an emerging research theme
4: Motivation to attend and travel to events involves both intrinsic (leisure) and extrinsic motivators; both seeking and escaping affect motives; there are typically a rage of generic (e.g. entertainment, novelty-seeking, escapism, socializing) and event-specific motives attracting people to the same events	– Event-tourist career theory needs considerable testing and refinement, including comparisons of the ways in which people get involved in leisure and work pursuits
5: Segmentation studies of event attendees are frequent, based on socio-demographic variables, resident versus tourist, expenditures, first-time versus repeat visitor, loyalty; (recently popular is the modelling of linked aspects of consumer behaviour (i.e. motivation, experience, satisfaction, as influences on future behaviour including loyalty)	– Loyalty versus novelty seeking is important to marketers and is insufficiently understood – Gender perspectives on events and tourism are an emerging theme – Post-event evaluations of experiences and effects on future intentions should be developed
6: Event-tourism behaviour is shaped in part by involvement and commitment (i.e. serious leisure, recreation specialization, sub-cultures and social worlds) and by event-tourist experience (e.g. higher levels of involvement generate event-tourist careers); personal and group constraints have to be overcome to pursue an event travel career	– how do different segments use the internet and social media to make decisions? – Longitudinal studies of event careers and constraint negotiation are absent – Both religious and secular pilgrimages appear to be on the rise – is this an important trend?
7: The decision-making and choices of convention and exhibition attendees and participants reflect a blend of extrinsic and intrinsic motivators; place is an important factor, alongside potential return on personal investment	– Although a well-established research topic, venue and location choice by professional event planners requires constant monitoring on light of changing economic and technological conditions
8: Professional event planners influence travel through their decisions on venues and locations, all designed to maximise attendance and return on corporate/association investment	

4.3. Planning and managing event tourism

Published studies exist on event tourism planning, development and marketing (e.g., Bramwell, 1997a, Getz, 2003a, Getz, 2003b, Gnoth and Anwar, 2000 and Higham, 2005), and this line of inquiry encompasses the organizations involved, stakeholder networks, policy making, goals and strategies, impacts and evaluations. Attention to event stakeholder management, partnerships and collaboration is a strong area of interest (e.g., Phi, Dredge and Whitford, 2014; Getz et al., 2007, Larson, 2002, Larson and Wikstrom, 2001, Long, 2000, Pappas, 2014, Parent and Sequin, 2007, Tkaczynski, 2013 and Ziakas, 2013), with research by Whitford, 2004a and Whitford, 2004b specifically taking a stakeholder perspective on event tourism policy making in Australia. Yaghmour and Scott (2011) also identified inter-organizational collaboration as a key success factor in the context of the Jeddah Festival in Saudi Arabia. Weed (2003) studied sport tourism policy in the context of stakeholder relationships, while the book by Weed and Bull (2009) further addresses these issues in sport tourism policy. Parent and Sequin (2007) used stakeholder theory in their study of major event failure. Irrational event planning is a topic seldom addressed (see Armstrong, 1985 and Bramwell, 1997b). It relates to the notion of civic or tourism 'boosterism' and the exercise of power.

Strategy for event tourism is a relatively new topic for scholars. Stokes, 2004, Stokes, 2006 and Stokes, 2008 and Stokes and Jago (2007) have examined this theme in Australia in relation to event-tourism strategy environment and processes. Baumann, Matheson, and Muroi (2009) addressed the effectiveness of sports-based tourism strategies in the Hawaiian context. Ford, Peeper, and Gresock (2009) examined stakeholder management in strategy-making. Pacione (2012) reviewed culture and event-led strategies being employed by cities looking for post-industrial prosperity, connected to the theme of long-term legacy building. Ziakas' (2013) analysis of event portfolios is of direct relevance, as many event development agencies and DMOs find themselves managing and marketing numerous events. Increasing attention has been given to the image-enhancement potential of events and their media

4. A framework for knowledge creation and theory development in event tourism

coverage, including how this might generate induced demand for the destination. Within a place-marketing and urban repositioning context, the role of events in creating or changing image now seems to be equal in importance to tourist attractiveness. Pertinent research has been mostly on how events might change destination image (e.g., Chalip et al., 2003, Ferreira and Donaldson, 2013, Hede, 2005, Kim et al., 2012, Kim et al., 2014, Li and Vogelsong, 2005, Mossberg, 2000a, Mossberg, 2000b, Ritchie et al., 2006, Shibli and the Sport Industry Research Centre, 2002 and Smith, 2005). Although it can be concluded from the evidence that events have an image-change effect, the measurement of media effects remains a difficult problem for evaluators. Advertising-equivalence measures are predominant, with the major shortcoming of only considering quantity and content, not impact. More attention will have to be given to the evaluation of media management as discussed by Getz and Fairley (2004). In this vein, Jutbring (2014) analysed exactly how brand values of the destination can be encoded in media coverage of events.

Co-branding events and destinations is a related topic (Chalip and Costa, 2006 and Jago et al., 2003). Arellano (2011) examined how the staging of the New France Festival related to branding of the Province of Quebec. The leveraging of events for additional benefits is a growing concern (Chalip and Leyns, 2002, Chalip and McGuirty, 2004, Gratton et al., 2000, Karadakis et al., 2010, Morse, 2001 and O'Brien, 2006). These topics also connect to the goal of generating a lasting event legacy (Dimanche, 1996, Hall, 1994, Mihalik, 1994 and Ritchie, 2000) and key research themes on planning and managing event tourism are outlined in Table 6.

4.4. Patterns and processes

The key themes are reviewed in Table 7.

Table 6. Research themes on planning and managing event tourism.

Major themes, concepts and terms: planning and managing event tourism	Future directions
1: The typical goals of event tourism are derived from a set of core propositions: attract tourists and stimulate new spending; combat seasonality; spread tourism spatially; generate positive images and co-brand with destination; be catalysts for development and enhanced marketing; generate a long-term, positive legacy.	– Conduct more case studies and cross-case analysis of event planning and destination strategies; encourage benchmarking among destinations – The dynamics and health of populations and portfolios of events (managed or not) is largely unknown; how is a sustainable niche defined and achieved for events and event-tourism destinations?
2: Demonstrating the public good arising from intervention is essential to the justification of event tourism	– A key question is how to increase rationality and professionalism in event development, bidding and hosting?
3: Civic/national boosterism and irrational decision-making are frequent explanations for bidding on or creating new events	– There is a need for measures and methods to evaluate long-term, cumulative and synergistic impacts of event tourism
4: Events can be classified on the basis of tourism-related functionality: destination event; mega, hallmark and iconic; regional and media events	– Determine how iconic events form and evolve (they must be symbolically important within social worlds and sub-cultures)
5: Events are often favoured because they offer a quicker and cheaper form of attraction development, however major events are dependent upon venues, especially purpose-built convention and exhibition centres and sport arenas and stadia; cities and large resorts consequently hold competitive advantages	– Assess how mass and social media influence perceptions of, and attitudes toward the costs and benefits of event tourism – Both practical and theoretical work is needed on event populations and the planning and managing of portfolios: what is a healthy population? (requires application of various theories from organizational ecology); success and sustainability measures are required; interactions of stakeholders in overlapping portfolios have not been studied
6: While it is desirable to sustain a comprehensive portfolio of permanent events, gaps can be filled through bidding within a specialized marketplace; these are called biddable, winnable, or one-time events	– For DMOs and other organizations examine stakeholder relations and power; how to foster bottom-up strategies and development is a key question (incorporating community development)
7: Planning for a long-term or permanent legacy is overtaking short-term measures of economic impact in justifying event tourism; halo effects relate to the short-term image boost; quantum leap means using events to accelerate growth; capacity building requires consideration of cumulative, sustainable benefits; repositioning stems from the exploitation of events in re-branding a destination; leveraging applies to a variety of methods intended to increase visitor spending and longer-term trade or development gains	– Measure the effectiveness of destination event-tourism strategies including co-branding and leveraging efforts – Study innovation processes and measures of success (achieving, defining and sustaining competitive advantages)

4. A framework for knowledge creation and theory development in event tourism

Table 7. Future research themes associated with dynamic processes in event tourism.

Major themes, concepts and terms: dynamic processes	Future directions
Spatial:	- Population dynamics and the health of event populations is a new area for researchers (start by conducting a census of events in a given area and looking for resource dependencies) - Future studies should be applied to event tourism, resulting in scenarios for strategy and theory-building purposes - For the life cycle, determine factors which factors most shape their evolution? - Capacity or saturation: do communities or destinations inevitably reach event and event tourism saturation? can populations be managed to ensure growth? - Adaptation strategies: how do events adapt to changes in their environment? (connect to niche theory in population ecology); - Why events fail is a largely unexplored issue
1: The distribution of festivals as linked to human–resource interactions such as agriculture and other resource exploitation	
2: Central-place theory applies (i.e., the concentration of events and event tourism in larger centres with venues and tourism infrastructure); core-periphery economics is a related construct as it is difficult to spread positive benefits away from main event venues	
3: Hierarchies of events appear to develop naturally, ranging from numerous local events through fewer regional events and a relatively small number with global attractiveness (i.e., the size pyramid)	
4: Distance-decay: travel distance, time and cost affects event-tourist demand and shapes market areas	
5: Events have zones of influence measured in newly generated travel, displacement effects, and short-term and long-term impacts	
6: Capacity (there are limits to event numbers or size in peak time periods and particular areas)	
7: Events play a role in fostering place identity; Hallmark events are dependent upon their place of origin — they are 'attached' or 'anchored'	
Temporal:	
1: Festivals tend to concentrate in good-weather and holiday periods (as in the peak festival season)	
2: Iconic events have the power to attract dedicated event tourists in the off-peak	
3: Event life cycles have been explored, raising concern for renewal or planned obsolescence	
4: The sustainability of events and event-tourism portfolios is of increasing importance	
5: Time switching is an essential consideration when attributing tourism demand and expenditure to events	

Policy:

1: Events and event tourism have been legitimized as policy instruments across economic, social, cultural and environmental domains; this has led to criticism of the 'festivalization' of urban policy and spaces and the generation of sameness (or loss of authenticity)
2: Justification of event tourism usually rests on purported tourism and economic benefits; longer-term legacy effects are of increasing significance
3: Events focus attention on how power, the interests of elite groups, and irrational decision-making shape policy and underscore decisions

- What are the ways in which stakeholders exercise power, and negotiate, to develop event tourism and related policy? who gets excluded or marginalised?
- How can an open, public discourse on costs and benefits of event tourism be fostered?
- Governance: how can public–private policymaking be made to work?
- Evaluation: how do we know when event tourism policies are effective and efficiently administered?
- Which justifications for public involvement in event tourism are supported, and why?
- What are the ideological foundations of event tourism policy?
- Each of the core propositions of event tourism should lead to interdisciplinary theory development
- Ontology: continued assessment of claims to knowledge, concepts and terms will help develop the field
- Epistemology: advances are required in adapting methodology from foundation disciplines and closely-related fields

Knowledge creation:

1: Many well-established lines of research support the core propositions of event tourism
2: There has been an over-emphasis of knowledge creation in the realm of economic impacts, although the discourse has been broadening quickly
3: Although fashionable, a consumerist perspective on event demand and decision making (i.e., modelling simple expressed motives with measures of satisfaction and future intentions), overlooks fundamental needs, event meanings, and complex antecedents
4: Interest in events continues to expand within all closely-related fields and many foundation disciplines, giving rise to new theoretical perspectives and useful methodologies

4.4.1. Spatial

The geography of events is fertile ground for both researchers and marketers. Getz (2004a) reviewed the pertinent literature, while Higham and Hinch (2006) provided an overview of the geography of sport and tourism. Janiskee, 1994 and Janiskee, 1996) ground-breaking contributions to event geography have to be acknowledged although his papers mostly examine the spatial and temporal distribution of festivals and what caused these patterns, not travel to events. Janiskee demonstrated the connection to resources, as in agricultural products that gave rise to festival themes. He also addressed the question of whether or not a region or a time-spot could reach its capacity in terms of event numbers. More theoretical research on the notion of event spaces and how they are transforming the everyday to something special remain a fertile area for research developing the interconnections of anthropology, sociology, urban studies and behavioural geography. Supply-demand interactions are fertile ground for event geographers (e.g. see Hall & Page, 2012). Analysis and forecasting of demand for a particular event or a region's events will in part depend on population distribution, competition, and intervening opportunities. Along these lines, Bohlin (2000) used a traditional tool of geographers, the distance-decay function, to exam festival-related travel in Sweden. He found that attendance decreased with distance, although recurring and well-established events have greater drawing power. Market potential for events was examined geographically by Wicks and Fesenmaier (1995). The market areas and tourist attractiveness of events have also been studied by Verhoven, Wall, and Cottrell (1998) employing demand mapping, and by Lee and Crompton (2003). Travel cost analysis as a measure of an event's economic value was addressed by Prabha, Rolfe, and Sinden (2006). Lee, Jee, Funk, and Jordan (2015) determined that a significant difference in expenditure patterns existed between first time and repeat attendees, as well as between long haul and short haul travellers at an annual event in Miami, USA.

Getz (1991) illustrated several models of potential event tourism patterns in a region. One option is clustering events in service centres, as opposed to dispersing them over a large, rural area. These are related to the concept of attractiveness and also have implications for the distribution of benefits and costs. Analysis of the zones of influence of events has been undertaken by Teigland (1996) specific to the Lillehammer (Norway) Winter Olympics, and this method has implications for event planning, especially regarding mega events with multiple venues. The elements of these zones of influence are the gateways, venue locations, tourist flows, transport management, and displacement of other activities. Daniels (2007) applied central-place theory to a sport event, finding that the economic benefits were quite different in the two adjacent counties that were co-hosts. The larger population centre received most of the economic benefits because it was able to accommodate the service requirements of visitors. Major events also motivate people to travel to one place as opposed to another, so that during the World's Fair (Expo) in Vancouver, Canada (1986) normal travel patterns were disrupted – Vancouver and British Columbia gained, but the rest of Canada lost traffic (Lee, 1987). The spatial distribution of costs and benefits is of particular interest in event geography, and so too are issues of social equity. Two very specific geographic questions are those of defining the region for which economic benefits are to be estimated, and measuring the spatial distribution of spending by visitors as examined by Connell and Page (2005). Sherwood (2007) obtained data for mapping travel to events in order to assess an event's 'energy footprint'. Recent geographic segmentation studies include those by Warnick, Bojanic, Mathur, and Ninan (2011) and Bojanic and Warnick (2012). Smith (2009) pointed out the need for conscious leveraging to spread benefits of event tourism from core to peripheral areas.

More recent emphasis on rural and regional development through tourism and event strategies has been of particular interest to scholars in Australia. Moscardo (2008) examined the roles of festivals and events in regional (i.e., rural) development. Her content analysis identified themes associated with the effectiveness of festivals and events in supporting regional development, including building social capital, enhancing community capacity, and supporting non-tourism-related products and services. A landmark book edited by Gibson and Connell (2011) documents the spread and nature of festivals and their roles in revitalising rural economies and towns, together with analysis of related issues. Gibson and Connell (2012) then focused on

4. A framework for knowledge creation and theory development in event tourism

music festivals in a separate book, including profiles of towns that had employed them as branding and development tools. Population-level analysis of festivals in three Norwegian counties revealed interesting processes of legitimation and growth of the sector, as well as spatial patterns (Andersson, Getz, & Mykletun, 2013). Andersson et al. (2013) found that large events are small in number compared to numerous small events, reflecting a clear hierarchy, and that there were more small events spread over remote, rural areas (on a per capita basis) compared to the density of events in cities.

Ecological-impact issues of event tourism can benefit from geographical analysis, such as how to tip the balance between private and transit-based event travel. Capacity to absorb tourism is a well-established theme, but capacity to absorb events is not. A starting point is population ecology and the inter-related issues of competition for resources, finding a market niche, and relationships with the host community and other stakeholders. This requires more complex analysis of an area's capacity for venues and other infrastructure, and how event tourism alters the landscape and ecological systems. Events help shape and define spaces, both private and public. Higham and Hinch (2006) considered how sport relates to place identity, attachment and dependence. Place identity and attachment linked to events have also been examined by De Bres and Davis, 2001, Derrett, 2003 and Robertson and Wardrop, 2012 and Laing and Frost (2013). Further research is needed in the area of place dependence, as in the context of or the permanence of hallmark events and why some places foster them while others do not to challenge the assumption that an event tourism strategy will automatically lead to a growth in visitors.

Technological advances have made it possible to track movements at the micro level, adding a new dimension to spatial studies. For example, Pettersson and Getz (2009) and Pettersson and Zillinger (2011) reported on how global positioning has been combined with other methods to gain insights on event-goer movements and the identification of experiential hot spots (also see Nilbe, Ahas and Silm, 2014 on the travel distances of visitors). Event tourists are now aided by a considerable number of social-media channels, formal and informal information sources, and fan-zone entertainment. Drawing upon the concepts of landscape, servicescape and experiencescape (see O'Dell and Billing, 2005) a relatively new line of research and theory-building is forming around the concept of festivalscape (or eventscape). In this context, Nelson (2009) considered how the event experience can be enhanced through design, and Yang, Gu, and Cen (2011) brought together emotion and perceived value in testing the moderating effect of festivalscape on behavioural intentions. Mason and Paggiaro (2012) applied the concept to food and wine events. As the event experience is shaped by interactions of place, setting, people and management, the implications for design and marketing are substantial.

4.4.2. Temporal issues

Seasonality of demand is the main temporal theme in event tourism, starting with the classic Ritchie and Beliveau (1974) research paper through to the detailed analysis by Connell et al. (2015). Events are one important way in which destinations can combat low tourist demand, yet as revealed by Janiskee, 1996 and Ryan et al., 1998, and others there is in most destinations a pronounced peaking of events in the high summer season, thereby presenting a challenge to the DMO. Yoon, Spencer, Holecek, and Kim (2000) undertook one of the few studies of the seasonality of the event tourism market, in Michigan. Higham and Hinch (2002) looked at sport events as an answer to seasonal tourist demand. Displacement of residents and other tourists is an occasionally researched temporal/spatial issue in event tourism (e.g., Brannas and Nordstrom, 2006 and Hultkrantz, 1998). This occurs when an event fills up available accommodation, or when publicity leads to the perception of crowding or high expense and this causes people to leave town or stay away. Obviously it is a major reason for bidding on or creating events in the off-peak tourist season. Displacement is also a critical consideration in estimating economic impacts of events.

The event life cycle, both in terms of changing market appeal and long-term sustainability or institutionalization, is an important temporal theme that has received attention by researchers (see Beverland et al., 2001, Frisby and Getz, 1989, Getz, 1993, Getz, 2000a, Getz, 2000b, Getz and Frisby, 1988,

4. A framework for knowledge creation and theory development in event tourism

Richards and Ryan, 2004, Sofield and Li, 1998, Sofield and Sivan, 2003 and Walle, 1994). Within a portfolio approach some thought has to be given to the image and freshness of events appealing to specific market segments, and the attractiveness of the overall mix of events. This relates to population ecology theory in the sense that the health of the portfolio is probably more important than the sustainability or appeal of individual events—but only in a strategic marketing sense, and not necessarily in terms of social and cultural factors. Why events fail (Carlsen et al., 2010 and Getz, 2003a) is a related line of research that is in need of progress, partially addressed by Connell et al. (2015). Time switching is an important issue in event tourism, being the propensity of people to alter the timing of their travel plans to take in an event. They are not necessarily attracted to travel because of the event and therefore their spending cannot be considered a benefit of the event (this is part of the general 'attribution problem' in event impact assessment; see for example Dwyer et al., 2000a and Dwyer et al., 2000b).

4.4.3. Policy for event tourism

Hall and Rusher (2004, p. 229) concluded that "there still remains relatively little analysis of the political context of events and the means by which events come to be developed and hosted within communities." However, interest in the policy dimension has grown rapidly, no doubt reflecting both the magnitude of the event sector and related controversy. Foley, McGillivray, and McPherson (2011) provide considerable insights to event-tourism policy, notably in the Scottish context where EventScotland and Creative Scotland both pursue development strategies, in effect generating overlapping festival and event portfolios. Getz (2009) argued for a new events policy paradigm combining sustainability and social responsibility, to which Dredge and Whitford (2010) responded and argued for a more nuanced approach. Dredge and Whitford (2011) subsequently examined governance issues, or how stakeholders participate, arguing for public-private decision-making in which stakeholders deliberate on and take action to achieve common goals. Policy and strategy is often formulated at the level of local authorities or municipalities, and Whitford, 2004a, Whitford,

2004b and Whitford, 2009 examined local authority policy towards events in Queensland, Australia, where it was mostly a top-down process. Pugh and Wood (2004) focused on the strategic use of events in UK local authorities, while Reid (2006) looked at the politics of city imaging surrounding an event, and Thomas and Wood (2004) discussed event-based tourism and local government in the UK. O'Sullivan, Pickernell, and Senyard (2009) discovered a gap between the expressed socio-cultural reasons for supporting festivals and events in Wales and the reliance on economic outcome performance measures.

The urban studies and policy literature has generated interesting perspectives such as Gotham (2002), examining Mardi Gras in New Orleans from the perspective of place marketing, commodification, spectacle, globalization and political economy. Gladstone (2012) referred to New Orleans events in discussing policy "regimes" and how they determine tourism strategies. Merkel (2013) indicates that the emergence of new lines of critical theory are now apparent. Weed's (2003) research in the UK revealed tensions between the two communities of sport and tourism including funding and resources, top-down policy-making, organization and professionalization, internal focus, and project-based liaison. Results showed how development of this policy network (i.e., sport plus tourism) can be made sustainable. Devine, Boyd, and Boyle (2010) also examined the sports tourism policy arena, highlighting the need for collaboration. Stakeholder analysis, networking and power relationships have been conceptualized as a political market square by Larson and Wikstrom (2001) and Larson (2002), while Larson (2009) employed the metaphors of the jungle, the park and the garden to highlight the varying dynamics of event networks. Robertson and Wardrop (2012) developed a conceptual model called the "spatial domain of politics and policy relating to festivals and other events" which notes the many interrelated policy aims connected to six dimensions: quality of life; place identity; culture; tourism; economy, and social capital. Scotland is used as a case in point, and EventScotland has also been profiled in Getz (2013a) as an exemplar of event-tourism strategy, organisation and development at the national level.

4. A framework for knowledge creation and theory development in event tourism

Research on social-cultural impacts and resident perceptions and attitudes has been evolving into research on resident support or opposition (e.g., Boo et al., 2011 and Richards et al., 2013). Open, public discourse on the costs and benefits of events and event tourism is rare, and generally ill-informed. The proponents of mega events in particular either shut down dissent or cynically manipulate debate through emphasis on intangibles like national pride. Tourism is often appropriated by the proponents as a major beneficiary, allegedly leading to jobs and prosperity, yet opportunity costs are ignored. Full accounting is absent after the event. Overall, the most crucial public-policy issue facing event tourism is that of justifying intervention, whether in the form of direct provision of events, building and subsidizing event venues, bidding on one-time events, or managing sustainable event portfolios. Unless 'public good' can be demonstrated through complete transparency and accountability on the part of all the agents, event tourism is likely to garner increasing opposition for its more obvious costs and perceived problems.

4.4.4. Knowledge creation

Knowledge creation in this field has largely been ad hoc and fragmented among diverse interest groups. Review articles like this one have as one of their main purposes the integration of pertinent literature, as do the growing number of textbooks. However, research on the process and actors in knowledge creation for event tourism is largely absent. Stokes (2004) examined knowledge networks in the Australian events sector. In advancing knowledge a number of important actors have to be involved, and perhaps some new collaborative processes developed. Event and tourism studies, like other immature fields of inquiry, are mostly multi-disciplinary in nature, drawing theory, knowledge, methodologies and methods from many established disciplines. It is also accomplished indirectly, by drawing on closely related professional fields like leisure studies. When two or more disciplinary foundations are applied to the problem we enter the realm of interdisciplinary research, with the long-term goal being to establish unique, interdisciplinary theory and knowledge. Anyone undertaking research on events should view the established disciplinary perspectives as a legitimate starting point. Even if the research problem is rooted in a policy or management need, it is highly possible that geography, economics, or another discipline already provides a solid foundation for conducting the research. However, within these disciplines the study of events and tourism is often incidental to a broader issue or theoretical problem and Table 7 outlines some of the future research themes that need to be addressed in terms of dynamic processes.

4.5. Outcomes and the impacted

Event tourism is primarily driven by the goal of economic benefits, but we need to examine outcomes and impacts at the personal and societal levels, and also in terms of cultural and environmental change. Event tourism should be viewed in an open-system perspective, identifying 'inputs' (what it takes to make events happen, including the costs of bidding, facility development and marketing), 'transforming processes' (events as agents of change), and 'outcomes' (desired and undesired impacts, including externalities). Depending on one's perspective, outcomes and change processes might be interpreted as a positive or negative impact. As so much research and applied work has been devoted to economic impacts, other outcomes were neglected for many years, along with development of suitable and convincing measures of intangible impacts and event value or worth. However, social and cultural outcomes are now fairly-well understood and a range of indicators are available. And although the environmental effects of events and tourism are also being addressed by researchers, incommensurability is hampering the utilization of a triple-bottom-line approach to event-tourism evaluation. In short, it is generally easier and often more politically effective to put outcomes into monetary terms.

Carlsen, Getz, and Soutar (2001) sought to establish broader measures of event impacts, and Sherwood et al., 2004 and Sherwood et al., 2005) and Sherwood (2007) advanced a triple bottom line approach to event sustainability. Fredline, Raybould, Jago, and Deery (2005) recommended use of the 'event footprint' as a concept of triple-bottom-line accounting. This graphical technique plots scores from key indicators on three dimensions. To make progress in both impact assessment and evaluation of worth, a

4. A framework for knowledge creation and theory development in event tourism

more comprehensive system will be needed, one that allows comparison of multiple dimensions without resorting to purely monetary measurement. This could be accomplished by a goal-attainment approach in which progress towards attaining specific outcomes, and broader sustainability and management goals is periodically undertaken.

The notion of event legacy planning has become firmly established, particularly with regard to mega events that cannot easily be justified in terms of tourist expenditure, project finances, or by reference to any short-term measures. Thomson, Schlenker, and Schulenkorf (2013, 111) stated the concept "has attracted limited critical analysis", and identified five key considerations: define the period of time early on, with a clear vision; plan the legacy from the concept stage; consider outcomes at all stages of the event life cycle; consider different perspectives on what is positive or negative; engage all stakeholders to ensure maximum reach for the desired outcomes. Clifton, O'Sullivan, and Pickernell (2012) took a different approach by linking a theme year to capacity building. Veitch (2013) explored mythology as a hallmark-event legacy, with reference to the lasting influence of the America's Cup on Fremantle, Western Australia. Many observers, and critics of mega events in particular, point to the credibility gap between what is promised and is actually delivered. For example, Minnaert (2012) argued the Olympics generally bring few benefits for socially excluded groups, although these supposed benefits are often important justifications.

4.5.1. Economic outcomes

As discussed earlier in the paper, the earliest journal articles were by Della Bitta et al. (1978), and Davidson and Schaffer (1980). The first truly comprehensive event impact research was conducted on the Adelaide Grand Prix (see Burns, Hatch, & Mules, 1986). Since then, a number of scholars have lamented the lack of consistency used in event impact studies (Dwyer et al., 2005, Sherwood, 2007 and Uysal and Gitelson, 1994), but there is now so much literature available that practitioners should be able to avoid the main pitfalls. Several noteworthy articles were published at the turn of the century, including state-of-the-art commentary and methodology for conducting economic impact assessments by Dwyer et al., 2000a and Dwyer et al., 2000b. These more or less laid to rest any debate on what needed to be done, and how to do it validly.

Projects funded through the Cooperative Research Centre for Sustainable Tourism in Australia resulted in the ENCORE toolkit, which in its final version adopted a triple-bottom-line approach. Jago and Dwyer (2006) documented all the basic methods and issues, but economists have subsequently debated the validity of using input–output as a foundation for estimating secondary and induced income. Dwyer et al. (2010) include state-of-the-art thinking on event tourism impacts and their assessment, within the broader tourism context. Although income and value-added multipliers are typically used when converting direct (in-scope) event tourism spending into gross economic impacts, others have used econometric modelling and most recently economists have been recommending use of General Equilibrium Models (Dwyer and Forsyth, 2009 and Dwyer et al., 2006).

Research concerning the economic impacts of single events dominates the literature (e.g., Peeters, Matheson and Szymanski, 2014;Saayman & Saayman, 2012) but various types of events are well studied. Grado et al. (1998) examined conferences and conventions, and Dwyer (2002) provided an overview of convention tourism impacts. Solberg, Andersson, & Shibli, 2002 examined 'business' travellers to events, notably the media and officials. Impacts of events on the public sector have been studied (Andersson and Samuelson, 2000) and it is especially noteworthy that tax benefits for all levels of government constitute one of the biggest benefits of event tourism (Turco, 1995). Fundamental to event impact assessment is detailed analysis of visitor spending, with appropriate methods reviewed by Case et al., 2010 and Case et al., 2013. Wicker and Hallmann (2013) continued this line of research by employing willingness to pay measures for marathon runners. Davies, Coleman, and Ramchandani (2013) argue for the Direct Expenditure Approach to event impact assessment, which is simply a calculation of direct tourist expenditure attributable to an event, minus first-round leakages and taking into account organizational spend and surplus or deficit.

4. A framework for knowledge creation and theory development in event tourism

A number of authors have called for more comprehensive cost and benefit evaluations (Burgan and Mules, 2001, Mules and Dwyer, 2006 and Whitson and Horne, 2006). As early as 1973 Cicarelli and Kowarsky conducted a cost-benefit evaluation of the Olympics. These are, very regrettably, seldom conducted, perhaps owing to the broad scope and the difficulty of comparing tangibles with intangibles. But there is also good reason to believe that full accounting of mega-project costs is not desired by proponents, and that after the event it is easier and more politically safe to let people believe that initial forecasts of costs and benefits were realized rather than to prove that they were. As the discourse broadens from economic to more comprehensive impact assessments, the more fundamental question has arisen of how society, politicians, and various other stakeholders value, or determine the worth of events and event-tourism portfolios. Within this debate are considerations of opportunity costs, gigantism (in terms of costs and incurred debt) and questionable sustainability (Hall, 2012, Preuss, 2007 and Preuss, 2009).

In order to avoid "black-box" models based on different assumptions and multipliers, and to achieve standardization in forecasts and post-event impact assessments, there is a move towards adopting a set of key performance indicators. Is it necessary for an event to forecast and account for every new dollar brought into the destination, or is it more important to know if local suppliers were used and their businesses enhanced? Do we need to prove that every event puts money into local government coffers, or is it better to measure the distributional effects of event-related expenditures? As the field matures, politicians demanding validity and standardization will increasingly pinpoint the economic and other outcomes they want measured and future research issues are outlined in Table 8.

Table 8. Future research issues associated with the economic effects of events.

Established concepts and terms: economic outcomes	Future directions
1: Numerous economic-impact studies of single events, and a few on multiple events have established how event-tourism changes consumption patterns, generates income/wealth, has a minor role in creating employment, and contributes to other forms of economic and urban/rural development	- Generalized Equilibrium Models are becoming preferred over multipliers, but more applications are needed
2: Analysis frequently reveals the dedicated event tourist to be a high-yield visitor with distinct consumption patterns; this varies, depending on the event and the target markets	- The distribution of costs and benefits among persons and groups and between cities and regions/countries remains a key issue
3: Taxation makes governments at all levels the primary beneficiaries of new event-tourism demand	- Full triple-bottom-line impact assessments are still in their infancy; commensurability remains a problem
4: The traditional reliance on multipliers (income, value, employment) to estimate total direct, indirect and induced economic impacts has received a lot of criticism for exaggerating benefits; furthermore, studies have revealed that direct event-tourist spending accounts for the vast majority of income benefits; multipliers are sometimes misused on purpose; their basis in Input-Output tables has been challenged owing to a propensity to exaggerate benefits	- Destinations and events always need fresh market intelligence on who are the high-yield event tourists, and how they should be attracted – this has theoretical implications in the context of event-tourist careers
5: Cost/benefit evaluation methods are rarely applied; many event supporters do not want a full accounting of costs; many costs are hidden (e.g., security, transport infrastructure) or are falsely claimed as benefits; externalities such as pollution or social problems are typically ignored	- In micro-economic terms, how does sustainability apply to the financial viability of the event and the event organization? of portfolios of events?
6: The attribution of new or incremental spending to dedicated event tourists (who travel because of the event) is the key to calculating tourism benefits; in-scope expenditure refers to both the specific event(s) and the geographical area for which costs and benefits are to be calculated; casual event tourists (those already in the area) must be discounted, unless they stay longer or spend more because of an event	- Long-term, cumulative economic impacts and legacy effects require study
7: Displacement effects (by types of visitor, by economic sector, and spatially) can be substantial and must be deducted from gross tourist income	- Opportunity costs are seldom built into feasibility and impact studies; the same goes for externalities such as amenity loss

4. A framework for knowledge creation and theory development in event tourism

4.5.2. Personal, social, cultural, and political outcomes

There are difficulties involved in distinguishing between personal, social, and cultural outcomes, all of which can have political repercussions. It is often impossible to prove cause and effect in these areas, moreover the impacts felt by people might be attributable to the holding of events, participation in them, or the tourism dimension. Media reports can greatly influence perceptions and attitudes. Tourism-related outcomes can be complex and hard to assess, starting with the travel dimension (e.g., it can impose congestion, pollution, noise, accidents), to host-guest interactions (such as servitude or expansion of social networks), or more intangible effects stemming from resident perceptions of their place in the world (e.g., their identity, place meanings and civic or national pride). Sharpley and Stone (2012) considered "social-cultural impacts" to be the simplest way of labelling this broad category, although they pointed out that social pertains to effects on people's day-to-day life and cultural impacts relate to beliefs, values, norms and traditions of groups. Research on social and cultural impacts of event tourism has its roots in anthropological studies (e.g. Greenwood, 1972), who raised the enduring issue of commodification of cultural traditions, and a consequent loss of authenticity. A conceptual overview was provided by Ritchie (1984), and a noteworthy piece of sociological research was conducted by Cunneen and Lynch (1988) who studied ritualized rioting at a sport event. The literature on resident perceptions has also grown substantially as Table 9 illustrates.

Social capital is a relatively new theme in the events literature (e.g., Arcodia and Whitford, 2007, Finkel, 2010, Schulenkorf et al., 2011 and Sharpley and Stone, 2012), and features prominently Richards et al. (2013). However it is sometimes unclear if it is an input necessary for establishing and sustaining events, or – if it is an outcome – whether it accrues from engagement with an event (as volunteer, organizer or participant) or can be a general effect of the networking and relationship building inherent in producing events. And where tourism exactly figures into the social-capital equation is somewhat of a mystery. Cultural change is usually less evident than social impacts, and therefore has been less subject to empirical research. Examples include

Garcia (2005) on the cultural effects of Glasgow's City of Culture year. Stevenson (2012) focused on the cultural legacy of the Olympics. More specific links between event tourism and culture can be found in the context of religious pilgrimage (Singh, 2006 and Timothy and Olsen, 2006), the events held by social worlds and sub cultures (Getz & Patterson, 2013), and of course arts events and their legacies (Quinn, 2009 and Quinn, 2010).

While the list of potential outcomes and positive or negative impacts felt by people and groups is long, there cannot be any certainty that event-tourism generates any one in particular. That is why it is necessary to understand the mechanisms by which change is initiated, and how people perceive the results (Kim, Jun, Walker, & Drane 2015). A number of theoretical perspectives are being taken to explain the socio-cultural impact processes and resident reactions to events or event-tourism. Exchange theory is the most widely cited. For example, Jackson (2008) found that residents are generally in favour of events that contribute socially and economically to the destination; residents are willing to cope with negatives as long as the perceived benefits are greater. Social representations, such as those communicated by media coverage, affect perceptions and attitudes. This has been explored in the event context by Fredline and Faulkner (2002b) and Cheng and Jarvis (2010). Other theoretical perspectives including personal and group identity and legitimation, place meanings, attachment and identity have been developed (Boyko, 2008 and De Bres and Davis, 2001). A feminist, critical-theory approach is rarely taken, but examples include studies by Eder et al., 1995 and Coughlin, 2010 and Fullagar and Pavlidis (2012), while Evans (2007) employed post-colonial theory in examining a film festival. There is a clear gap in the research on political outcomes of event tourism. These could include corruption, changes in government, the evolution of governance (i.e., new models involving stakeholders), or the politicization of decision-making about events and tourism. Henderson (2007) discussed the dilemma of how to deal with the question of demonstrations that are affected by prevailing political cultures, and it can be asked if growing opposition to mega-events (e.g., in Rio de Janeiro) will have any impact on praxis or politics and future research themes are outlined in Table 10.

4. A framework for knowledge creation and theory development in event tourism

Table 9. Resident and social impact studies on event tourism.

Residents' perceptions of, and attitudes towards events emerged as a major research and theoretical theme, although the tourism-specific dimensions have not been fully examined in this context. Key studies include:
- Soutar and McLeod (1993)
- Delamere (1997, 2001)
- Delamere, Wankel, and Hinch (2001)
- Fredline and Faulkner (1998, 2000, 2002a, 2002b)
- Mihalik (2001)
- Fredline, 2006; Fredline, Jago, and Deery (2003, 2005)
- Cegielski and Mules (2002)
- Small et al. (2005)
- Wood (2005)
- Ohmann, Jones, and Wilkes (2006)
- Xiao and Smith (2004a)
- Gursoy and Kendall (2006)
- Lim and Lee (2006)
- Rollins and Delamare (2007)
- Bull and Lovell (2007)
- Reid (2007)
- Jackson (2008)
- Small (2007)
- Wang and Pfister (2008)
- Zhou and Ap (2009)
- Zhou (2010)
- Chen (2011)
- Lorde, Greenidge, and Devonish (2011)
- Balduck, Maes, and Buelens (2011)
- Pranic, Petric, and Cetinic (2012)
- Prayag et al. (2013).

A number of specific social impact studies have been conducted including:
- Barker, Page, and Meyer's (2002a, 2002b, 2003) papers on event-related crime and perceptions of safety during an event. Even so Schroeder and Pennington-Grey (2014) argued that perceptions of crime at the Olympics lacked research despite the major security exercises surrounding such events.
- Woosnam, Van Winkle, and An (2013) sought to confirm their festival social impact attitude scale in Texas.
- Di Giovine (2009) considered how a destination event facilitates resident communitas and urban regeneration.
- Robertson, Rogers, and Leask (2009) sought to develop a set of indicators for socio-cultural festival impact assessments.
- Deery and Jago (2010) dealt with anti-social behaviour at events.
- Kaplanidou et al. (2013) tested perceived satisfaction with quality of life by residents in South Africa before and after the World Cup.
- Focussing on how publicity in advance of the Olympics shaped resident support, Chien, Ritchie, Shipway, and Henderson (2012) concluded that attitudes were shaped by perceived fairness of the coverage, taking into account imputed benefits and forecast costs or negative impacts.

Table 10. Future research themes on the personal, social and cultural outcomes of event tourism.

Major themes, concepts and terms: personal, social, and cultural outcomes	Future directions
1: Many resident-perceived impacts (both positive and negative) have been identified at the personal and community levels	- More is needed on personal development through event engagement and participation (e.g., how do people describe and explain why event tourism experiences are satisfying, memorable or transforming? what are the personal and social consequences of negative event tourism experiences?)
2: Exchange theory helps explain why many people are supportive of event-tourism, or not (because they perceive benefits or costs accrue to them)	- Evaluation tools and measures are needed for intangible effects and long-term, cumulative social/cultural legacies
3: Proximity effects: being close to events and venues is potentially an important explanatory factor	- Establishing cause and effect in social and cultural change is always problematic (e.g., does commodification of an event cause loss of tradition or authenticity? does gigantism and mega-event costs/debt generate social discontent?)
4: Social representation through media coverage affects perceptions and attitudes	- Compare discourses on costs and benefits (e.g., post-colonial, feminist, power and politics, stakeholder interactions)
5: Commodification through event tourism is a threat to cultural authenticity	- How are social representations of events formed and communicated?
6: Events can help preserve traditions, foster civic and national pride develop participation in and support for the arts	- How does the nature and extent of community involvement influence event tourism success and outcomes?
7: Events are sometimes platforms for protests, demonstrations, anti-social behaviour; these effects are often connected to the extent of media coverage	- What strategies work best for maximizing community benefits?
8: Legitimation and identity building for groups occurs through organizing or participating in events	- The politics of event tourism and response to perceived impacts requires study
9: Voluntarism and other forms of engagement fosters personal development and group identity	- Explore the process of how events contribute to place identity and attachment
10: Production of, and engagement with events can create social and cultural capital – especially through increased personal and institutional networking	

4.5.3. Environmental

The environmental impacts of events and tourism have remained a largely neglected area of academic research. For example, Sherwood (2007) examined 85 event economic impact studies prepared in Australia, and while the economic impact assessment was inconsistent but well established, social and cultural event impacts were being given more and more attention, but there was still a great need for advancing environmental impact assessment. According to Sherwood, only two published papers by May (1995) and Harris and Huyskens (2002) had dealt explicit with the environmental impacts of events, and only one of 85 event impact studies actually employed a triple bottom line approach. The research literature on environmental outcomes has been expanding slowly. Ahmed, Moodley, and Sookrajh (2008) looked at the environmental impacts of beach sport tourism in South Africa surrounding a surfing event, including the potential for environmental education. Collins and Flynn (2008) and Collins et al., 2009 and Collins et al., 2012 have tested different ways of measuring the environmental sustainability of a major sporting event. Recent additions to the literature on green and sustainable events include books by Raj and Musgrave, 2009 and Jones, 2010 and Goldblatt and Goldblatt (2012). Case (2013) covers relationships between events and the environment comprehensively, while Pernecky and Luck (2012) examine event sustainability in its many dimensions, including case studies. Case (2012) chronicled the Olympic movement and environmental issues and profiled the 2012 London Olympics in the context of sustainability. Theodoraki (2009) studied official communications on the impacts of the Athens 2004 Olympic Games, finding that at the bid stage the efforts were directed at building a positive image for the games, while accounts of post-game economic benefits were unsophisticated and unverifiable. In an effort to overcome the problem of incommensurability in triple-bottom-line approaches to impact assessment, festival-related research by Andersson, Armbrecht and Lundberg (2012) and Andersson and Lundberg (2013) placed a monetary value on the use and non-use values expressed through willingness to pay by event tourists and residents, and on the ecological footprint of the event. Footprint analysis, which focuses on consumption and waste, including carbon, offers the prospect of making it possible to compare events systematically and to monitor their sustainability efforts.

In terms of standards and certification, the main development is that of ISO, 2012 which "provides the framework for identifying the potentially negative social, economic and environmental impacts of events by removing or reducing them, and capitalizing on more positive impacts through improved planning and processes." Standards like these impose management and reporting systems but do not answer the question 'what is a sustainable event'? Standards and practices for green events and venues are now widely implemented, but the literature does not provide any comparative evidence of results. As argued by Case (2013), who considered the entire set of relationships between planned events and the environment, it is necessary to consider resources consumed, and the micro and macro environmental outcomes. To focus on tourism we need to question the additional resources consumed, and outcomes generated by tourists, as well as any modifications introduced by the tourism dimension. In particular, this systems approach leads to the critical issue of transportation and energy consumption, but might also focus attention on the peculiar eating and purchasing habits of event tourists, different patterns of destination activities, and preferences for particular forms of accommodation. It might also be the case that event tourists bring different values and attitudes to a place, resulting in conflicts with residents.

As Table 11 demonstrates on concepts and future research issues, there are few established concepts and terms specific to event tourism and the environment. The location and setting of events certainly has an effect, with purpose-built venues permanently changing the landscape and generating on-going costs and resource demands. More attention should be given to achieving positive environmental outcomes, including the role of event tourism in nature conservation through education and interpretation. In this context the notion of ecotourism events is relevant; that is events that are designed to attract ecotourists who expect to make a positive contribution and a model for ecotourism events has been developed by Getz (2013b). However, more generically the literature on future studies and events has remained a largely undeveloped area and so it is pertinent to briefly review this area and its contribution to the long-term development of event tourism research.

4. A framework for knowledge creation and theory development in event tourism

Table 11. Future research issues on environmental outcomes and event tourism.

Major themes, concepts and terms: environmental outcomes	Future directions
1: Event-tourism is a major consumer of energy and other resources, generating high ecological and carbon footprints	- There is an on-going need to advance environmental impact evaluation methods (e.g. ecological footprint; carbon footprint) - Cumulative, long-term impacts and ecological sustainability of event populations are unexplored
2: Event types and settings influence environmental impacts (e.g., indoor versus outdoor, festival versus sport); reliance on private automobile access is a major issue; some events drastically alter consumption patterns, such as spending on travel, food, accommodation	- Compare event types, formats and locations in terms of propensity to harm or benefit the environment - Evaluate the effects of green and sustainability certification and standards - Are ecotourism events suitable for sensitive environments? - Advance interpretation and environmental education at, and through events
3: Practices and standards have been established for the "greening" of events and sustainable venues	

4.5.4. Future studies and event tourism: a neglected area for further research?

There is a growing body of knowledge emerging within tourism management on the value of future studies and their contribution to understanding the future changes to demand and supply issues and how they may impact tourism. The extant literature in event studies is still in its infancy with much of the tourism and broader management literature providing the foundations for the analysis of event futures. As Page, Yeoman, Munro, Connell, and Walker (2006) outlined, the UK Cabinet Office Performance and Innovation Unit (2001) suggested that when examining futures issues there are three principal questions we need to pose: what may happen (possible futures), what is the most likely to happen (probable futures) and what would we prefer to happen (preferable futures). For this reason, a number of distinct approaches exist which can be employed in futures research as outlined by Page et al. (2006) based on the growth in this area within management science from softer (often qualitative techniques) through to much harder techniques. In fact Cornish's (2004) overview of futures techniques illustrates the breadth of futures research highlighted which comprise environmental scanning, trend analysis, trend monitoring, trend projection, scenarios, polling, brainstorming, modelling, gaming, historical analysis and visioning (i.e. looking more than 10 years ahead). This underlines the importance of the type of futures research question one wishes to ask – and whether you are looking for certainty and ambiguity, with the much harder techniques lending themselves to a greater element of certainty. In terms of the harder techniques that principally focus on forecasting, they were grouped by Calentone, Benedetto, and Bojanic (1987) into four types: exploratory forecasting based on extrapolating past trends using regression and similar techniques; normative, integrative and speculative forecasting (see Song & Li, 2008 for a review of their use in tourism). In terms of the qualitative area, Lin and Song (2014) examined the comparative neglect of the use of Delphi techniques as a way of understanding futures using more qualitative tools while several uses of

4. A framework for knowledge creation and theory development in event tourism

scenario planning have been employed in the tourism literature (see Page et al., 2006 and Page et al., 2010) for a detailed review) which document the emergence and use to understand possible futures to pose the 'what if' question to managers.

Within the field of event tourism, there are a wide range of challenges facing managers which futures research will help in a greater understanding of what are the key trends and factors affecting those trends (e.g. drivers of change) and the main issues which might occur in an event setting (e.g. the occurrence of random events such an extreme event such as a heatwave through to more atypical events such as terrorism through to more commonplace safety and security risks). Recent studies such as Yeoman, Robertson, McMahon-Beattie, Smith, and Backer (2014) map out some of the issues which future event tourism research will have to face and a useful summary of the key drivers of consumer behaviour in event tourism are outlined in Table 12 based on Yeoman (2013).

What Table 12 illustrates is the scope of probable drivers of change that managers will need to understand in terms of demand although from a supply perspective a number of other challenges exist as outlined by Adema and Roehl, 2010 and Mair, 2011 and Yeoman et al. (2014) where the following drivers of event tourism may include (as also highlighted in Table 2, Table 3, Table 4, Table 5, Table 6, Table 7, Table 8, Table 9, Table 10 and Table 11):

- Environmental and green issues
- The impact of climate change and the need for venues and event stagers to consider the capacity to adapt to and mitigate the impacts
- Security and safety issues
- Globalization and the global audiences for event tourism.

Table 12. A summary of consumer trends shaping future events and festivals.

Trend term	Summary
Everyday exceptional	An increase in celebration and the transformation of the everyday experience into some more extraordinary and exceptional events
Magic nostalgia	A greater focus on reminiscence and celebration of the past in events and festivals
Leisure upgrade	The aspiration for leisure participation increases with affluence and events offer a new form of social capital where participation is celebrated as an experience
Mobile living	We are living in more connected societies and living more connected lives which also transcends our leisure lives in which events (and non-leisure events) occur
Performative leisure	We are increasingly witnessing people celebrating their involvement in events and enjoyment through sharing the experiences via social media and mobile technology
Authentic experience	Consumers are seeking to accumulate more authentic leisure experiences and events and festivals offer one way to do this, increasingly through co-creation
Affluence	Consumers are becoming more demanding in terms of their needs and consumption within the experience economy
Ageless society	The rising age of the population in the developed world, due to greater life expectancy, has transformed the participation in events and festivals
Consuming with ethics	Consumers are starting to recognise the challenge of green issues and their own carbon footprint in everyday life and this may start to shape leisure consumption in the future around participation in event tourism
Accumulation of social capital	Consumers want to celebrate their achievements and participation in key events and festivals and this is part of the desire to accumulate experiences as part of their social capital repertoire

Source: Developed from Yeoman (2013).

5. CONCLUSIONS

Event tourism is a sub-field at the nexus of tourism and event studies as demonstrated in Fig. 1 and Fig. 2, and its growing importance as an economic activity with various development roles has resulted in substantial global competition. It is instrumentalist in nature, meaning that events and tourism are valued for many purposes, both in the public and private sectors. A set of core propositions have been identified (see Section 1.2), and these

5. Conclusions

collectively explain why event tourism is considered to be important, as well as generating goals for its development. The pertinent literature is vast. Indeed it is too big and diverse to thoroughly cover, but in this systematic summary we have attempted to signpost key developments and areas of research that have evolved and areas for further development. One of the central challenges is that many other disciplines and fields have taken an interest in event-related phenomena, contributing to many discourses and research topics that link events and tourism in an interdisciplinary manner. The review process has been shaped by a framework (see Fig. 5) that can also be used by managers and policy makers to shape their overall understanding and approach to event tourism. The tables in the paper summarize knowledge in this field by identifying major themes, concepts and terms being employed in each of the elements of the framework: the core phenomenon (experience and meanings), antecedents, outcomes, planning and management, and dynamic patterns and processes.

While a complete ontological mapping of event tourism remains to be attempted, this review has identified major claims to knowledge that delimit event tourism, with insights on how this knowledge has been determined through various research methods. This process has also provided a research agenda by suggesting new and emerging topics and specific methods that can prove useful. To make progress ontologically will require a systematic analysis of all the pertinent literature, both from the research journals and from praxis, with the aim of pinpointing all claims to knowledge (including all concepts being employed), and an effort made to codify and standardize terminology.

5.1. Trends

By focussing on changes in the literature a number of hot topics and trends can be identified. First, strategy for event tourism is attracting more attention from researchers and theorists, and in particular a new emphasis has emerged on developing and managing portfolios of events. Most of the literature pertains to single events, their planning, management and impact; but growth in the events sector combined with global competitiveness means that most cities and destinations possess numerous events with long-term and synergistic effects. Application of portfolio concepts from the investment field, and testing of various propositions from population and organizational ecology will advance praxis and foster interdisciplinary theory for event tourism. There is a need to study the ever-expanding roles of events as they have become legitimized as instruments of varied government policies, corporate marketing, and industry strategy. While the impetus for event tourism in most cases was that of tourist attractiveness, related to overcoming seasonality and generating economic impacts. The image enhancement and co-branding roles of events have become equally or more valued. A major source of concern for events and destinations is how to measure real effects of media coverage and whether or not image enhancement translates into future demand or growth. Greater attention to media management is warranted, specifically how images and brand values can be conveyed to target audiences through the planned-event medium.

The roles of events in urban regeneration and re-positioning go well beyond place marketing and image effects. Links between events, travel and health are under-researched and this should become a priority research topic – especially directed at youth. How do events, and the development of event travel careers, engage youth and keep people active throughout their lives? Do owners and organizers of events establish health goals and evaluate outcome? The oft-claimed effects of sport-events that they generate increased participation have been challenged and longitudinal research is therefore required. Who exactly participates more as a consequence of events, and can we distinguish between the effects of publicity (e.g., merely watching an event on TV) actual participation as a volunteer or athlete, and other forms of engagement such as educational programmes attached to events? Among key themes for future development are the following.

Special-Interest Event Tourism: The range of special interests generating events and tourism continues to expand, from yoga and food to ever-diversifying participation sports. How involvement is established and leads to event-tourist careers is of direct relevance to both understanding this trend and to taking advantage of it for competitive reasons. While traditions and the institutionalization process give rise to permanent hallmark events,

5. Conclusions

special interests generate numerous targeted events and a number of Iconic events that hold special meaning for the highly involved.

Understanding the travel experience has always been a theme, advanced most recently through ethnographic accounts, but now experience design is the very popular topic.

Increasingly it will be necessary to 'custom-design' highly targeted event experiences, and this has to be based on greater knowledge of the planned event experience in all its dimensions (by type of event, setting and management systems). A variety of research approaches and many comparisons will be required, from evaluations of those attending events to qualitative studies of what people are looking for, meanings they attach to their experiences, and influences on future attitudes and behaviour. In this demand-side approach, market intelligence equates with more and better research and theory development on the roles of event tourism in social worlds connected to leisure, sport and lifestyle.

Evaluation and Impact Assessment: Inevitably the rise of event tourism is generating a greater need for accountability, transparency, and comprehensiveness in evaluation of policies, strategies, investments and interventions. This applies to bidding, developing a comprehensive portfolio, and construction or replacement of venues. Researchers have made great strides in developing theory and methods for non-economic impact topics, most recently by stressing social capital, use and non-use values, footprint calculations and taking a multistakeholder approach. But full cost and benefit evaluations are rare and it is exceptional to see proper consideration of opportunity costs or key externalities like security and infrastructure costs for mega events. Long-term evaluation of leveraging and legacy effects is needed. Key indicators are being developed to both reflect triple-bottom-line thinking and encourage standardization of impact assessments and forecasts.

A primary need is to focus attention on the bigger evaluation questions of what an event is worth, how to value events within a portfolio, and the relative value of permanent versus one-time events.

Running in parallel to developments in evaluation and impact assessment has been a rise in critical discourse related to events and event tourism (e.g. see Merkel, 2013). This reflects the growing scholarly interest in contested meanings and the worth of events, how ideology, power and politics shape event tourism, events as propaganda, protests, and the distribution of costs and benefits.

Sustainability: Certainly this is not a new theme, but its relevance will not diminish. While there is no single definition or approach to achieving sustainability that is accepted by all, there has been recognition that it is much more than a "greening of events" and in this sense the discourse is closely related to evaluation of worth, justifications for public-sector intervention, portfolios and populations, and to various lines of argument within critical discourse. Many critics flatly reject the notion that mega events can ever be green or sustainable.

Futurism. Tourism forecasting is a well-established theme, but we are now seeing more interest in future studies applied to the events sector (e.g. Yeoman et al. (2014) highlighting the significance of trend analysis.

Finally it is pertinent to conclude with a focus on education for event tourism. Event management education is now well established in many countries, but it is inevitable that event tourism will find a place of its own. This is in part a function of the increasing number of jobs specific to event tourism indicative of the potential drop in student demand for tourism and value in combining tourism and closely related studies like events more closely. The synergistic effects of these new (or perhaps re-invented) mergers offer students more choices and bring applied management fields closer together as advocated by Getz (2014).

REFERENCES

1. Adema, K. L., & Roehl, W. S. (2010). Environmental scanning the future of event design. *International Journal of Hospitality Management, 29*(2), 199e207.

2. Ahmed, Z. U. (1992). Islamic pilgrimage (Hajj) to Ka'aba in Makkah (Saudi Arabia): an important international tourism activity. *Journal of Tourism Studies, 3*(1), 35e43. Ahmed, F., Moodley, V., & Sookrajh, R. (2008). The environmental impacts of beach sport tourism events: a case study of the Mr. Price Pro surfing event, Durban, South Africa. *Africa Insight, 38*(3), 73e85.
3. AIEST. (1987). *The role and impact of mega-events and attractions on regional and national tourism* (Vol. 28). St. Gallen, Switzerland: Editions AIEST.
4. Alexandris, K., & Kaplanidou, K. (2014). Marketing sport event tourism: sport tourist behaviors and destination provisions. *Sport Marketing Quarterly, Special Issue: Marketing Sport Event Tourism, 23*(3), 125e126.
5. Andersson, T., Armbrecht, J., & Lundberg, E. (2012). Estimating use and non-use values of a music festival. *Scandinavian Journal of Hospitality and Tourism, 12*(3), 215e231.
6. Andersson, T., & Getz, D. (2009). Tourism as a mixed industry: differences between private, public and not-for-profit festivals. *Tourism Management, 30*(6), 847e856.
7. Andersson, T., Getz, D., & Mykletun, R. (2013). The festival size pyramid. *Convention and Event Tourism, 14*(2), 81e103.
8. Andersson, T. D., & Lundberg, E. (2013). Commensurability and sustainability: triple impact assessments of a tourism event. *Tourism Management, 37*, 99e109.
9. Andersson, T., Persson, C., Sahlberg, B., & Strom, L. (Eds.). (1999). *The impact of mega events*. Ostersund, Sweden: European Tourism Research Institute.
10. Andersson, T., & Samuelson, L. (2000). Financial effects of events on the public sector. In L. Mossberg (Ed.), *Evaluation of Events: Scandinavian experiences* (pp. 86e103). New York: Cognizant.
11. Anwar, S., & Sohail, S. (2004). Festival tourism in the United Arab Emirates: first- time visitors versus repeat visitor perceptions. *Journal of Vacation Marketing, 10*(2), 161e170.

12. Arcodia, C., & Whitford, M. (2007). Festival attendance and the development of social capital. *Journal of Convention & Event Tourism, 8*(2), 1e18.
13. Arellano, A. (2011). A history of Quebec e branded: the staging of the New France festival. *Event Management, 15*(1), 1e12.
14. Armstrong, J. (1985). International events: the real tourism impact. In *Conference proceedings of the Canada chapter* (pp. 9e37). Edmonton: Travel and Tourism Research Association.
15. Ashworth, G., & Page, S. J. (2011). Urban tourism research: recent progress and current paradoxes. *Tourism Management, 32*(1), 1e15.
16. Avgousti, K. (2012). The use of events in development of the tourism industry: the case of Cyprus. *Event Management, 16*(3), 203e221.
17. Backman, K., Backman, S., Uysal, M., & Sunshine, K. (1995). Event tourism: an ex- amination of motivations and activities. *Festival Management and Event Tourism, 3*(1), 15e24.
18. Backman, K., Hsu, C.-H., & Backman, S. (2011). Tao Residents' Perceptions of Social and Cultural Impacts of Tourism in Lan-Yu, Taiwan. *Event Management, 15*, 121e136.
19. Baker, K., & Draper, J. (2013). Importanceeperformance analysis of the attributes of a cultural festival. *Journal of Convention & Event Tourism, 14,*(2), 104e123.
20. Balduck, A., Maes, M., & Buelens, M. (2011). The social impact of the Tour de France: comparisons of residents' pre- and post-event perceptions. *European Sport Management Quarterly, 11*(2), 91e113.
21. Barker, M., Page, S. J., & Meyer, D. (2002a). Evaluating the impact of the 2002 America's Cup on Auckland, New Zealand. *Event Management, 7*(2), 79e92.
22. Barker, M., Page, S. J., & Meyer, D. (2002b). Modelling tourism crime: the 2000 America's Cup. *Annals of Tourism Research, 29*(3), 762e782.
23. Barker, M., Page, S. J., & Meyer, D. (2003). Urban visitor perceptions of safety during a special event. *Journal of Travel Research, 41*, 355e361.

24. Barron, P., & Leask, A. (2012). Events management education. In S. Page, & J. Connell (Eds.), *Routledge handbook of events*. London: Routledge.
25. Basu, P. (2005). Macpherson country: genealogical identities, spatial histories and the Scottish diasporic clanscape. *Cultural Geographies, 12*(2), 123e150.
26. Bauer, T., Law, R., Tse, T., & Weber, K. (2008). Motivation and satisfaction of mega- business event attendees: the case of ITU Telecom World 2006 in Hong Kong. *International Journal of Contemporary Hospitality Management, 20*(2), 228e234.
27. Baum, T., Deery, M., Hanlon, C., Lockstone, L., & Smith, K. (Eds.). (2009). *People and work in events and conventions*. Wallingford, UK: CABI.
28. Baumann, R., Matheson, V., & Muroi, C. (2009). Bowling in Hawaii: examining the effectiveness of sports-based tourism strategies. *Journal of Sport Economics, 10*(1), 107e123.
29. Benckendorff, P., & Pearce, P. (2012). The psychology of events. In S. Page, & J. Connell (Eds.), *Routledge handbook of events*. London: Routledge.
30. Benedict, B., & Dobkin, M. M. (1983). *The anthropology of World's fairs: San Fran- cisco's Panama Pacific International Exposition of 1915*. CA: Lowie Museum of Anthropology Berkeley.
31. Benson, A. M., Dickson, T. J., Terwiel, F. A., & Blackman, D. A. (2014). Training of Vancouver 2010 volunteers: a legacy opportunity? *Contemporary Social Science, 9*(2), 210e226.
32. Bernini, C. (2009). Convention industry and destination clusters: evidence from Italy. *Tourism Management, 30*(6), 878e889.
33. Berridge, G. (2010). Event pitching: the role of design and creativity. *International Journal of Hospitality Management, 29*(2), 208e215.
34. Berridge, G. (2014). The Gran Fondo and sportive experience: an exploratory look at cyclists' experiences and professional event staging. *Event Management, 18*(1), 75e88.

35. Beverland, M., Hoffman, D., & Rasmussen, M. (2001). The evolution of events in the Australasian wine sector. *Tourism Recreation Research, 26*(2), 35e44.
36. Blesic, I., Pivac, T., Stamenkovic, I., & Besermenji, S. (2013). Motives of visits to ethno music festivals with regard to gender and age structure of visitors. *Event Management, 17*(2), 145e154.
37. Blichfeldt, B. S., & Halkier, H. (2014). Mussels, tourism and community develop- ment: a case study of place branding through food festivals in Rural North Jutland, Denmark. *European Planning Studies, 22*(8), 1587e1603.
38. Bohlin, M. (2000). Traveling to events. In L. Mossberg (Ed.), *Evaluation of events: Scandinavian experiences* (pp. 13e29). New York: Cognizant.
39. Bojanic, D., & Warnick, R. (2012). The role of purchase decision involvement in a special event. *Journal of Travel Research, 51*(3), 357e366.
40. Bolan, P. (2014). A perspective on the near future: mobilizing events and social media. In I. Yeoman, et al. (Eds.), *The future of events and festivals* (pp. 200e208). Abingdon: Routledge.
41. Boo, S., & Busser, J. (2006). Impact analysis of a tourism festival on tourists' desti- nation images. *Event Management, 9*(4), 223e237.
42. Boorstin, D. (1961). *The image: A guide to pseudo-events in America.* New York: Harper & Row.
43. Boo, S., Koh, Y., & Jones, D. (2008). An Exploration of Attractiveness of Convention Cities
44. Based on Visit Behavior. *Journal of Convention & Event Tourism, 9*(4), 239e257.
45. Boo, S., Wang, Q., & Yu, L. (2011). Residents' support of mega-events: a reexami- nation. *Event Management, 15*(3), 215e232.
46. Bos, H. (1994). The importance of mega-events in the development of tourism demand. *Festival Management and Event Tourism, 2*(1), 55e58.
47. Bowdin, G., Allen, J., Harris, R., McDonnell, I., & O'Toole, W. (2012). *Events man- agement.* Routledge.

48. Boyko, C. (2008). Are you being served? The impacts of a tourist hallmark event on the place meanings of residents. *Event Management, 11*(4), 161e177.
49. Bracalente, B., Chirieleison, C., Cossignani, M., Ferrucci, L., Gigliotti, M., & Ranalli, M. (2011). The economic effects of cultural events: the Pintoricchio exhibition in Perugia. *Event Management, 15*(2), 137e149.
50. Bramwell, B. (1997a). Strategic planning before and after a mega-event. *Tourism Management, 18*(3), 167e176.
51. Bramwell, B. (1997b). A sport mega-event as a sustainable tourism development strategy. *Tourism Recreation Research, 22*(2), 13e19.
52. Brannas, K., & Nordstrom, J. (2006). Tourist accommodation effects of festivals. *Tourism Economics, 12*(2), 291e302.
53. Breiter, D., & Milman, A. (2006). Attendees' needs and service priorities in a large convention center: application of the importanceeperformance theory. *Tourism Management, 27*(6), 1364e1370.
54. Brewster, M., Connell, J., & Page, S. J. (2009). The Scottish Highland Games: evolu- tion, development and role as a community event. *Current Issues in Tourism, 12*(3), 271e293.
55. Brida, J. G., Disegna, M., & Scuderi, R. (2014). Segmenting visitors of cultural events: the case of Christmas Market. *Expert Systems with Applications, 41*(10), 4542e4553.
56. Brown, G. (2002). Taking the pulse of Olympic sponsorship. *Event Management, 7*(3), 187e196.
57. Brown, K. (2010). Come on home: visiting friends and relatives e the Cape Breton Experience. *Event Management, 14*(4), 309e318.
58. Bryan, H. (1977). Leisure value systems and recreation specialization: the case of trout fisherman. *Journal of Leisure Research, 9*(3), 174e187.
59. Bryan, H. (2000). Recreation specialization revisited. *Journal of Leisure Research, 32*(1), 18e21.
60. Buck, R. (1977). Making good business better: a second look at staged tourist at- tractions. *Journal of Travel Research, 15*(3), 30e31.
61. Bull, C., & Lovell, J. (2007). The impact of hosting major sporting events on local residents. an analysis of the views and perceptions of

Canterbury residents in relations to the Tour de France 2007. *Journal of Sport & Tourism, 12*(3e4), 229e248.
62. Burgan, B., & Mules, T. (2001). Reconciling costebenefit and economic impact assessment for event tourism. *Tourism Economics, 7*(4), 321e330.
63. Burns, J., Hatch, J., & Mules, T. (Eds.). (1986). *The Adelaide Grand Prix: The impact of a special event*. Adelaide: The Centre for South Australian Economic Studies.
64. Burns, J., & Mules, T. (1989). An economic evaluation of the Adelaide grand prix. In G. Syme, B. Shaw, D. Fenton, & W. Mueller (Eds.), *The planning and evaluation of hallmark events* (pp. 172e185). Aldershot: Gower Publishing Company.
65. Burr, S., & Scott, D. (2004). Application of the recreational specialization framework to understanding visitors to the Great Salt Lake Bird Festival. *Event Management, 9*(1/2), 27e37.
66. Business Tourism Partnership. (2004). *Better business tourism in Britain*. http:// www.businesstourismpartnership.com/research-and-publications/research/ category/4-bvep-research.
67. Buzinde, C., Kalavar, J., Kohli, N., & Manuel-Navarrete, D. (2014). Emic un- derstandings of Kumbh Mela pilgrimage experiences. *Annals of Tourism Research, 49*(November), 1e18.
68. Cahir, D., & Clark, I. (2010). 'An edifying spectacle': a history of 'tourist corroborees' in Victoria, Australia, 1835e1870. *Tourism Management, 31*(3), 412e420.
69. Calentone, R., Benedetto, A., & Bojanic, D. (1987). A comprehensive review of tourism forecasting literature. *Journal of Travel Research, 36*(3), 79e84.
70. Carlsen, J., Andersson, T., Ali-Knight, J., Jaeger, K., & Taylor, R. (2010). Festival management innovation and failure. *International Journal of Event and Festival Management, 1*(2), 120e131.
71. Carlsen, J., & Getz, D. (2006). Strategic planning for a regional wine festival: the Margaret River wine region festival. In J. Carlsen, & S. Charters (Eds.), *Global wine tourism: Research, management and marketing* (pp. 209e224). Wallingford: CABI. Carlsen, J., Getz, D., &

Soutar, G. (2001). Event evaluation research. *Event Manage- ment,* 6(4), 247e257.
72. Carlsen, J., & Taylor, A. (2003). Mega-events and urban renewal: the case of the Manchester 2002 Commonwealth Games. *Event Management,* 8(1), 15e22.
73. Carmichael, B. (2002). Global competitiveness and special events in cultural tourism: the example of the Barnes Exhibit at the Art Gallery of Ontario, Tor- onto. *The Canadian Geographer,* 46(4), 310e324.
74. Carnegie, E., & McCabe, S. (2008). Re-enactment events and tourism: meaning, authenticity and identity. *Current Issues in Tourism,* 11(4), 349e368.
75. Case, R. (2012). Event impacts and environmental sustainability. In S. Page, &
76. J. Connell (Eds.), *Routledge handbook of events.* London: Routledge. Case, R. (2013). *Events and the environment.* London: Routledge.
77. Case, R., Dey, T., Hobbs, S., Hoolachan, J., & Wilcox, A. (2010). An examination of sporting event direct-spending patterns at three competitive levels. *Journal of Convention & Event Tourism,* 11(2), 119e137.
78. Case, R., Dey, T., Lu, J., Phang, J., & Schwanz, A. (2013). Participant spending at sporting events: an examination of survey methodologies. *Journal of Convention & Event Tourism,* 14(1), 21e41.
79. Cavicchi, A., & Santini, C. (2014). *Food and wine events in Europe.* Abingdon: Rout- ledgeAbingdon. Routledge.
80. Cegielski, M., & Mules, M. (2002). Aspects of residents' perceptions of the GMC 400dCanberra's V8 supercar race. *Current Issues in Tourism,* 5(1), 54e70.
81. Chalip, L., & Costa, C. (2006). Building sport event tourism into the destination brand: foundations for a general theory. In H. Gibson (Ed.), *Sport tourism: Concepts and theories* (pp. 86e105). London: Routledge.

82. Chalip, L., Green, C., & Hill, B. (2003). Effects of sport media on destination image and intentions to visit. *Journal of Sport Management, 17,* 214e234.
83. Chalip, L., & Leyns, A. (2002). Local business leveraging of a sport event: managing an event for economic benefit. *Journal of Sport Management, 16,* 132e158.
84. Chalip, L., & McGuirty, J. (2004). Bundling sport events with the host destination. *Journal of Sport Tourism, 9*(3), 267e282.
85. Chang, S., & Mahadevan, R. (2014). Fad, fetish or fixture: contingent valuation of performing and visual arts festivals in Singapore. *International Journal of Cul- tural Policy, 20*(3), 318e340.
86. Chang, W., & Yuan, J. (2011). A taste of tourism: visitors' motivations to attend a food festival. *Event Management, 15*(1), 13e23.
87. Che, D. (2008). Sports, music, entertainment and the destination branding of post- Fordist Detroit. *Tourism Recreation Research, 33*(2), 195e206.
88. Chen, P. (2006). The attributes, consequences, and values associated with event sport tourists' behaviour: a means-end chain approach. *Event Management, 10*(1), 1e22.
89. Chen, S. (2011). Residents' perceptions of the impact of major annual tourism events in Macao: cluster analysis. *Journal of Convention & Event Tourism, 12*(2), 106e128.
90. Cheng, E., & Jarvis, N. (2010). Residents' perception of the social-cultural impacts of the 2008 Formula 1 Singtel Singapore Grand Prix. *Event Management, 14*(2), 91e106.
91. Chhabra, D. (2005). Defining authenticity and its determinants: toward an authenticity flow model. *Journal of Travel Research, 44*(1), 64e73.
92. Chhabra, D., Healy, R., & Sills, E. (2003). Staged authenticity and heritage tourism. *Annals of Tourism Research, 30*(3), 702e719.
93. Chien, M., Ritchie, B. W., Shipway, R., & Henderson, H. (2012). I Am having a dilemma: factors affecting resident support of event development in the community. *Journal of Travel Research, 51*(4), 451e463

94. Cicarelli, J., & Kowarsky, D. (1973). The economics of the Olympic games. *Business and Economic Dimensions*, (September/October), 1e5.
95. Clark, D. (2006). What are cities really committing to when they build a convention center? *Journal of Convention & Event Tourism, 8*(4), 7e27.
96. Clifton, N., O'Sullivan, D., & Pickernell, D. (2012). Capacity building and the
97. contribution of public festivals: evaluating "Cardiff 2005". *Event Management, 16*(1), 77e91.
98. Coates, D., & Humphreys, B. (2008). Do economists reach a conclusion on subsidies for sports Franchises, Stadiums, and mega-events? *Econ Journal Watch, 5*(3), 294e315. Sep.
99. Coughlan, A., & Filo, K. (2013). Using constant comparison method and qualitative data to understand participants' experiences at the nexus of tourism, sport and charity events. *Tourism Management, 35*, 122e131.
100. Cohen, E. (1988). Authenticity and commoditization in tourism. *Annals of Tourism Research, 15*(3), 371e386.
101. Cohen, E. (2007). The "postmodernization" of a mythical event: Naga Fireballs on the Mekong River. *Tourism Culture & Communication, 7*(3), 169e181.
102. Collıns, A., & Flynn, A. (2008). measuring the environmental sustainability of a major sporting event: a case study of the FA Cup final. *Tourism Economics, 14*(4), 751e768.
103. Collins, A., Jones, C., & Munday, M. (2009). Assessing the environmental impacts of mega sporting events: two options? *Tourism Management, 30*(6), 828e837.
104. Collins, A., Munday, M., & Roberts, A. (2012). Environmental consequences of tourism consumption at major events: an analysis of the UK stages of the 2007 Tour de France. *Journal of Travel Research, 51*(5), 577e590.

105. Connell, J., & Page, S. J. (2005). Evaluating the economic and spatial effects of an event: the case of the world medical and health games. *Tourism Geographies, 7*(1), 63e85.
106. Connell, J., Page, S. J., & Meyer, D. (2015). Visitor attractions and events: responding to seasonality. *Tourism Management, 46*, 283e298.
107. Convention Industry Council. (2012). *The economic significance of meetings to the U.S. Economy.* http://www.conventionindustry.org/ResearchInfo/EconomicSignific anceStudy.aspx.
108. Coopers and Lybrand Consulting Group. (1989). *NCR 1988 festivals study final report* (Vol. 1). Ottawa: Report for the Ottawa-Carleton Board of Trade. Coppock, J. (1977). *Tourism as a tool for regional development. Tourism: A tool for regional development* (pp. 1.2e1.15). Edinburgh: Tourism Recreation Research Unit.
109. Cornish, E. (2004). *Futuring: The exploration of the future.* Bethesda, MD: World Future Society.
110. Costa, X. (2002). Festive identity: personal and collective identity in the fire carnival of the 'Fallas' (Vale ncia, Spain). *Social Identities, 8*(2), 321e345.
111. Coughlin, M. (2010). The spectacle of piety on the Brittany Coast. *Event Management, 14*(4), 287e300.
112. Cowie, I. (1985). Housing policy options in relation to the America's Cup. *Urban Policy and Research, 3*, 40e41.
113. Crompton, J. (1979). Motivations for pleasure vacation. *Annals of Tourism Research, 6*(4), 408e424.
114. Crompton, J. (1999). *Measuring the economic impact of visitors to sports tournaments and special events.* Ashburn, Virginia: Division of Professional Services, National Recreation and Park Association.
115. Crompton, J., & McKay, S. (1994). Measuring the economic impact of festivals and events: some myths, misapplications and ethical dilemmas. *Festival Manage- ment and Event Tourism, 2*(1), 33e43.
116. Crompton, J., & McKay, S. (1997). Motives of visitors attending festival events. *Annals of Tourism Research, 24*(2), 425e439.

117. Crouch, G., & Ritchie, J. R. B. (1997). Convention site selection research: a review, conceptual model, and propositional framework. *Journal of Convention & Exhi- bition Management, 1*(1), 49e69.
118. Csikszentmihalyi, M. (1990). *Flow: The psychology of optimal experience.* New York: Harper Perennial.
119. Csikszentmihalyi, M., & Csikszentmihalyi, I. (1988). *Optimal experience: Psychologi- cal studies of flow on consciousness.* Cambridge, MA: Cambridge University Press. Cunneen, C., & Lynch, R. (1988). The social meanings of conflict in riots at the Australian Grand Prix Motorcycle Races. *Leisure Studies, 7*(1), 1e19.
120. Daniels, M. (2007). Central place theory and sport tourism impacts. *Annals of Tourism Research, 34*(2), 332e347.
121. Daniels, M., & Norman, W. (2003). Estimating the economic impacts of seven reg- ular sport tourism events. *Journal of Sport Tourism, 8*(4), 214e222.
122. Dann, G. (1977). Anomie, ego-enhancement and tourism. *Annals of Tourism Research, 4*(4), 184e194.
123. Dann, G. (1981). Tourist motivation an appraisal. *Annals of Tourism Research, 8*(2), 187e219.
124. Dansero, E., & Puttilli, M. (2010). Mega-events tourism legacies: the case of the Torino 2006 Winter Olympic Gamesea territorialisation approach. *Leisure Studies, 29*(3), 321e341.
125. Davıdson, R. (2003). Adding pleasure to business: conventions and tourism. *Journal of Convention and Exhibition Management, 5*(1), 29e39.
126. Davidson, R., & Cope, B. (2003). *Business travel: Conferences, incentive travel, exhi- bitions.* Corporate Hospitality and Corporate Travel.
127. Davidson, R., & Rogers, T. (2006). *Marketing destinations and venues for conferences, conventions and business events.* Oxford: Butterworth-Heinemann.
128. Davidson, L., & Schaffer, W. (1980). A discussion of methods employed in analyzing the impact of short-term entertainment events. *Journal of Travel Research, 28*(3), 12e16.

129. Davidson, R., Holloway, C., & Humphreys, C. (2009). *The business of tourism.* Prentice Hall.
130. Davies, L., Coleman, R., & Ramchandani, G. (2013). Evaluating event economic impact: rigour versus reality? *International Journal of Event and Festival Man- agement, 4*(1), 31e42.
131. De Bres, K., & Davis, J. (2001). Celebrating group and place identity: a case study of a new regional festival. *Tourism Geographies, 3,* 326e337.
132. Deery, M., & Jago, L. (2010). Social impacts of events and the role of anti-social behaviour. *International Journal of Event and Festival Management, 1*(1), 8e28.
133. Deery, M., Jago, L., & Fredline, E. (2004). Sport tourism or event tourism: are they one and the same? *Journal of Sport Tourism, 9*(3), 235e246.
134. Delamere, T. (1997). Development of scale items to measure the social impact of community festivals. *Journal of Applied Recreation Research, 22*(4), 293e315.
135. Delamere, T. (2001). Development of a scale to measure resident attitudes toward the social impacts of community festivals, Part 2: verification of the scale. *Event Management, 7*(1), 25e38.
136. Delamere, T., Wankel, L., & Hinch, T. (2001). Development of a scale to measure resident attitudes toward the social impacts of community festivals, Part 1: item generation and purification of the measure. *Event Management, 7*(1), 11e24.
137. Della Bitta, A., Loudon, D., Booth, G., & Weeks, R. (1978). Estimating the economic impact of a short-term tourist event. *Journal of Travel Research, 16,* 10e15.
138. Deng, Q., & Li, M. (2014). A model of event-destination image transfer. *Journal of Travel Research, 53*(1), 69e82.
139. Derrett, R. (2003). Making sense of how festivals demonstrate a community's sense of place. *Event Management, 8*(1), 49e58.
140. Destination Marketing Association International (www.destinationmarketing.org/). Devine, A., Boyd, S., & Boyle, E. (2010). Unravelling the complexities of inter- organisational

relationships within the sports tourism policy arena. *Journal of Policy Research in Tourism, Leisure and Events, 2*(2), 93e112.

141. Dewar, K., Meyer, D., & Li, W. M. (2001). Harbin, lanterns of ice, sculptures of snow.*Tourism Management, 22*(5), 523e532.

142. Di Giovine, M. (2009). Revitalization and counter-revitalization: tourism, heritage, and the Lantern Festival as catalysts for regeneration in Hoi An, Viet Nam. *Journal of Policy Research in Tourism, Leisure and Events, 1*(3), 208e230.

143. Díaz-Barriga, M. (2003). Materialism and sensuality: visualizing the devil in the festival of our lady of Urkupina. *Visual Anthropology, 16*(2e3), 245e261.

144. Dimanche, F. (1996). Special events legacy: the 1984 Louisiana world fair in New Orleans. *Festival Management and Event Tourism, 4*(1), 49e54.

145. DiPietro, R., Breiter, D., Rompf, P., & Godlewsk, M. (2008). An exploratory study of differences among meeting and exhibition planners in their destination selec- tion criteria. *Journal of Convention & Event Tourism, 9*(4), 258e276.

146. Dodd, T., Yuan, J., Adams, C., & Kolyesnikova, N. (2006). Motivations of young people for visiting wine festivals. *Event Management, 10*(1), 23e33.

147. Dolles, H. (2013). In D. Getz (Ed.), *Event tourism*. Abingdon: Routledge. Dolles, H., & S€derman, S. (2008). Mega-sporting events in Asiadimpacts on soci- ety, business and management: an introduction. *Asian Business & Management, 7*(2), 147e162.

148. Donovan, A., & Debres, K. (2006). Foods of freedom: Juneteenth as a culinary tourist attraction. *Tourism Review International, 9*(4), 379e389.

149. Dredge, D., & Whitford, M. (2010). Policy for sustainable and responsible festivals and events: institutionalization of a new paradigm e a response. *Journal of Policy Research in Tourism, Leisure and Events, 2*(1), 1e13.

150. Dredge, D., & Whitford, M. (2011). Event tourism governance and the public sphere. *Journal of Sustainable Tourism, 19*(4e5), 479e499.

151. Du Cros, H., & Jolliffe, L. (2014). *The arts and events*. London: Routledge.
152. Dungan, T. (1984). *How cities plan special events* (pp. 83e89). The Cornell H.R.A. Quarterly (May).
153. Duvignaud, J. (1976). Festivals-sociological approach. *Cultures, 3*(1), 14e25.
154. Dwyer, L. (2002). Economic contribution of convention tourism: conceptual and empirical issues. In K. Weber, & K. Chon (Eds.), *Convention tourism: International research and industry perspectives* (pp. 21e35). New York: Haworth.
155. Dwyer, L., & Forsyth, P. (2009). Public sector support for special events. *Eastern Economic Journal, 35*(4), 481e499.
156. Dwyer, L., Forsyth, P., & Dwyer, W. (2010). *Tourism economics and policy*. Bristol: Channel View.
157. Dwyer, L., Forsyth, P., & Spurr, R. (2005). Estimating the impacts of special events on an economy. *Journal of Travel Research, 43*(4), 351e359.
158. Dwyer, L., Forsyth, P., & Spurr, R. (2006). Assessing the economic impacts of events: a computable general equilibrium approach. *Journal of Travel Research, 45*(1), 59e66. Dwyer, L., Mellor, R., Mistillis, N., & Mules, T. (2000a). A framework for assessing 'tangible' and 'intangible' impacts of events and conventions. *Event Manage-ment, 6*(3), 175e189.
159. Dwyer, L., Mellor, R., Mistillis, N., & Mules, T. (2000b). Forecasting the economic impacts of events and conventions. *Event Management, 6*(3), 191e204.
160. Easto, P. C., & Truzzi, M. (1973). Towards an ethnography of the carnival social system. *The Journal of Popular Culture, 6*(3), 550e566.
161. Eder, D., Staggenborg, S., & Sudderth, L. (1995). The National Women's Music Festival: collective identity and diversity in a lesbian-feminist community. *Journal of Contemporary Ethnography, 23*(4), 485e515.
162. Edinburgh Festivals Impact Study. (online at www. eventscotland. org/funding-and- resources/downloads/get/56).

163. Edvinsson, T. N. (2014). Before the sunshine: organising and promoting the Olympic games in Stockholm 1912. *International Journal of the History of Sport, 31*(5), 563e580.
164. Elston, K., & Draper, J. (2012). A review of meeting planner site selection criteria research. *Journal of Convention & Event Tourism, 13*(3), 203e220.
165. EMBOK (the event management body of knowledge e see www.embok.org/).
166. Emery, P. (2001). Bidding to host a major sports event. In C. Gratton, & I. Henry (Eds.), *Sport in the city: The role of sport in economic and social regeneration* (pp. 91e108). London: Routledge.
167. ENCORE Event Evaluation Toolkit (www.sustainabletourismonline.com).
168. Evans, O. (2007). Border exchanges: the role of the European Film Festival. *Journal of Contemporary European Studies, 15*(1), 23e33.
169. Fairley, S., & Gammon, S. (2006). Something lived, something learned: Nostalgia's expanding role in sport tourism. In H. Gibson (Ed.), *Sport tourism: Concepts and theories* (pp. 50e65). London: Routledge.
170. Falassi, A. (1987). *Time out of time: Essays on the festival.* Albuquerque: University of New Mexico Press.
171. Faulkner, B., Chalip, L., Brown, G., Jago, L., March, R., & Woodside, A. (2000). Monitoring the tourism impacts of the Sydney 2000 Olympics. *Event Manage- ment, 6*(4), 231e246.
172. Fenich, G. (2005). *Meetings, expositions, events, and conventions: An introduction to the industry.* Upper Saddle River NJ: Pearson.
173. Fenich, G., Halsell, S., Ogbeide, G.-C., & Hashimoto, K. (2014). What the Millennial generation from around the world prefers in their meetings, conventions, and events. *Journal of Convention & Event Tourism, 15*(3), 236e241.
174. Ferdinand, N., & Williams, N. L. (2013). International festivals as experience pro-duction systems. *Tourism Management, 34*, 202e210.
175. Ferreira, S., & Donaldson, R. (2013). Global imaging and branding: source market newspaper reporting of the 2010 Fifa World Cup. *Tourism Review International, 1*, 253c265.

176. Filo, K., Chen, N., & Funk, D. (2013). Sport tourists' involvement with a destination a stage-based examination. *Journal of Hospitality & Tourism Research, 37*(1), 100e124.
177. McFarland. In Findling, J. E., & Pelle, K. D. (Eds.), *Encyclopaedia of World's fairs and expositions.*
178. Finkel, R. (2010). Dancing around the ring of fire: social capital, tourism resistance, and gender dichotomies at up Helly Aa in Lerwick, Shetland. *Event Management, 14*(4), 275e285.
179. Foley, M., McGillivray, D., & McPherson, G. (2011). *Event policy: From theory to strategy.* London: Routledge.
180. Foley, M., McGillivray, D., & McPherson, G. (2012). Policy pragmatism: qatar and the global events circuit. *International Journal of Event and Festival Management, 3*(1), 101e115.
181. Foley, C., Schlenker, K., Edwards, D., & Lewis-Smith, L. (2013). Determining business event legacies beyond the tourism spend: an Australian case study approach. *Event Management, 17*(3), 311e322.
182. Ford, R. C., & Peeper, W. C. (2007). The past as prologue: predicting the future of the convention and visitor bureau industry on the basis of its history. *Tourism Management, 28*(4), 1104e1114.
183. Ford, R., Peeper, W., & Gresock, A. (2009). Friends to grow and foes to know: using a stakeholder matrix to identify management strategies for convention and vis- itors bureaus. *Journal of Convention & Event Tourism, 10*(3), 166e184.
184. Formica, S. (1998). The development of festivals and special events studies. *Festival Management and Event Tourism, 5*(3), 131e137.
185. Formica, S., & Murrmann, S. (1998). The effects of group membership and moti- vation on attendance: an international festival case. *Tourism Analysis, 3*(3/4), 197e207.
186. Formica, S., & Uysal, M. (1998). Market segmentation of an international cultur- alehistorical event in Italy. *Journal of Travel Research, 36*(4), 16e24.
187. Foster, K., & Robinson, P. (2010). A critical analysis of the motivational factors that influence event attendance in family groups. *Event Management, 14*(2), 107e125.

188. Fredline, E. (2006). Host and guest relations and sport tourism. In H. Gibson (Ed.), *Sport tourism: Concepts and theories* (pp. 131e147). London: Routledge.
189. Fredline, E., & Faulkner, B. (1998). Resident reactions to a major tourist event: the Gold Coast Indy car race. *Festival Management and Event Tourism, 5*(4), 185e205.
190. Fredline, E., & Faulkner, B. (2000). Host community reactions: a cluster analysis.*Annals of Tourism Research, 27*(3), 763e784.
191. Fredline, E., & Faulkner, B. (2002a). Residents' reactions to the staging of major motorsport events within their communities: a cluster analysis. *Event Man- agement, 7*(2), 103e114.
192. Fredline, E., & Faulkner, B. (2002b). Variations in residents' reactions to major motorsport events: why residents perceive the impacts of events differently. *Event Management, 7*(2), 115e125.
193. Fredline, E., Jago, L., & Deery, M. (2003). The development of a generic scale to measure the social impacts of events. *Event Management, 8*(1), 23e37.
194. Fredline, E., Raybould, M., Jago, L., & Deery, M. (2005). *Triple bottom line event evaluation: A proposed framework for holistic event evaluation*. Paper presented at the third international event management research conference, Sydney. University of Technology, Sydney.
195. Frew, E. A., & Williams, K. M. (2014). Dancesport in Australia: sport tourists' behaviour and patterns. *Managing Leisure, 19*(3), 171e187.
196. Frisby, W., & Getz, D. (1989). Festival management: a case study perspective. *Journal of Travel Research, 28*(1), 7e11.
197. Fullagar, S., & Pavlidis, A. (2012). "It's all about the journey": women and cycling events. *International Journal of Event and Festival Management, 3*(2), 149e170.
198. Funk, D. (2012). *Consumer behaviour in sport and events*. Abingdon: Routledge. Funk, D. (2008). *Consumer behaviour in sport and events: Marketing action*. Oxford: Elsevier.

199. Funk, D., Alexandris, K., & Ping, Y. (2009). To go or stay home and watch: exploring the balance between motives and perceived constraints for major events: a case study of the 2008 Beijing Olympic Games. *International Journal of Tourism Research, 11*(1), 41e53.
200. Funk, D., & James, J. (2001). The psychological continuum model: a conceptual framework for understanding an individual's psychological connection to sport. *Sport Management Review, 4*(2), 119e150.
201. Funk, D., & James, J. (2006). Consumer loyalty: the meaning of attachment in the development of sport team allegiance. *Journal of Sport Management, 20*(2),189e217. Funk, D., Mahony, D., Nakazawa, M., & Hirakawa, S. (2001). Development of the Sports Interest Inventory(SII): implications for measuring unique consumer motives at sporting events. *International Journal of Sports Marketing & Spon-sorship, 3*.
202. Funk, D., Ridinger, L., & Moorman, A. (2003). Understanding consumer support: extending the Sport Interest Inventory (SII) to examine individual differences among women's professional sport consumers. *Sport Management Review, 6*(1), 1e31.
203. Funk, D., Toohey, K., & Bruun, T. (2007). International sport event participation: prior sport involvement; destination image; and travel motives. *European Sport Management Quarterly, 7*(3), 227e248.
204. Gammon, S. (2004). Secular pilgrimage and sport tourism. In B. Ritchie, & D. Adair (Eds.), *Sport tourism: Interrelationships, impacts and issues*. Clevedon: Channel View.
205. Gammon, S., & Kurtzman, J. (2002). *Sport tourism: Principles and practice*. East-bourne: Leisure Studies Association Publications.
206. Garcia, B. (2005). Deconstructing the city of culture: the long-term cultural legacies of Glasgow 1990. *Urban Studies, 42*(5e6), 841e868.
207. Gartner, W., & Holocek, D. (1983). Economic impact of an annual tourism industry exposition. *Annals of Tourism Research, 10*(2), 199e212.

References

208. Gelder, G., & Robinson, P. (2009). A critical comparative study of visitor motivations for attending music festivals: a case study of Glastonbury and V Festival. *Event Management, 13*(3), 181e196.
209. Getz, D. (1989). Special events: defining the product. *Tourism Management, 10*(2), 135e137.
210. Getz, D. (1991). *Festivals, special events, and tourism*. New York: Van Nostrand Rheinhold.
211. Getz, D. (1993). Corporate culture in not-for-profit festival organizations. *Festival Management and Event Tourism, 1*(1), 11e17.
212. Getz, D. (1997). *Event management and event tourism* (1st ed.). New York: Cognizant Communications Corporation.
213. Getz, D. (2000a). Festivals and special events: life cycle and saturation issues. In W. Garter, & D. Lime (Eds.), *Trends in outdoor recreation, leisure and tourism* (pp. 175e185). Wallingford, UK: CABI.
214. Getz, D. (2000b). Developing a research agenda for the event management field. In J. Allen, et al. (Eds.), *Events beyond 2000: Setting the agenda, pro- ceedings of conference on event evaluation, research and education* (pp. 10e21). Sydney: Australian Centre for Event Management, University of Technology. Getz, D. (2003a). Why festivals fail. *Event Management, 7*(4), 209e219.
215. Gctz, D. (2003b). Sport event tourism: planning, development, and marketing. In S. Hudson (Ed.), *Sport and adventure tourism* (pp. 49e88). New York: Haworth. Getz, D. (2004a). Geographic perspectives on event tourism. In A. Lew, M. Hall, & Williams (Eds.), *A companion to tourism* (pp. 410e422). Oxford: Blackwell Publishing.
216. Getz, D. (2004b). Bidding on events: critical success factors. *Journal of Convention and Exhibition Management, 5*(2), 1e24.
217. Getz, D. (2005). *Event management and event tourism* (2nd ed.). New York: Cognizant.
218. Getz, D. (2007). *Event studies: Theory, research and policy for planned events*. Oxford: Elsevier.

219. Getz, D. (2008). Event tourism: definition, evolution, and research. *Tourism Man- agement, 29*, 403e428.
220. Getz, D. (2009). Policy for sustainable and responsible festivals and events: insti- tutionalization of a new paradigm. *Journal of Policy Research in Tourism, Leisure and Events, 1*(1), 61e78.
221. Getz, D. (2011). The nature and scope of festival studies. *International Journal of Event Management Research, 5*(1) (online).
222. Getz, D. (2012a). *Event studies: Theory, research and policy for planned events* (2nd ed.). Abingdon: Routledge.
223. Getz, D. (2012b). Event Studies. In S. Page, & J. Connell (Eds.), *Routledge Handbook of Events*. London: Routledge.
224. Getz, D. (2013a). *Event tourism: Concepts, international case studies, and research.* New York: Cognizant.
225. Getz, D. (2013b). Ecotourism events. In R. Ballantyne (Ed.), *International handbook on ecotourism*. Edward Elgar Press.
226. Getz, D. (2014). Event studies, management, and tourism: three discourses and their relevance to curriculum. In M. Gross, D. Dredge, & D. Airey (Eds.), *The Routledge handbook of tourism and hospitality education*. Abingdon: Routledg- eAbingdon. Routledge.
227. Getz, D., & Andersson, T. (2008). Sustainable festivals: on becoming an institution.*Event Management, 12*(1), 1e17.
228. Getz, D., & Andersson, T. (2010). The event-tourist career trajectory: a study of high- involvement Amateur distance runners. *Scandinavian Journal of Tourism and Hospitality, 19*(4), 468e491.
229. Getz, D., Anderson, D., & Sheehan, L. (1998). Roles, issues and strategies for convention and visitors bureaux in destination planning and product devel- opment: a survey of Canadian bureaux. *Tourism Management, 19*(4), 331e340.
230. Getz, D., Andersson, T., & Larson, M. (2007). Festival stakeholder roles: concepts and case studies. *Event Management, 10*(2/3), 103e122.
231. Getz, D., Andersson, T., & Mykletun, R. (2013). Sustainable festivals: an organiza- tional ecology approach. *Tourism Analysis, 18*(3/4).
232. Getz, D., & Cheyne, J. (2002). Special event motives and behaviour. In C. Ryan (Ed.),*The tourist experience* (2nd ed.). (pp. 137e155). London:

Continuum.Getz, D., & Fairley, S. (2004). Media management at sport events for destination promotion. *Event Management, 8*(3), 127e139.

233. Getz, D., & Frisby, W. (1988). Evaluating management effectiveness in community- run festivals. *Journal of Travel Research, 27*(1), 22e27.
234. Getz, D., & McConnell, A. (2011). Serious sport tourism and event travel careers.*Journal of Sport Management, 25*, 326e338.
235. Getz, D., & McConnell, A. (2014). Comparing trail runners and mountain bikers: motivation, involvement, portfolios, and event-tourist careers. *Journal of Convention & Event Tourism, 15*(1), 69e100.
236. Getz, D., O'Neill, M., & Carlsen, J. (2001). Service quality evaluation at events through service mapping. *Journal of Travel Research, 39*(4), 380e390.
237. Getz, D., & Patterson, I. (2013). Social worlds as a framework for examining event and travel careers. *Tourism Analysis, 18*(5), 485e501.
238. Getz, D., Robinson, R., Andersson, T., & Vujicic, S. (2014). *Foodies and food tourism*.Oxford: Goodfellow.
239. Getz, D., Svensson, B., Pettersson, R., & Gunnervall, A. (2012). Hallmark events: definition, goals and planning process. *International Journal of Event Manage- ment Research, 7*(1/2) (online).
240. Gibson, H. (1998). Sport tourism: a critical analysis of research. *Sport Management Review, 1*, 45e76.
241. Gibson, H. (Ed.). (2006). *Sport tourism: Concepts and theories*. London: Routledge. Gibson, C., & Connell, J. (Eds.). (2011). *Festival places: Revitalising rural Australia*.Bristol: Channel View.
242. Gibson, C., & Connell, J. (2012). *Music festivals and rural development in Australia*.Farnham, England: Ashgate Publishing.
243. Gibson, H., Walker, M., Thapa, B., Kaplanidou, K., Geldenhuys, S., & Coetzee, W. (2014). Psychic income and social capital among host nation residents: a pre- post analysis of the 2010 FIFA World Cup in South Africa. *Tourism Manage- ment, 44*, 113e122.
244. Gilbert, D., & Lizotte, M. (1998). Tourism and the performing arts. *Travel & Tourism Analyst, 1*, 82e96.

245. Giovanardi, M., Lucarelli, A., & Decosta, P. (2014). Co-performing tourism places: the "Pink Night" festival. *Annals of Tourism Research, 44*, 102e115.
246. Gladstone, D. (2012). Event-based urbanization and the New Orleans tourist regime: a conceptual framework for understanding structural change in US tourist cities. *Journal of Policy Research in Tourism, Leisure and Events, 4*(3), 221e248.
247. Glos. (2005). *The economics of staging the Olympics. A comparison of the games 1972e2008*. Cheltenham, UK: Edward Elgar Publishing.
248. Gnoth, J., & Anwar, S. (2000). New Zealand bets on event tourism. *Cornell Hotel and Restaurant Administration Quarterly*, (August), 72e83.
249. Goldblatt, J. (1990). *Special events: The art and science of celebration*. New York: Van Nostrand Rheinhold. Goldblatt, S., & Goldblatt, J. (2012). *The complete guide to greener meetings and events*. New York: Wiley.
250. Gold, J. R., & Gold, M. M. (2010). *Olympic cities: City agendas, planning, and the World's Games, 1896e2016*. Taylor & Francis.
251. Gotham, K. (2002). Marketing Mardi Gras: commodification, spectacle and the political economy of tourism in New Orleans. *Urban Studies, 39*(10), 1735e1756. Grado, S., Strauss, C., & Lord, B. (1998). Economic impacts of conferences and con-ventions. *Journal of Convention and Exhibition Management, 1*(1), 19e33.
252. Gratton, C., Dobson, N., & Shibli, S. (2000). The economic importance of major sports events: a case study of six events. *Managing Leisure, 5*, 14e28.
253. Gratton, C., & Kokolakakis, T. (1997). *Economic impact of sport in England 1995*.London: The Sports Council.Green, B. (2008). Leveraging subculture and identity to promote events. In M. Weed (Ed.), *Sport and tourism: A reader* (pp. 362e376). London: Routledge.
254. Green, B., & Chalip, L. (1998). Sport tourism as the celebration of subculture. *Annals of Tourism Research, 25*(2), 275e291.
255. Greenwood, D. (1972). Tourism as an agent of change: a Spanish Basque case study.*Ethnology, 11*, 80e91.

References

256. Gripsrud, G., Nes, E., & Olsson, U. (2010). Effects of hosting a mega-sport event on country image. *Event Management, 14*, 193e204.
257. Grix, J. (2012). 'Image' leveraging and sports mega-events: Germany and the 2006 FIFA World Cup. *Journal of Sport & Tourism, 17*(4), 289e312.
258. de Groote, P. (2005). A multidisciplinary analysis of world fairs (14 Expos) and their effects. *Tourism Review, 60*(1/3), 12e19.
259. Gunn, C. (1979). *Tourism planning*. New York: Crane Russak.
260. Gunn, C., & Wicks, B. (1982). *A study of visitors to Dickens on the strand*. Galveston, Texas: Galveston Historical Foundation.
261. Gursoy, D., & Kendall, K. (2006). Hosting mega events: modelling locals' support. *Annals of Tourism Research, 33*(3), 603e623.
262. Hall, C. M. (1989). The definition and analysis of hallmark tourist events. *GeoJournal, 19*(3), 263e268.
263. Hall, C. M. (1992). *Hallmark tourist events: Impacts, management and planning*. London: Belhaven.
264. Hall, C. M. (1994). Mega-events and their legacies. In P. Murphy (Ed.), *Quality management in urban tourism: Balancing business and environment* (pp. 109e123). University of Victoria.
265. Hall, C. M. (2012). Sustainable mega-events: beyond the myth of balanced ap- proaches to mega-event sustainability. *Event Management, 16*(2), 119e131.
266. Hall, C. M., & Page, S. J. (2012). Geography and the study of events. In S. J. Page, & J. Connell (Eds.), *Routledge handbook of events*. London: Routledge.
267. Hall, C. M., & Rusher, K. (2004). Politics, public policy and the destination. In Yeoman, et al. (Eds.), *Festival and events management* (pp. 217e231). Oxford: Elsevier.
268. Hanly, P. (2012). Measuring the economic contribution of the international asso- ciation conference market: an Irish case study. *Tourism Management, 33*(6), 1574e1582.
269. Hannam, K., & Halewood, C. (2006). European Viking themed festivals: an expression of identity. *Journal of Heritage Tourism, 1*(1), 17e31.

270. Harris, R. (2014). The role of large-scale sporting events in host community edu- cation for sustainable development: an exploratory case study of the Sydney 2000 Olympic Games. *Event Management, 18,*(3), 207e230.
271. Harris, R., & Huyskens, M. (2002). Public events: can they make a contribution to ecological sustainability?. In *CAUTHE 2002: Tourism and Hospitality on the Edge; Proceedings of the 2002 CAUTHE conference* (p. 319) Edith Cowan University Press.
272. Harris, R., Jago, L., Allen, J., & Huyskens, M. (2001). Towards an Australian event research agenda: first steps. *Event Management, 6*(4), 213e221.
273. Hatten, A. (1987). *Economic impact of expo '86*. Victoria, British Columbia: Ministry of Finance and Corporate Relations.
274. Havitz, M., & Dimanche, F. (1999). Leisure involvement revisited: drive properties and paradoxes. *Journal of Leisure Research, 31*(2), 122e149.
275. Hawkins, D., & Goldblatt, J. (1995). Event management implications for tourism education. *Tourism Recreation Research, 20*(2), 42e45.
276. Hede, A. (2005). Sports-events, tourism and destination marketing strategies: an Australian case study of Athens 2004 and its media telecast. *Journal of Sport Tourism, 10*(3), 187e200.
277. Hede, A., Jago, L., & Deery, M. (2003). An agenda for special event research: lessons from the past and directions for the future. *Journal of Hospitality and Tourism Management, 10*(Suppl.), 1e14.
278. Hede, A., Jago, L., & Deery, M. (2004). Segmentation of special event attendees using personal values: relationships with satisfaction and behavioural in- tentions. *Journal of Quality Assurance in Hospitality and Tourism, 5*(2/3/4), 33e55.
279. Hektner, J. M., & Csikszentmihalyi, M. (2002). The experience sampling method: measuring the context and content of lives. In R. Bechtel, & A. Churchman (Eds.), *Handbook of environmental psychology* (pp. 233e243). New York: Wiley. Henderson, J. (2007).

Hosting major meetings and accompanying protestors: Singapore 2006. *Current Issues in Tourism, 10*(6), 543e557.
280. Higham, J. (1999). Commentary: sport as an avenue of tourism development: an analysis of the positive and negative impacts of sport tourism. *Current Issues in Tourism, 2*(1), 82e90.
281. Higham, J. (Ed.). (2005). *Sport tourism destinations: Issues, opportunities and analysis.* Oxford: Elsevier.
282. Higham, J., & Hinch, T. (2002). Tourism, sport and seasons: the challenges and potential of overcoming seasonality in the sport and tourism sectors. *Tourism Management, 23*(2), 175e185.
283. Higham, J., & Hinch, T. (2006). Sport and tourism research: a geographic approach. *Journal of Sport & Tourism, 11*(1), 31e49.
284. Hiller, H. (2000a). Mega-events, urban boosterism and growth strategies: an analysis of the objectives and legitimations of the Cape Town 2004 Olympic bid. *International Journal of Urban and Regional Research, 24*(2), 439e458.
285. Hiller, H. (2000b). Toward an urban sociology of mega-events. *Research in Urban Sociology, 5*, 181e205.
286. Hiller, H. (2012). *Host cities and the Olympics: An interactionist approach.* London: Routledge.
287. Hiller, H. H., & Wanner, R. A. (2011). Public opinion in host Olympic cities: the case of the 2010 Vancouver Winter Games. *Sociology, 45*(5), 883e899.
288. Hinch, T., & Delamere, T. (1993). Native festivals as tourism attractions: a commu- nity challenge. *Journal of Applied Recreation Research, 18*(2), 131e142.
289. Hinch, T., & Higham, J. (2011). *Sport tourism development* (2nd. ed). Bristol: Channel View.
290. Hinch, T., Jackson, E., Hudson, S., & Walker, G. (2006). Leisure constraint theory and sport tourism. In H. Gibson (Ed.), *Sport tourism: Concepts and theories* (pp. 10e31). London: Routledge.
291. Hsu, C., & Gartner, W. (2012). *The Routledge handbook of tourism research.* Abingdon: RoutledgeAbingdon. Routledge.

292. Hudson, S. (Ed.). (2002). *Sport and adventure tourism*. New York: Haworth.
293. Hudson, S., Roth, M., Madden, T., & Hudson, R. (2015). The effects of social media on emotions, brand relationship quality, and word of mouth: an empirical study of music festival attendees. *Tourism Management, 47*, 68e76.
294. Hughes, H. (1993). Olympic tourism and urban regeneration. *Festival Management and Event Tourism, 1*(4), 157e162.
295. Hughes, H. (2000). *Arts, entertainment and tourism*. Oxford: Butterworth- Heinemann.
296. Hultkrantz, L. (1998). Mega-event displacement of visitors: the World Champion- ship in Athletics, Goteborg 1995. *Festival Management and Event Tourism, 5*(1/2).Indiana University e Purdue University Indianapolis (http://petm.iupui.edu/academics/tcem/degrees.php.ISO. (2012). Event sustainability management systems. www.iso.org/iso/sustainable_ events_iso_ 2012.pdf.
297. Iso-Ahola, S. (1980). *The social psychology of leisure and recreation*. Dubuque, IA: Brown.
298. Iso-Ahola, S. (1983). Towards a social psychology of recreational travel. *Leisure Studies, 2*(1), 45e57.
299. Jackson, P. (1988). Street life: the politics of carnival. *Environment and Planning D: Society and Space, 6*(2), 213e227.
300. Jackson, E. (Ed.). (2005). *Constraints on leisure*. State College PA: Venture Publishing. Jackson, L. (2008). Residents' perceptions of the impacts of special event tourism.*Journal of Place Management and Development, 1*(3), 240e255.
301. Jackson, E., Crawford, D., & Godbey, G. (1992). Negotiation of leisure constraints. *Leisure Sciences, 15*, 1e12.
302. Jaeger, K., & Mykletun, R. (2013). Festivals, identities, and belonging. *Event Man- agement, 17*(3), 213e226.
303. Jafari, J. (1987). Tourism models: the sociocultural aspects. *Tourism Management, 8*(2), 151e159.

304. Jago, L., Chalip, L., Brown, G., Mules, T., & Shameem, A. (2003). Building events into destination branding: insights from experts. *Event Management, 8*(1), 3e14.
305. Jago, L., & Deery, M. (2005). Relationships and factors influencing convention de- cision-making. *Journal of Convention & Event Tourism, 7*(1), 23e41.
306. Jago, L., & Dwyer, L. (2006). *Economic evaluation of special events: A practitioner's guide*. Gold Coast Australia: Cooperative Research Centre for Sustainable Tourism.
307. James, J., & Ross, S. (2004). Comparing sport consumer motivations across multiple sports. *Sport Marketing Quarterly, 13*(1), 17e25.
308. Jamieson, N. (2014). Sport tourism events as community buildersdhow social capital helps the "locals". *Cope Journal of Convention & Event Tourism, 15*(1), 57e68.
309. Janiskee, R. (1994). Some macroscale growth trends in America's community festival industry. *Festival Management and Event Tourism, 2*(1), 10e14.
310. Janiskee, R. (1996). The temporal distribution of America's community festivals. *Festival Management and Event Tourism, 3*(3), 129e137.
311. Jepson, A., & Clarke, A. (2014). *Exploring community festivals and events*. Abingdon: Routledge.
312. Jepson, A., Clarke, A., & Ragsdell, G. (2014). Investigating the application of the motivation-opportunity-ability model to reveal factors which facilitate or inhibit inclusive engagement within local community festivals. *Scandinavian Journal of Hospitality and Tourism, 14*(3), 331e348.
313. Jin, X., & Weber, K. (2013). Developing and testing a model of exhibition brand preference: the exhibitors' perspective. *Tourism Management, 38*, 94e104.
314. Jones, M. (2010). *Sustainable event management: A practical guide*. London: Earthscan.
315. Jones, C. (2012). Events and festivals: fit for the future? *Event Management, 16*(2), 107e118.

316. Jones, I., & Green, C. (2006). Serious leisure, social identity and sport tourism. In
317. H. Gibson (Ed.), *Sport Tourism: Concepts and theories* (pp. 32e49). London: Routledge.
318. Junge, B. (2008). Heterosexual attendance at gay events: the 2002 Parada Livre Festival in Porto Alegre, Brazil. *Sexuality & Culture, 12*(2), 116e132.
319. Jutbring, H. (2014). Encoding destination messages in media coverage of an inter- national event: a case study of the European athletics indoor championships. *Journal of Destination Marketing & Management, 3*(1), 29e36.
320. Kang, Y., & Perdue, R. (1994). Long term impact of a mega-event on international tourism to the host country: a conceptual model and the case of the 1988 Seoul Olympics. In M. Uysal (Ed.), *Global tourist behaviour* (pp. 205e225). Interna- tional Business Press.
321. Kaplanidou, K., Karadakis, K., Gibson, H., Thapa, B., Walker, M., Geldenhuys, S., & Coetzee, W. (2013). Quality of life, event impacts, and mega-event sup- port among residents before and after the event: the case of the 2010 FIFA World Cup in South Africa. *Journal of Travel Research September, 52*(5), 631e645.
322. Karadakis, K., Kaplanidou, K., & Karlis, G. (2010). Event leveraging of mega sport events: a SWOT analysis approach. *International Journal of Event and Festival Management, 1*(3), 170e185.
323. Kasimati, E. (2003). Economic aspects and the Summer Olympics: a review of related research. *International Journal of Tourism Research, 5*(6), 433e444.
324. Kay, P. (2004). Cross-cultural research issues in developing international tourist markets for cultural events. *Event Management, 8*(4), 191e202.
325. Kennelly, M., & Toohey, K. (2014). Strategic alliances in sport tourism: national sport organisations and sport tour operators. *Sport Management Review, 17*(4), 407e418.

326. Kim, S., Ao, Y., Lee, H., & Pan, S. (2012). A study of motivations and the image of Shanghai as perceived by foreign tourists at the Shanghai EXPO. *Journal of Convention & Event Tourism, 13*(1), 48e73.
327. Kim, J., Boo, S., & Kim, Y. (2013). Patterns and trends in event tourism study topics over 30 years. *International Journal of Event and Festival Management, 4*(1), 66e83.
328. Kim, H., Borges, M., & Chon, J. (2006). Impacts of environmental values on tourism motivation: the case of FICA, Brazil. *Tourism Management, 27*(5), 957e967.
329. Kim, J., Kang, J., & Kim, Y.-K. (2014). Impact of mega sport events on destination image and country image. *Sport Marketing Quarterly, 23*(3), 161e175.
330. Kim, W., Jun, H., Walker, M., & Drane, D. (2015). Evaluating the perceived social impacts of hosting large-scale sport tourism events: scale development and validation. *Tourism Management, 48*, 21e32.
331. Kim, S., Scott, D., & Crompton, J. (1997). An exploration of the relationships among social psychological involvement, behavioural involvement, commitment, and future intentions in the context of birdwatching. *Journal of Leisure Research, 29*(3), 320e341.
332. Kim, S. S., Yoon, S., & Kim, Y. (2011). Competitive positioning among international convention cities in the East Asian region. *Journal of Convention & Event Tourism, 12*(2), 86e105.
333. Koo, G. Y., & Hardin, R. (2008). Difference in interrelationship between spectators' motives and behavioural intentions based on emotional attachment. *Sport Marketing Quarterly, 17*(1), 30e43.
334. Kruger, M., & Saayman, M. (2012a). Show me the band and I will show you the market. *Journal of Convention & Event Tourism, 13*(4), 250e269.
335. Kruger, M., & Saayman, M. (2012b). Listen to your heart: motives for attending Roxette live. *Journal of Convention and Event Tourism, 13*(3), 181e202.
336. Kruger, S., Saayman, M., & Ellis, S. (2014). The influence of travel motives on visitor happiness attending a wedding expo. *Journal of Travel and Tourism Marketing, 31*(5), 649e665.

337. Krugman, C., & Wright, R. R. (2006). *Global meetings and exhibitions*. Hoboken NJ: Wiley.Kyle, G., & Chick, G. (2002). The social nature of leisure involvement. *Journal of Leisure Research, 34*(4), 426e448.

338. Kyle, G., Absher, J., Norman, W., Hammit, W., & Jadice, L. (2007). A modified involvement scale. *Leisure Studies, 26*(4), 399e427.

339. Ladkin, A., & Weber, K. (2010). Career aspects of convention and exhibition pro- fessionals in Asia. *International Journal of Contemporary Hospitality Management, 22*(6), 871e886.

340. Lai, K. (2015). Destination images penetrated by mega-events: a behaviorist study of the 2008 Beijing Olympics. *Asia Pacific Journal of Tourism Research, 20*(4), 378e398.

341. Lai, K., & Li, Y. (2014). Image impacts of planned special events: literature review and research agenda. *Event Management, 18*(2), 111e126.

342. Laing, J., & Frost, W. (2013). Food, wine. Heritage, identity? Two case studies of Italian Diaspora Festivals in Regional Victoria. *Tourism Analysis, 18*(3), 323e334. Lamont, M., & Jenkins, J. (2013). Segmentation of cycling event participants: a two- step cluster method utilizing recreation specialization. *Event Management, 17*(4), 391e407.

343. Lamont, M., & Kennelly, M. (2011). I can't do everything! Competing priorities as constraints in triathlon event travel careers. *Tourism Review International, 14*(2), 85e97.

344. Lamont, M., & Kennelly, M. (2012). A qualitative exploration of participant motives among committed amateur triathletes. *Leisure Sciences, 34*(3), 236e255.

345. Lamont, M., Kennelly, M., & Wilson, E. (2012). Competing priorities as constraints in event travel careers. *Tourism Management, 33*(5), 1068e1079.

346. Lang Research. (2007). *Attending rock concerts and recreational dancing*. http:// publications.gc.ca/pub?id¼348066&sl¼0.

347. Larson, M. (2002). A political approach to relationship marketing: case study of the Storsjoyran festival. *International Journal of Tourism Research, 4*(2), 119e143.
348. Larson, M. (2009). Joint event production in the jungle, the park, and the garden: metaphors of event networks. *Tourism Management, 30*(3), 393e399.
349. Larson, R., & Csikszentmihalyi, M. (1983). The experience sampling method. In H. Reis (Ed.), *Naturalistic approaches to studying social interaction* (pp. 41e56). San Francisco, CA: Jossey-Bass.
350. Larson, M., & Wikstrom, E. (2001). Organising events: managing conflict and consensus in a political market square. *Event Management, 7*(1), 51e65.
351. Lashua, B., Spracklen, K., & Long, P. (2014). Introduction to the special issue: music and Tourism. *Tourist Studies, 14*(1), 3e9.
352. Lavenda, R. H. (1980). The festival of progress: the globalizing world-system and the transformation of the caracas carnival. *The Journal of Popular Culture, 14*(3), 465e475.
353. Lee, J. (1987). The impact of Expo' 86 on British Columbia markets. In P. Williams, J. Hall, & M. Hunter (Eds.), *Tourism: Where is the client. Conference papers of the travel and tourism research association*. Canada chapter.
354. Lee, I., Arcodia, C., & Lee, T. (2012). Benefits of visiting a multicultural festival: the case of South Korea. *Tourism Management, 33*(2), 334e340.
355. Lee, S., Arcodia, C., & Lee, T. (2012). Multicultural festivals: a niche tourism product in South Korea. *Tourism Review, 67*(1), 34e41.
356. Lee, J., & Back, K. (2005). A review of convention and meeting management research. *Journal of Convention and Event Tourism, 7*(2), 1e19.
357. Lee, S., & Crompton, J. (2003). The attraction power and spending impact of three festivals in Ocean City, Maryland. *Event Management, 8*(2), 109e112. Lee, S., Harris, J., & Lyberger, M. (2010). The economic impact of college sporting events: a case study of division I-a football games. *Event Management, 14*, 157e165.

358. Lee, S., Jee, W., Funk, D., & Jordan, J. (2015). Analysis of attendees' expenditure patterns to recurring annual events: examining the joint effects of repeat attendance and travel distance. *Tourism Management, 46*, 177e186.
359. Lee, C., & Kim, J. (1998). International tourism demand for the 2002 World Cup Korea: a combined forecasting technique. *Pacific Tourism Review, 2*(2), 157e166.
360. Lee, S., Kim, J., & Parrish, C. (2012). Are you ready for the extra inning? An exploratory study of the evaluation of professional sport teams' websites as marketing tools to prospective meeting/event customers. *Journal of Convention & Event Tourism, 13*(4), 270e289.
361. Lee, S., & Krohn, B. (2013). A study of psychological support from local residents for hosting mega-sporting events: a case of the 2012 Indianapolis Super Bowl XLVI. *Event Management, 17*(4), 361e376.
362. Lee, J., & Kyle, G. (2012). Recollection consistency of festival consumption emotions. *Journal of Travel Research, 51*, 178.
363. Lee, J., & Kyle, G. (2014). Segmenting festival visitors using psychological commit- ment. *Journal of Travel Research, 53*(5), 656e669.
364. Lee, J., Kyle, G., & Scott, D. (2012). The mediating effect of place attachment on the relationship between festival satisfaction and loyalty to the festival hosting destination. *Journal of Travel Research, 51*(6), 754e767.
365. Lee, M. J., & Lee, S. (2014). Subject areas and future research agendas in exhibition research: visitors' and organizers' perspectives. *Event Management, 18*(3), 377e386.
366. Lee, C., Lee, Y., & Wicks, B. (2004). Segmentation of festival motivation by nation- ality and satisfaction. *Tourism Management, 25*(1), 61e70.
367. Lee, C.-K., Mjelde, J., Kim, T.-K., & Lee, H.-M. (2014). Estimating the inten- tionebehavior gap associated with a mega event: the case of the Expo 2012 Yeosu Korea. *Tourism Management, 41*, 168e177. Lee, D., & Palakurthi, R. (2013). Marketing strategy to increase exhibition

attendance through controlling and eliminating leisure constraints. *Event Management, 17*(4), 323e336.

368. Lee, M. J., Yeung, S., & Dewald, B. (2010). An exploratory study examining the de- terminants of attendance motivations as perceived by attendees at Hong Kong exhibitions. *Journal of Convention & Event Tourism, 11*(3), 195e208.

369. Leibold, M., & van Zyl, C. (1994). The Summer Olympic Games and its tourismmarketing: city tourism marketing experiences and challenges with specific reference to Cape Town, South Africa. In P. Murphy (Ed.), *Quality management in urban tourism: Balancing business and the environment, proceedings* (pp. 135e151). University of Victoria.

370. Leiper, N. (1979). The framework of tourism: towards a definition of tourism, tourist, and the tourist industry. *Annals of Tourism Research, 6*(4), 390e407.

371. Leiper, N. (1990). Tourist attraction systems. *Annals of Tourism Research, 17*(3), 367e384.

372. Leiper, N. (2008). Why 'the tourism industry'is misleading as a generic expression: the case for the plural variation,'tourism industries'. *Tourism Management, 29*(2), 237e251.

373. Ley, D., & Olds, K. (1988). Landscape as spectacle: world fairs and the culture of heroic consumption. *Environment and Planning D: Society and Space, 6*(2), 191e212.

374. Lillywhite, J., Simonsen, J., & Wilson, B. (2013). A portrait of the US fair sector. *Event Management, 17*, 241e255.

375. Lim, S., & Lee, J. (2006). Host population perceptions of the impact of mega-events.*Asia Pacific Journal of Tourism Research, 11*(4), 407e421.

376. Lin, V. S., & Song, H. (2014). A review of Delphi forecasting research in tourism.*Current Issues in Tourism* (in press).

377. Li, R., & Petrick, J. (2006). A review of festival and event motivation studies. *Event Management, 9*(4), 239e245.

378. Li, X., & Vogelsong, H. (2005). Comparing methods of measuring image change: a case study of a small-scale community festival. *Tourism Analysis, 10*(4), 349e360.Liu, D., Broom, D., & Wilson, R.

(2014). Legacy of the Beijing Olympic Games: a non- host city perspective. *European Sport Management Quarterly.* in press.

379. Lockstone-Binney, L., Whitelaw, P., Robertson, M., Junek, O., & Michael, I. (2014). The motives of ambassadors in bidding for international association meetings and events. *Event Management, 18*(1), 65e74.

380. Long, P. (2000). After the event: perspectives on organizational partner- ship in the management of a themed festival year. *Event Management, 6*(1), 45e59.

381. Lorde, T., Greenidge, D., & Devonish, D. (2011). Local residents' perceptions of the impacts of the ICC Cricket World Cup 2007 on Barbados: comparisons of pre- and post-games. *Tourism Management, 32*(2), 349e356.

382. Lu, T., & Cai, L. (2011). An analysis of image and loyalty in convention and exhibition tourism in China. *Event Management, 15*(1), 37e48.

383. MacCannell, D. (1976). *The tourist: A new theory of the leisure class.* Berkeley: University of California Press.

384. Mackellar, J. (2006a). Conventions, festivals, and tourism: exploring the network that binds. *Journal of Convention and Event Tourism, 8*(2), 45e56.

385. Mackellar, J. (2006b). Special interest events in a regional destinationdexploring differences between specialist and generalist events. In C. Arcodia, M. Whitford,& C. Dickson (Eds.), *Global events congress proceedings* (pp. 176e195). Brisbane: University of Queensland.

386. Mackellar, J. (2009). An examination of serious participants at the Australian Wintersun Festival. *Leisure Studies, 28*(1), 85e104.

387. Mackellar, J. (2013). *Event audiences and expectations.* Abingdon: RoutledgeA- bingdon. Routledge.

388. Maennig, W., & Zimbalist, A. S. (Eds.). (2012). *International handbook on the eco- nomics of mega sporting events.* Cheltenham: Edward Elgar Publishing.

389. Mair, J. (2011). Events and climate change: an Australian perspective. *International Journal of Event and Festival Management, 2*(3), 245e253.

390. Mair, J. (2012). A review of business events literature. *Event Management, 16*(2), 133e141.
391. Mair, J. (2013). *Conferences and conventions: A research perspective.* Abingdon: Routledge.
392. Mair, J., & Thompson, K. (2009). The UK association conference attendance decision- making process. *Tourism Management, 30*(3), 400e409.
393. Mair, J., & Whitford, M. (2013). An exploration of events research: event topics, themes and emerging trends. *International Journal of Event and Festival Man- agement, 4*(1), 6e30.
394. Mannell, R. C., & Iso-Ahola, S. E. (1987). Psychological nature of leisure and tourism experience. *Annals of Tourism Research, 14*(3), 314e331.
395. Marshall, G. (1998). Exchange theory. In *A Dictionary of sociology*. Retrieved December 24, 2009 from Encyclopedia.com http://www.encyclopedia.com/doc/ 1O88-exchangetheory.html.
396. Maslow, A. (1954). *Motivation and personality* (2nd ed). NY: Harper and Row.
397. Mason, M., & Paggiaro, A. (2012). Investigating the role of festivalscape in culinary tourism: the case of food and wine events. *Tourism Management, 33*(6), 1329e1336.
398. Masterman, G. (2009). *Strategic sports event management.* Abingdon: Routledge. Matheson, C. (2005). Festivity and sociability: a study of a Celtic music festival.*Tourism Culture & Communication, 5*(3), 149e163.
399. Matheson, C., Rimmer, R., & Tinsley, R. (2014). Spiritual attitudes and visitor moti- vations at the Beltane Fire Festival, Edinburgh. *Tourism Management, 44*, 16e33. May, V. (1995). Environmental implications of the 1992 Winter Olympic Games.*Tourism Management, 16*(4), 269e275.
400. Mayfield, T., & Crompton, J. (1995). The status of the marketing concept among festival organizers. *Journal of Travel Research*, (Spring), 14e22.

401. MBECS (Meeting and Business Event Competency Standards e see www.mpiweb. org/MBECS).
402. McCabe, V. (2008). Strategies for career planning and development in the convention and exhibition industry in Australia. *International Journal of Hospi- tality Management, 27*(2), 222e231.
403. McCabe, V. (2012). Developing and sustaining a quality workforce: lessons from the convention and exhibition Industry. *Journal of Convention & Event Tourism, 13*(2), 121e134.
404. McGehee, N., Yoon, Y., & Cardenas, D. (2003). Involvement and travel for recrea- tional runners in North Carolina. *Journal of Sport Management, 17*(3), 305e324.
405. McKercher, B., & du Cros, H. (2002). *Cultural tourism.* New York: Haworth. McKercher, B., Mei, W., & Tse, T. (2006). Are short duration festivals tourist at-tractions? *Journal of Sustainable Tourism, 14*(1), 55e66.
406. Mehmetoglu, M., & Ellingsen, K. (2005). Do small-scale festivals adopt "market orientation" as a management philosophy? *Event Management, 9*(3), 119e132.
407. Mehus, I. (2005). Sociability and excitement motives of spectators attending entertainment sport events: spectators of soccer and ski-jumping. *Journal of Sport Behavior, 28*(4), 333.
408. Melbourne Convention and Exhibition Centre. (2010). *Melbourne convention dele- gate study 2010.* www.melbournecb.com.au/.
409. Mendell, R., MacBeth, J., & Solomon, A. (1983). The 1982 world's fair-a synopsis. Leisure Today. *Journal of Physical Education, Recreation and Dance,* (April), 48e49. Merkel, U. (Ed.). (2013). *Power, politics and international events: Socio-cultural ana-lyses of festivals and spectacles.* Abingdon: Routledge.
410. Mihalik, B. (1994). Mega-event legacies of the 1996 Atlanta Olympics. In P. Murphy (Ed.), *Quality management in urban tourism: Balancing business and environment.* Proceedings (pp. 151e162). University of Victoria.

References

411. Mihalik, B. (2001). Host population perceptions of the 1996 Atlanta Olympics: support, benefits and liabilities. *Tourism Analysis, 5*(1), 49e53.
412. Mihalik, B. J., & Ferguson, M. F. (1995). Visitor profile analysis: a pilot market research study of an American State Fair. *Journal of Travel & Tourism Marketing, 3*(4), 85e103.
413. Mihalik, B., & Wing-Vogelbacher, A. (1992). Travelling art expositions as a tourism event: a market research analysis for Ramesses the Great. *Journal of Travel and Tourism Marketing, 1*(3), 25e41.
414. Miller, A. (2012). Understanding the event experience of active sports tourists. In *International sports events: Impacts, experiences and identities* (pp. 99e112). London: Routledge.
415. Mill, R., & Morrison, A. (1985). *The tourism system*. Englewood Cliffs, NJ: Prentice-Hall. Mills, B., & Rosentraub, M. (2013). Hosting mega-events: a guide to the evaluation of development effects in integrated metropolitan regions. *Tourism Management,34*, 238e246.
416. Milne, G., & McDonald, M. (1999). Motivations of the sport consumer. In G. Milne, &
417. M. McDonald (Eds.), *Sport marketing: Managing the exchange process* (pp. 21e38). Sudbury MA: Jones and Bartlett.
418. Milner, L., Jago, L., & Deery, M. (2004). Profiling the special event nonattendee: an initial investigation. *Event Management, 8*(3), 141e150.
419. Minnaert, L. (2012). An Olympic legacy for all? The non-infrastructural outcomes of the Olympic Games for socially excluded groups (Atlanta 1996eBeijing 2008). *Tourism Management, 33*(2), 361e370.
420. Mitchell, C., & Wall, G. (1986). Impacts of cultural festivals on Ontario communities.*Recreation Research Review, 13*(1), 28e37.
421. Mitchell, C. J., & Wall, G. (1989). The arts and employment: a case study of the Stratford Festival. *Growth and Change, 20*(4), 31e40.
422. Mohan, L., & Thomas, L. (2012). Effect of identification on attendance at team sporting events. *Event Management, 16*(4), 341e349.

423. Mohr, K., Backman, K., Gahan, L., & Backman, S. (1993). An investigation of festival motivations and event satisfaction by visitor type. *Festival Management and Event Tourism, 1*(3), 89e97.
424. Morse, J. (2001). The Sydney 2000 Olympic Games: how the Australian Tourist Commission leveraged the Games for tourism. *Journal of Vacation Marketing, 7*(2), 101e107.
425. Moscardo, G. (2008). Analyzing the role of festivals and events in regional devel- opment. *Event Management, 11*(1e2), 1e2.
426. Mossberg, L. (Ed.). (2000a). *Evaluation of events: Scandinavian experiences*. New York: Cognizant Communication Corp.
427. Mossberg, L. (2000b). Effects of events on destination image. In L. Mossberg (Ed.), *Evaluation of events: Scandinavian experiences* (pp. 30e46). New York: Cognizant Communication Corp.
428. Mules, T., & Dwyer, L. (2006). Public sector support for sport tourism events: the role of costebenefit assessment. In H. Gibson (Ed.), *Sport tourism: Concepts and theories* (pp. 206e223). London: Routledge.
429. Mules, T., & McDonald, S. (1994). The economic impact of special events: the use of forecasts. *Festival Management and Event Tourism, 2*(1), 45e53.
430. Muller, D., & Pettersson, R. (2006). Sami heritage at the winter festival in Jokkmokk, Sweden. *Scandinavian Journal of Hospitality and Tourism, 6*(01), 54e69.
431. Nelson, R. (2006). Public financing of headquarter hotels in the United States. *Journal of Convention & Event Tourism, 8*(4), 29e46.
432. Nelson, K. (2009). Enhancing the Attendee's experience through creative design of the event environment: applying Goffman's dramaturgical perspective. *Journal of Convention & Event Tourism, 10*(2), 120e133.
433. Nelson, R., Baltin, B., & Feighner, B. (2012). Public-private financing Structures used in the United States to develop convention hotels. *Journal of Convention & Event Tourism, 13*(2), 135e146.
434. Neuenfeldt, K. W. (1995). The Kyana corroboree: cultural production of indigenous ethnogenesis. *Sociological Inquiry, 65*(1), 21e46.

435. New Zealand Tourist and Publicity Department. (1987). *New Zealand tourism report no. 38 (November)*.
436. Ngamsom, B., & Beck, J. (2000). A pilot study of motivations, inhibitors, and facili- tators of association members in attending international conferences. *Journal of Convention and Exhibition Management, 2*(2/3), 97e111.
437. Nicholson, R., & Pearce, D. (2001). Why do people attend events: a comparative analysis of visitor motivations at four South Island events. *Journal of Travel Research, 39*, 449e460.
438. Nilbe, K., Ahas, R., & Silm, S. (2014). Evaluating the travel distances of events visitors and regular visitors using mobile positioning data: the case of estonia. *Journal of Urban Technology, 21*(2), 91e107.
439. Nishio, T. (2013). The impact of sports events on inbound tourism in New Zealand. *Asia Pacific Journal of Tourism Research, 18*(8), 934e946.
440. Noël, F. (2008). Old home week celebrations as tourism promotion and commemoration: North Bay, Ontario, 1925 and 1935. *Urban History Review, 37*(1), 36e47.
441. Nolan, M. L., & Nolan, S. (1992). Religious sites as tourism attractions in Europe. *Annals of Tourism Research, 19*(1), 68e78.
442. Nordvall, A., Pettersson, R., Svensson, B., & Brown, S. (2014). Designing events for social interaction. *Event Management, 18*(2), 141e151.
443. Nurse, K. (2004). Trinidad carnival: festival tourism and cultural industry. *Event Management, 8*(4), 223e230.
444. O'Dell, T., & Billing, P. (Eds.). (2005). *Experiencescapes: Tourism, culture and economy*. Copenhagen Business School Press.
445. O'Sullivan, D., Pickernell, D., & Senyard, J. (2009). Public sector evaluation of festivals and special events. *Journal of Policy Research in Tourism, Leisure and Events, 1*(1), 19e36.
446. O'Toole, W. (2011). *Events feasibility and development: From strategy to operations*. Oxford: Butterworth-Heinemann.
447. Oh, M., & Lee, J. (2012). How local festivals affect the destination choice of tourists. *Event Management, 16*(1), 1e9.

448. Ohmann, S., Jones, I., & Wilkes, K. (2006). The perceived social impacts of the 2006 Football World Cup on Munich residents. *Journal of Sport Tourism, 11*(2), 129e153.
449. Oppermann, M., & Chon, K. (1997). Convention participation decision- making process. *Annals of Tourism Research, 24*(1), 178e191.
450. O'Brien, D. (2006). Event business leveraging: the Sydney 2000 Olympic Games. *Annals of Tourism Research, 33*(1), 240e261.
451. Pacione, M. (2012). The role of events in urban regeneration. In S. J. Page, &J. Connell (Eds.), *Routledge handbook of events* (pp. 385e400). London: RoutledgePage, S. J. (1990). Arena development in the UK and urban regeneration in London Docklands. *Sport Place: An International Journal of Sports Geography, 4*(1), 2e15.
452. Page, S. J., & Connell, J. (2010a). *Leisure: An introduction*. Harlow: Pearson. Page, S. J., & Connell, J. (Eds.). (2010b). *Tourism, 6 vols*. Los Angeles: Sage.
453. Page, S. J., & Connell, J. (Eds.). (2012). *Routledge handbook of events*. London: Routledge.
454. Page, S. J., Yeoman, I., Connell, J., & Greenwood, C. (2010). Scenario planning as a tool to understand uncertainty in tourism: the example of transport and tourism in Scotland in 2025. *Current Issues in Tourism, 13*(2), 99e137.
455. Page, S. J., Yeoman, I., Munro, C., Connell, J., & Walker, L. (2006). A case study of best practice e VisitScotland's prepared response to an influenza pandemic. *Tourism Management, 27*(3), 361e393.
456. van Panhuis, W. G., Hyun, S., Blaney, K., Marques, E. T. A., Jr., Coelho, G. E., et al. (2014). Risk of dengue for tourists and teams during the World Cup 2014 in Brazil. *PLoS Neglected Tropical Diseases, 8*(7).
457. Pappas, N. (2014). Hosting mega events: Londoners' support of the 2012 Olympics.
458. *Journal of Hospitality and Tourism Management, 21*, 10e17.
459. Parent, M., & Sequin, B. (2007). Factors that led to the drowning of a world championship organising committee: a stakeholder approach. *European Sport Management Quarterly, 7*(2), 187e212.

460. Park, S. (2009). Segmentation of boat show attendees by motivation and charac- teristics: a case of New York National Boat Show. *Journal of Convention & Event Tourism, 10*(1), 27e49.
461. Park, J., Wu, B., Ye, S., Morrison, A., & Kong, y (2014). The great halls of China? Meeting planners' perceptions of Beijing as an international convention desti- nation. *Journal of Convention & Event Tourism, 15*(4), 244e270.
462. Patterson, I., & Getz, D. (2013). At the nexus of leisure and event studies. *Event Management, 17*(3), 227e240.
463. Pearce, P. (2005). *Tourist behaviour: Themes and conceptual schemas.* Clevedon: Channel View.
464. Pechlaner, H., Dal Bc , G., & Pichler, S. (2013). Differences in perceived destination image and event satisfaction among cultural visitors: the case of the Euro- pean Biennial of Contemporary Art "Manifesta 7". *Event Management, 17*(2), 123e133.
465. Peeters, T., Matheson, V., & Szymanski, S. (2014). Tourism and the 2010 World Cup: lessons for developing countries. *Journal of African Economies, 23*(2), 290e320. Pegg, S., & Patterson, I. (2010). Rethinking music festivals as a staged event: gaining insights from understanding visitor motivations and the experiences they seek. *Journal of Convention and Event Tourism, 11*(2), 85c99.
466. *People and Physical Environment Research, Conference Proceedings.* (1987). Perth WA: Center for Urban Research University of Western Australia.
467. Pernecky, T., & Luck, M. (Eds.). (2012). *Events, society and Sustainability: Critical and contemporary approaches.* Oxon: Routledge.
468. Persson, C. (2002). The Olympic games site decision. *Tourism Management, 23*(1), 27e36. Petrick, J., Bennett, G., & Yosuke. (2013). Development of a scale for measuring event attendees' evaluations of a sporting event to determine loyalty. *Event Man-agement, 17*(2), 97e110.
469. Pettersson, R., & Getz, D. (2009). Event experiences in time and space. *Scandinavian Journal of Hospitality and Tourism, 9*(2/3), 308e326.

470. Pettersson, R., & Zillinger, M. (2011). Time and space in event behaviour: tracking visitors by GPS. *Tourism Geographies, 13*(1), 1e20.
471. Phi, G., Dredge, D., & Whitford, M. (2014). Understanding conflicting perspectives in event planning and management using Q method. *Tourism Management, 40*, 406e415.
472. Picard, D., & Robinson, M. (Eds.). (2006a). *Festivals, tourism and social change: Remaking worlds*. Clevedon: Channel View.
473. Picard, D., & Robinson, M. (2006b). Remaking worlds: festivals, tourism and change. In D. Picard, & M. Robinson (Eds.), *Festivals, tourism and social change: Remaking worlds* (pp. 1e31). Clevedon: Channel View.
474. Pike, S., & Page, S. J. (2014). Destination marketing organizations and destination mar- keting: a narrative analysis of the literature. *Tourism Management, 41*, 202e227.
475. Pine, B., & Gilmore, J. (1999). *The experience economy: Work is theatre and every business a stage*. Boston: Harvard Business School Press.
476. Pitts, B. (1999). Sports tourism and niche markets: Identification and analysis of the growing lesbian and gay sports tourism industry. *Journal of Vacation Marketing, 5*(1), 31e50.
477. Ploner, J., & Robinson, M. (2012). Tourism at the Olympic Games: visiting the world. *Journal of Tourism and Cultural Change, 10*(2), 99e104. Special Issue: Tourism at The Olympics.
478. Potwarka, L. R., Nunkoo, R., & McCarville, R. E. (2014). Understanding television viewership of a mega event: the case of the 2010 Winter olympics. *Journal of Hospitality Marketing and Management, 23*(5), 536e563.
479. Poynter, G., & MacRury, I. (Eds.). (2009). *Olympic cities: 2012 and the remaking of London*. Ashgate Publishing, Ltd.
480. Prabha, P., Rolfe, J., & Sinden, J. (2006). A travel cost analysis of the value of special events: gemfest in Central Queensland. *Tourism Economics, 12*(3), 403e420.
481. Pranic, L., Petric, L., & Cetinic, L. (2012). Host population perceptions of the social impacts of sport tourism events in transition countries:

evidence from Croatia. *International Journal of Event and Festival Management, 3*(3), 236e256.

482. Prayag, G., & Grivel, E. (2014). Motivation, satisfaction, and behavioral intentions: segmenting youth participants at the Interamnia World Cup 2012. *Sport Marketing Quarterly, 23*(3), 148e160.

483. Prayag, G., Hosany, S., Nunkoo, R., & Alders, T. (2013). London residents' support for the 2012 Olympic Games: the mediating effect of overall attitude. *Tourism Management, 36*, 629e640.

484. Presenza, A., & Sheehan, L. (2013). Planning tourism through sporting events. *International Journal of Event and Festival Management, 4*(2), 125e139.

485. Preuss, H. (2004). *The economics of staging the olympics: A comparison of the games 1972e2008*. Edward Elgar Publishing.

486. Preuss, H. (2005). The economic impact of visitors at major multi-sport events. *European Sport Management Quarterly, 5*(3), 281e301.

487. Preuss, H. (Ed.). (2007). *The impact and evaluation of major sporting events*. London: Routledge.

488. Preuss, H. (2009). Opportunity costs and efficiency of investments in mega sport events. *Journal of Policy Research in Tourism, Leisure and Events, 1*(2), 131e140.

489. Pugh, C., & Wood, E. (2004). The strategic use of events within local government: a study of London Borough councils. *Event Management, 9*(1/2), 61e71.

490. Pyo, S., Cook, R., & Howell, R. (1988). Summer Olympic tourist marketdLearning from the Past. *Tourism Management, 9*(2), 137e144.

491. Quinn, B. (2006). Problematising 'festival tourism': arts festivals and sustainable development in Ireland. *Journal of Sustainable Tourism, 14*(3), 288e306.

492. Quinn, B. (2009). The European capital culture initiative and cultural legacy: an analysis of the cultural sector in the aftermath of Cork 2005. *Event Management, 13*, 249e264.

493. Quinn, B. (2010). Arts festivals, urban tourism and cultural policy. *Journal of Policy Research in Tourism, Leisure and Events, 2*(3), 264e279.

494. Raj, R., & Musgrave, J. (Eds.). (2009). *Event management and sustainability*. Wall- ingford UK: CABI.
495. Ramirez, D., Laing, J., & Mair, J. (2013). Exploring intentions to attend a convention: a gender perspective. *Event Management, 17*(2), 165e178.
496. Ray, N., McCain, G., Davis, D., & Melin, T. (2006). Lewis and Clark and the corps of discovery: re-enactment event tourism as authentic heritage travel. *Leisure Studies, 25*(4), 437e454.
497. Reid, G. (2006). The politics of city imaging: a case study of the MTV Europe Music Awards in Edinburgh 03. *Event Management, 10*(1), 35e46.
498. Reid, S. (2007). Identifying social consequences of rural events. *Event Management, 11*, 89e98.
499. Richards, G. (Ed.). (1996). *European cultural tourism: Trends and future prospects*.Wallingford, UK: CABI.
500. Richards, G. (Ed.). (2007). *Cultural tourism: Global and local perspectives*. New York: Haworth.
501. Richards, G., de Brito, M., & Wilks, L. (2013). *Exploring the social impacts of events*.Abingdon: Routledge.Richards, G., & Palmer, R. (2010). *Eventful Cities: Cultural management and urban revitalization*. Oxford: Butterworth-Heinemann.
502. Richards, P., & Ryan, C. (2004). The Aotearoa Traditional Maori performing Arts festival 1972e2000: a case study of cultural event maturation. *Journal of Tourism and Cultural Change, 2*(2), 94e117.
503. Ritchie, J. R. B. (1984). Assessing the impacts of hallmark events: conceptual and research issues. *Journal of Travel Research, 23*(1), 2e11.
504. Ritchie, J. R. B. (2000). Turning 16 days into 16 years through Olympic legacies. *Event Management, 6*(2), 155e165.
505. Ritchie, B., & Adair, D. (2004). *Sport tourism: Interrelationships, impacts and issues*.Clevedon: Channel View.
506. Ritchie, J. R. B., & Beliveau, D. (1974). Hallmark events: an evaluation of a strategic response to seasonality in the travel market. *Journal of Travel Research, 14*, 14e20.
507. Ritchie, B., Sanders, D., & Mules, T. (2006). Televised events: shaping destination images and perceptions of capital cities from

the couch. In C. Arcodia, M. Whitford, & C. Dickson (Eds.), *Global events congress, proceedings* (pp. 286e299). Brisbane: University of Queensland.
508. Ritchie, B., Shipway, R., & Chien, M. (2010). The role of the media in influencing residents' support for the 2012 Olympic Games. *International Journal of Event and Festival Management, 1*(3), 202e219.
509. Ritchie, J. R. B., & Smith, B. (1991). The impact of a mega-event on host region awareness: a longitudinal study. *Journal of Travel Research, 30*(1), 3e10.
510. Rittichainuwat, B., Beck, J., & LaLopa, J. (2001). Understanding motivations, in- hibitors, and facilitators of association members in attending international conferences. *Journal of Convention and Exhibition Management, 3*(3), 45e62.
511. Rittichainuwat, B., & Mair, J. (2012). Visitor attendance motivations at consumer travel exhibitions. *Tourism Management, 33*(5), 1236e1244.
512. Robertson, M., Rogers, P., & Leask, A. (2009). Progressing socio-cultural impact evaluation for festivals. *Journal of Policy Research in Leisure, Tourism and Events, 1*(2), 156e169.
513. Robertson, M., & Wardrop, K. (2012). Festivals and events, government and spatial governance. In S. J. Page, & J. Connell (Eds.), *The handbook of events* (pp. 489e506). London: Routledge.
514. Robinson, R., & Clifford, C. (2012). Authenticity and festival foodservice experiences. *Annals of Tourism Research, 39*(2), 571e600.
515. Robinson, T., & Gammon, S. (2004). A question of primary and secondary motives: revisiting and applying the sport tourism framework. *Journal of Sport Tourism, 9*(3), 221e223.
516. Robinson, M., Picard, D., & Long, P. (2004). Festival tourism: producing, translating, and consuming expressions of culture(s). *Event Management, 8*(4), 187e189. Roche, M. (2000). *Mega-events and modernity: Olympics and expos in the growth of global culture*. London: Routledge.
517. Roche, M. (2006). Mega-events and modernity revisited: globalization and the case of the Olympics. *Sociological Review, 54*, 25e40.

518. Rollins, R., & Delamare, T. (2007). Measuring the social impact of festivals. *Annals of Tourism Research, 34*(3), 805e808.
519. Rooney, J. (1988). *Mega sports events as tourism attractions: A geographical analysis.*Paper presented at the TTRA annual conference, Montreal.
520. Rosentraub, M. (2009). *Major league winners: Using sports and cultural centers as tools for economic development.* Boca Raton FL: CRC Press.
521. Rosentraub, M., & Joo, M. (2009). Tourism and economic development: which in- vestments produce gains for regions? *Tourism Management, 30*(5), 759e770.
522. Rozin, S. (2000). *The amateurs who saved Indianapolis.* Business Week. April 10.
523. Ruback, R., Pandey, J., & Kohli, N. (2008). Evaluations of a sacred place: role and religious belief at the Magh Mela. *Journal of Environmental Psychology, 28*(2), 174e184.
524. Ryan, C. (Ed.). (2002). *The tourist experience* (2nd ed.). London: Continuum.
525. Ryan, C., & Lockyer, T. (2002). Masters' GamesdThe nature of competitors' involvement and requirements. *Event Management, 7*(4), 259e270.
526. Ryan, C., Smee, A., Murphy, S., & Getz, D. (1998). New Zealand events: a temporal and regional analysis. *Festival Management and Event Tourism, 5*(1/2), 71e83.
527. Saayman, M., & Saayman, A. (2012). The economic impact of the Comrades mara- thon. *International Journal of Event and Festival Management, 3*(3), 220e235.
528. Sadd, D. (2014). The future is virtual. In I. Yeoman, et al. (Eds.), *The future of events and festivals* (pp. 209e218). Abingdon: RoutledgeAbingdon. Routledge.
529. Saleh, F., & Ryan, C. (1993). Jazz and knitwear: factors that attract tourists to fes- tivals. *Tourism Management, 14*(4), 289e297.
530. Sant, S.-L., Mason, D., & Hinch, T. (2013). Conceptualising Olympic tourism legacy: destination marketing organisations and Vancouver 2010. *Journal of Sport & Tourism, 18*(4), 287e312.

531. Santos, J. M. C. M. (2014). Brazil: an emerging power establishing itself in the world of international sports mega-events. *International Journal of the History of Sport, 31*(10), 1312e1327.
532. Santos-Lewis, R., & Moital, M. (2013). Constraints to attend events across speciali- zation levels. *International Journal of Event and Festival Management, 4*(2), 107e124.
533. Savinovic, A., Kim, S., & Long, P. (2012). Audience members' motivation, satisfaction, and intention to re-visit an ethnic minority cultural festival. *Journal of Travel & Tourism Marketing, 29*(7), 682e694.
534. Schlentrich, U. (2008). The MICE industry: meetings, incentives conventions and exhibitions. In B. Brotherton, & R. Woods (Eds.), *The Sage handbook of hospitality management* (pp. 400e420). London: Sage.
535. Schneider, I., & Backman, S. (1996). Cross-cultural equivalence of festival motiva- tions: a study in Jordan. *Festival Management and Event Tourism, 4*(3e4), 3e4.
536. Schulenkorf, N., Thomson, A., & Schlenker, K. (2011). Intercommunity sport events: vehicles and catalysts for social capital in divided societies. *Event Management, 15*(2), 105e119.
537. Scotinform Ltd. (1991). *Edinburgh Festivals study 1990e91: Visitor survey and eco- nomic impact assessment.* Final report. Edinburgh: Scottish Tourist Board.
538. Scott, D. (1996). A comparison of visitors' motivations to attend three urban festi- vals. *Festival Management and Event Tourism, 3*(3), 121e128.
539. Scott, D., Steiger, R., Rutty, M., & Johnson, P. (2014). The future of the Olympic Winter Games in an era of climate change. *Current Issues in Tourism.*
540. Severt, K., Fjelstul, J., & Breiter, D. (2009). A comparison of motivators and inhibitors for association meeting attendance for three generational cohorts. *Journal of Convention & Event Tourism, 10*(2), 105e119.

541. Severt, D., Wang, Y., Chen, P., & Breiter, D. (2007). Examining the motivation, perceived performance, and behavioural intentions of convention attendees: evidence from a regional conference. *Tourism Management, 28*(2), 399e408.
542. Sharpley, R., & Stone, P. (2012). Socio-cultural impacts of events: meanings, authorised transgression and social capital. In S. Page, & J. Connell (Eds.), *The Routledge handbook of events*. London: Routledge.
543. Shaw, B. (1985). Freemantle, W. A. and the America's Cup... the spectre of devel- opment. *Urban Policy and Research, 3*(2), 38e40.
544. Sherwood, P. (2007). *A triple bottom line evaluation of the impact of special events: The development of indicators*. Unpublished Doctoral dissertation. Melbourne: Victoria University.
545. Sherwood, P., Jago, L., & Deery, M. (2004). Sustainability reporting: an application for the evaluation of special events. In C. Cooper, et al. (Eds.), *Proceedings of the annual conference of the Council for Australian University Tourism and Hospitality Education*. Brisbane: University of Queensland.
546. Sherwood, P., Jago, L., & Deery, M. (2005). Unlocking the triple bottom line of special event evaluations: what are the key impacts? In J. Allen (Ed.), *Proceedings of the third international event management research conference, Sydney* Sydney: Uni- versity of Technology.
547. Shibli, S., & The Sport Industry Research Centre. (2002). *The 2002 embassy world snooker championship, an evaluation of the economic impact, place marketing effects, and visitors' perceptions of Sheffield*. For Sheffield City Council.
548. Shinde, K. (2010). Managing Hindu festivals in pilgrimage sites: emerging trends, opportunities, and challenges. *Event Management, 14*, 53e67.
549. Shipway, R., & Fyall, A. (Eds.). (2012). *International sports events: Impacts, experiences and identities*. Abingdon: RoutledgeAbingdon. Routledge.

550. Shipway, R., & Jones, I. (2007). Running away from home: understanding visitor experiences and behaviour at sport tourism events. *International Journal of Tourism Research, 9*(5), 373e383.
551. Shipway, R., & Jones, I. (2008). The great suburban Everest: an 'insiders' perspective on experiences at the 2007 Flora London Marathon. *Journal of Sport and Tourism, 1,* 61e77.
552. Singh, R. (2006). Pilgrimage in Hinduism: historical context and modern perspec- tives. In D. Timothy, & D. Olsen (Eds.), *Tourism, religion and spiritual journeys* (pp. 220e236). London, New York: Routledge.
553. Small, K. (2007). Social dimensions of community festivals: an application of factor analysis in the development of the social impact perception (SIP) scale. *Event Management, 11*(1/2), 45e55.
554. Small, K., Edwards, D., & Sheridan, L. (2005). A flexible framework for socio-cultural impact evaluation of a festival. *International Journal of Event Management Research, 1*(1), 66e77.
555. Smith, M. (2005). Spotlight events, media relations, and place promotion: a case study. *Journal of Hospitality and Leisure Marketing, 12*(1/2), 115e134.
556. Smith, A. (2009). Spreading the positive effects of major events to peripheral areas.*Journal of Policy Research in Tourism, Leisure and Events, 1*(3), 231e246.
557. Smith, A. (2012). *Events and urban regeneration: The strategic use of events to revi- talise cities.* Abingdon: Routledge.
558. Sofield, T. (1991). Sustainable ethnic tourism in the South Pacific: some principles.*Journal of Tourism Studies, 1*(3), 56e72.
559. Sofield, T., & Li, F. (1998). Historical methodology and sustainability: an 800-year- old festival from China. *Journal of Sustainable Tourism, 6*(4), 267e292.
560. Sofield, T., & Sivan, A. (2003). From cultural festival to international sportdThe Hong Kong Dragon boat Races. *Journal of Sport Tourism, 81,* 9e20.

561. Solberg, H., Andersson, T., & Shibli, S. (2002). An exploration of the direct economic impacts from business travellers at world championships. *Event Management, 7*(3), 151e164.
562. Song, H., & Li, G. (2008). Tourism demand modelling and forecastingdA review of recent research. *Tourism Management, 29*(2), 203e220.
563. Soutar, G., & McLeod, P. (1993). Residents' perceptions on impact of the America's Cup. *Annals of Tourism Research, 20*(3), 571e582.
564. Spiller, J. (2002). History of convention tourism. In K. Weber, & K. Chon (Eds.), *Convention tourism: International research and industry perspectives* (pp. 3e20). New York: Haworth.
565. Spilling, O. (1998). Beyond intermezzo? On the long-term industrial impacts of mega-events: the case of Lillehammer 1994. *Festival Management and Event Tourism, 5*(3), 101e122.
566. Sports Business Market Research Handbook (2000). Sports Business Market Research Inc.SportBusiness Ultimate Sport Cities (www.sportbusiness.com).
567. Standeven, J., & De Knop, P. (1999). *Sport tourism*. Champaign, IL: Human Kinetics. Sustainable Tourism Cooperative Research Centre (STCRC)(www.sustainabletourismonline.com).
568. Stebbins, R. (1982). Serious leisure: a conceptual statement. *Pacific Sociological Re- view, 25*(2).
569. Stebbins, R. (2006). *Serious leisure: a perspective for our time*. Somerset,NJ: Aldine Transaction Publications.
570. Stein, A., & Evans, B. (2009). *An introduction to the entertainment Industry*. New York: Peter Lang.
571. Stevenson, N. (2012). Culture and the 2012 Games: creating a tourism legacy?*Journal of Tourism and Cultural Change, 10*(2), 137e149.
572. Stokes, R. (2004). A framework for the analysis of events-tourism knowledge net- works. *Journal of Hospitality and Tourism Management, 11*(2),108e123.
573. Stokes, R. (2006). Network-based strategy making for events tourism. *European Journal of Marketing, 40*(5/6), 682e695.

574. Stokes, R. (2008). Tourism strategy making: Insights to the events tourism domain. *Tourism Management, 29*, 252e262.
575. Stokes, R., & Jago, L. (2007). Australia's public sector environment for shaping event tourism strategy. *International Journal of Event Management Research, 3*(1), 42e53. Syme, G., Shaw, B., Fenton, D., & Mueller, W. (Eds.). (1989). *The planning and eval-uation of hallmark events*. Aldershot: Gower.
576. Tanford, S., Montgomery, R., & Nelson, K. (2012). Factors that influence attendance, satisfaction, and loyalty for conventions. *Journal of Convention & Event Tourism, 13*(4), 290e318.
577. Taylor, P., & Gratton, C. (1988). The Olympic games: an economic analysis. *Leisure Management, 8*(3), 32e34.
578. Teigland, J. (1996). *Impacts on tourism from mega-events: The case of winter Olympic games*. Sogndal: Western Norway Research Institute.
579. The National Endowment for the Arts. Survey of public Participation in the Arts. Washington D.C. (http://arts.gov/news/2013/national-endowment-arts-presents-highlights-2012-survey-public-participation-arts#sthash.cqCz23dy.dpuf).
580. Theodoraki, E. (2009). Organizational communication on the intended and achieved impacts of the Athens 2004 Olympic Games. *Journal of Policy Research in Tourism, Leisure and Events, 1*(2), 141e155.
581. Thomas, R., & Wood, E. (2004). Event-based tourism: a survey of local authority strategies in the UK. *Local Governance, 29*(2), 127e136.
582. Thomson, A., Schlenker, K., & Schulenkorf, N. (2013). Conceptualizing sport event legacy. *Event Management, 17*(2), 111e122.
583. Thrane, C. (2002). Music quality, satisfaction, and behavioural intentions within a jazz festival context. *Event Management, 7*(3), 143e150.
584. Tikkanen, I. (2008). Internationalization process of a music festival: case Kuhmo Chamber music festival. *Journal of Euromarketing, 17*(2), 127e139.

585. Timothy, D., & Olsen, D. (Eds.). (2006). *Tourism, religion and spiritual journeys*. London, New York: Routledge.
586. Tkaczynski, A. (2013). A stakeholder approach to attendee segmentation: a case study of an Australian Christian music festival. *Event Management, 17*, 283e298.
587. Tkaczynski, A., & Rundle-Thiele, S. (2011). Event segmentation: a review and research agenda. *Tourism Management, 32*(2), 426e434.
588. Tkaczynski, A., & Toh, Z. H. (2014). Segmentation of visitors attending a multicul- tural festival: an Australian Scoping Study. *Scandinavian Journal of Hospitality and Tourism, 14*(3), 296e314.
589. Tomljenovic, R., & Weber, S. (2004). Funding cultural events in Croatia: tourism- related policy issues. *Event Management, 9*(1/2), 51e59.
590. Toohey, K., & Veal, T. (2007). *The olympic games: A social science perspective*. Wall- ingford: CABI.
591. Trail, G., & James, J. (2001). The motivation scale for sport consumption: assessment of the scale's psychometric properties. *Journal of Sport Behaviour, 24*(1), 108e127.
592. Trauer, B. (2006). Conceptualizing special interest tourismdframeworks for anal- ysis. *Tourism Management, 27*(2), 183e200.
593. Travel Industry Association of America (TIAA). (1999). *Profile of travellers who attend sports events*. Washington, DC: TIAA.
594. Travel Industry Association of America, and Smithsonian Magazine. (2003). *The historic/cultural traveller*.
595. Travel and Tourism Research Association (TTRA). (1986). In *Proceedings of the Canada chapter conference*.
596. Trost, K., & Milohnic, I. (2012). Management attitudes towards event impacts in the tourist destination: the case of Istria, Croatia. *Event Management, 16*(1), 51e64. Turco, D. (1995). Measuring the tax impacts of an international festival: justification for government sponsorship. *Festival Management and Event Tourism, 2*(3/4), 191e195.
597. Turco, D., Riley, R., & Swart, K. (2002). *Sport tourism*. Morgantown, WV: Fitness Information Technology Inc.

598. Turner, V. (1969). *The ritual process: Structure and anti-structure.* New York: Aldine de Gruyter.
599. Turner, V. (1974). Liminal to liminoid, in play, flow and ritual: an essay in comparative symbology. In E. Norbeck (Ed.), *The anthropological study of human play* (vol. 60, pp. 53e92). Rice University Studies.
600. Turner, V. (Ed.). (1982). *Celebration: Studies in festivity and ritual.* Washington: Smithsonian Institution Press.
601. UFI. (2011). *Global exhibition industry statistics, 2011.* The Global Association of the Exhibition Industry. retrieved from http://www.ufi.org/Medias/pdf/thetradefairsector/2011_exhibiton_industry_statistics.pdf.
602. UFI. (2012). *The global exhibition barometer e Survey of the exhibition industry.* The Global Association of the Exhibition Industry. Retrieved from http://www.ufi.org/Medias/pdf/thetradefairsector/surveys/ufi_global_exhibition_barometer_july_2012.pdf.
603. UK Cabinet Office Performance and Innovation Unit. (2001). *A futurist's toolbox: Methodologies in futures work, strategic futures team.* London: UK Cabinet Office Performance and Innovation Unit.
604. The Contribution of Music Festivals and Major Concerts to Tourism in the UK by UK Music (www.ukmusic.org).
605. Universidad Europea de Madrid. (iede.universidadeuropea.es/en/academic-programs/... tourism.../data).
606. Unruh, D. (1980). The nature of social worlds. *The Pacific Sociological Review, 23*(3), 271e296.
607. US National Association of Sport Commissions (www.sportscommissions.org/).
608. Uysal, M., Gahan, L., & Martin, B. (1993). An examination of event motivations: a case study. *Festival Management and Event Tourism, 1*(1), 5e10.
609. Uysal, M., & Gitelson, R. (1994). Assessment of economic impacts: festivals and special events. *Festival Management and Event Tourism, 2*(1), 3e9.do Valle, P. O., Mendes, J., & Guerreiro, M. (2012). Residents'

participation in events, events image, and destination image: a correspondence analysis. *Journal of Travel & Tourism Marketing, 29*(7), 647e664.

610. Van Gennep, A. (1909). *The rites of passage (1960 translation by M. Vizedom and G. Coffee)*. London: Routledge and Kegan Paul.

611. Van Zyl, C., & Botha, C. (2004). Motivational factors of local residents to attend the Aardklop national arts festival. *Event Management, 8*(2), 213e222.

612. Vargo, S. L., & Lusch, R. F. (2004). The four service marketing myths: remnants of a goods-based, manufacturing model. *Journal of Service Research, 6*(4), 324e335. Vargo, S. L., & Lusch, R. F. (2008). From goods to Service(s): divergences and con-vergences of logics. *Industrial Marketing Management, 37*, 254e259.

613. Vaughan, R. (1979). *Does a festival pay? A case study of the Edinburgh Festival in 1976. Tourism Recreation Research Unit*. Working paper 5. University of Edinburgh.

614. Veitch, S. (2013). Mythology as hallmark event legacy: the endurance of America's Cup mythology in the city of Fremantle. *Journal of Sport & Tourism*,1e14 (ahead-of-print). Verhoven, P., Wall, D., & Cottrell, S. (1998). Application of desktop mapping as a mar- keting tool for special events planning and evaluation: a case study of the Newport News Celebration in Lights. *Festival Management and Event Tourism, 5*(3),123e130.

615. Waitt, G. (2003). Social impacts of the Sydney Olympics. *Annals of Tourism Research, 30*(1), 194e215.

616. Waitt, G. (2004). A critical examination of Sydney's 2000 Olympic games. In I. Yeoman, et al. (Eds.), *Festivals and events management* (pp. 391e408). Oxford: Elsevier.

617. Wallace, T. (2007). Went the day well: scripts, glamour and performance in war- weekends. *International Journal of Heritage Studies, 13*(3), 200e223.

618. Walle, A. (1994). The festival life cycle and tourism strategies: the case of the Cowboy Poetry Gathering. *Festival Management and Event Tourism, 2*, 85e94.

619. Wan, P. (2011). Assessing the strengths and weaknesses of Macao as an attractive meeting and convention destination: perspectives of key informants. *Journal of Convention & Event Tourism, 12*(2), 129e151.
620. Wang, Y., & Pfister, R. (2008). Residents' attitudes toward tourism and perceived personal benefits in a rural community. *Journal of Travel Research, 47*, 84e93.
621. Wang, C., & Yu, L. (2015). Managing student volunteers for mega events: motivation and psychological contract as predictors of sustained volunteerism. *Asia Pacific Journal of Tourism Research, 20*(3), 338e357.
622. Wann, D. (1995). preliminary validation of the sport fan motivation scale. *Journal of Sport and Social Issues, 19*(4), 377e396.
623. Wann, D., Grieve, F., Zapalac, R., & Pease, D. (2008). Motivational profiles of sport fans of different sports. *Sport Marketing Quarterly, 17*(1), 6e19.
624. Wann, D. L., Schrader, M. P., & Wilson, A. M. (1999). Sport fan motivation: ques- tionnaire validation, comparisons by sport, and relationship to athletic moti- vation. *Journal of Sport Behavior, 22*(1), 114e139.
625. Warnick, R. B., Bojanic, D. C., Mathur, A., & Ninan, D. (2011). Segmenting event at- tendees based on travel distance, frequency of attendance, and involvement measures: a cluster segmentation technique. *Event Management, 15*(1), 77e90.
626. Weber, K., & Chon, K. (2002). *Convention tourism: International research and industry perspectives.* New York: Haworth.
627. Weber, K., & Ladkin, A. (2004). Trends affecting the convention industry in the 21st century. *Journal of Convention and Event Tourism, 6*(4), 47e63.
628. Weber, K., & Ladkin, A. (2009). Career Anchors of Convention and Exhibition In- dustry Professionals in Asia. *Journal of Convention & Event Tourism Volume, 10*(4), 243e255.
629. Weed, M. (2003). Why the two won't tango! Explaining the lack of integrated policies for sport and tourism in the UK. *Journal of Sport Management, 17*(3), 258e283.

630. Weed, M. (2006). Sports tourism research 2000e2004: a systematic review of knowl- edge and a meta-evaluation of methods. *Journal of Sport and Tourism, 11*(1), 5e30.
631. Weed, M. (2009). *Sport and tourism: A reader.* London: Routledge. Weed, M. (Ed.). (2012a). *Olympic tourism* (2nd ed.). London: Routledge.
632. Weed, M. (2012b). Towards an interdisciplinary events research agenda across sport, tourism, leisure and health. In S. J. Page, & J. Connell (Eds.), *Routledge handbook of events.* London: Routledge.
633. Weed, M., & Bull, C. (2009). In *Sports tourism: Participants, policy and providers* (2nd. ed). Oxford: Elsevier.
634. Westerbeek, H., Turner, P., & Ingerson, L. (2002). Key success factors in bidding for hallmark sporting events. *International Marketing Review, 19*(3), 303e322.
635. Whitfield, J., Dioko, L. D. A. N., Webber, D., & Zhang, L. (2014). Attracting convention and exhibition attendance to complex MICE venues: emerging data from Macao. *International Journal of Tourism Research, 16*(2), 169e179.
636. Whitford, M. (2004a). Regional development through domestic and tourist event policies: Gold Coast and Brisbane, 1974e2003. *UNLV Journal of Hospitality, Tourism and Leisure Science, 1*, 1e24.
637. Whitford, M. (2004b). Event public policy development in the northern sub- regional organisation of councils, Queensland Australia: rhetoric or realisa- tion? *Journal of Convention and Event Tourism, 6*(3), 81e99.
638. Whitford, M. (2009). A framework for the development of event public policy: facilitating regional development. *Tourism Management, 30*(5), 674e682.
639. Whitford, M., & Dunn, A. (2014). Papua new guinea's indigenous cultural festivals: cultural tragedy or triumph? *Event Management, 18*(3), 265e283.
640. Whitson, D., & Horne, J. (2006). Underestimated costs and overestimated benefits? Comparing the outcomes of sports mega-events in Canada and Japan. *Socio- logical Review, 54*(2), 71e89.

641. Whitson, D., & Macintosh, D. (1996). The global circus: international sport, tourism, and the marketing of cities. *Journal of Sport and Social Issues, 20*(3), 275e295.

642. Wicker, P., & Hallmann, K. (2013). Estimating consumer's willingness-to-pay for participation in and traveling to Marathon events. *Event Management, 17*(3), 271e282.

643. Wicks, B., & Fesenmaier, D. (1995). Market potential for special events: a mid- western case study. *Festival Management and Event Tourism, 3*(1), 25e31.

644. Williams, K., Laing, J., & Frost, W. (2013). *Fashion, design and events*. Abingdon: Routledge.

645. Wood, E. (2005). Measuring the economic and social impacts of local authority events. *International Journal of Public Sector Management, 18*(1), 37e53.

646. Woosnam, K., Van Winkle, C., & An, S. (2013). Confirming the festival social impact attitude scale in the context of a rural Texas cultural festival. *Event Management, 17*(3), 257e270.

647. Wooten, M., & Norman, W. (2008). Differences in arts festival visitors based on level of past experience. *Event Management, 11*(3), 109e120.

648. Xiao, H., & Smith, S. (2004a). Residents' perceptions of Kitchener-Waterloo Okto- berfest: an inductive analysis. *Event Management, 8*(3), 161e175.

649. Xiao, P., & Smith, S. (2004b). Improving forecasts for world's fair attendance: incorporating income effects. *Event Management, 6*(1), 15e23.

650. Xie, P. (2003). The Bamboo-beating dance in Hainan, China. Authenticity and commodification. *Journal of Sustainable Tourism, 11*(1), 5e16.

651. Xie, P. (2004). Visitors' perceptions of authenticity at a rural heritage festival: a case study. *Event Management, 8*(3), 151e160.

652. Xing, X., Chalip, L., & Green, B. C. (2014). Marketing a social experience: how celebration of subculture leads to social spending during a sport event. *Sport Marketing Quarterly, 23*(3).

653. Yaghmour, S., & Scott, N. (2011). Inter-organizational collaboration characteristics and outcomes: a case study of the Jeddah Festival. *Journal of Policy Research in Tourism, Leisure and Events, 1*(2), 115e130.
654. Yang, J., Gu, Y., & Cen, J. (2011). Festival tourists' emotion, perceived value, and behavioural intentions: a test of the moderating effect of festivalscape. *Journal of Convention & Event Tourism, 12*(1), 25e44.
655. Yeoman, I. (2013). A futurist's thoughts on consumer trends shaping future festivals and events. *International Journal of Event and Festival Management, 4*(3), 249e260.
656. Yeoman, I., Robertson, M., McMahon-Beattie, U., Smith, K., & Backer, E. (Eds.). (2014). *The future of events and festivals*. Abingdon: Routledge.
657. Yoo, J., & Chon, K. (2010). Temporal changes in factors affecting convention participation decision. *International Journal of Contemporary Hospitality Man- agement, 22*(1), 103e120.
658. Yoon, S., Spencer, D., Holecek, D., & Kim, D. (2000). A profile of Michigan's festival and special event tourism market. *Event Management, 6*(1), 33e44.
659. Yoo, J., & Weber, K. (2005). Progress in convention tourism research. *Journal of Hospitality and Tourism Research, 29*(2), 194e222.
660. Yuan, J., Cai, L., Morrison, A., & Linton, S. (2005). An analysis of wine festival at- tendees' motivations: a synergy of wine, travel and special events? *Journal of Vacation Marketing, 11*(1), 41e58.
661. Yuan, J., & Jang, S. (2008). The effects of quality and satisfaction on awareness and behavioural intentions: exploring the role of a wine festival. *Journal of Travel Research, 46*(3), 279e288.
662. Yuan, J., Morrison, A., Cai, L., & Linton, S. (2008). A model of wine tourist behaviour: a festival approach. *International Journal of Tourism Research, 10*(3), 207e219.
663. Zhou, Y. (2010). Resident perceptions toward the impacts of the Macao grand prix. *Journal of Convention & Event Tourism, 11*(2), 138e153.

664. Zhou, Y., & Ap, J. (2009). Residents' perceptions towards the impacts of the Beijing 2008 Olympic games. *Journal of Travel Research, 48*(1), 78e91.
665. Ziakis, V. (2010). Understanding an event portfolio: The uncovering of in- terrelationships, synergies and leveraging opportunities. *Journal of policy Research in Tourism, Leisure and Events, 2*(2), 144e164.
666. Ziakas, V. (2013). *Event portfolio planning and management: A holistic approach*. Abingdon: Routledge.
667. Ziakas, V., & Boukas, N. (2012). A neglected legacy: examining the challenges and potential for sport tourism development in post-Olympic Athens. *International Journal of Event and Festival Management, 3*(3), 292e316.

CHAPTER 6

Destination Marketing Organizations and Climate Change—The Need for Leadership and Education

Rachel Dodds

Ted Rogers School of Hospitality & Tourism Management, Ryerson University, 350 Victoria Street, Toronto, ON, M5B 2K3, Canada

ABSTRACT

: Destination Marketing Organizations (DMOs) operate at many levels ranging from the national to the municipal and have evolved over the years to respond to the geographical and political realities that are associated with tourism supply. Alongside providing information to potential visitors, DMOs work to make a destination attractive by showcasing its unique aspects and attractions. As the appeal of destinations, cost of doing business and the destination brand may be affected by the possible effects of climate change, this study aims to identify opportunities and threats to municipal and provincial/territorial DMOs and their members as well as identify measures they are undertaking to address the potential impacts. A study conducted of Canada's provincial and municipal large DMOs was conducted in 2009. This research found that awareness of climate change in Canada's tourism industry is increasing, but more efforts must be undertaken to mitigate climate change. To address climate change and tourism, this paper suggests doing three things: (a) DMOs need to demonstrate leadership about climate change education and mitigation to all their members; (b) government policy and action are needed to provide incentives for industry to address climate change; and (c) industry members require further education to take the steps necessary mitigate risk and to adapt. The internet has changed the DMOs' roles and how they provide information to the consumer; as such, they have been presented with an opportunity to take on new roles as educational and marketing providers. This paper will outline in the current shifts among Canadian DMOs and will discuss the key issues that are applicable to DMOs worldwide.

KEYWORDS

destination marketing organizations; climate change; leadership; destination appeal

1. INTRODUCTION

The global scope of the tourism industry and its reliance on the mobilization of almost one billion people each year make the industry an important contributor of greenhouse gases and to climate change. The World Economic Forum (WEF) estimates that the tourism sector is responsible for 5% of global anthropogenic greenhouse gas emissions [1] Due to its size and dependence on its natural environment, tourism is particularly vulnerable to the impacts of climate change [2]. This study does not seek to detail or provide examples of specific climate scenarios but instead aims to understand opportunities and barriers facing tourism destinations with regard to climate change. Destination Marketing Organizations were the focus of this study as it is these organizations that represent much of a country's tourism product.

The Intergovernmental Panel on Climate Change (IPCC) defines climate change as

1. Introduction

"…a change in the state of the climate that can be identified (e.g., using statistical tests) by changes in the mean and/or the variability of its properties, and that persists for an extended period, typically decades or longer. Climate change may be due to natural internal processes or external forcings, or to persistent anthropogenic changes in the composition of the atmosphere or in land use".[3]

Although there is a natural cycle of climate changes [4], the extreme changes that are now facing the planet are agreed to be directly related to the rapid technological advances and population growth in the last century [4]. The UNEP/UNWTO [2] attempts to quantify the links between tourism and climate change and concludes that carbon dioxide emissions from tourism activities (transport, accommodation, activities) account for an estimated 4–6% of total emissions worldwide. If mitigation measures are not implemented, tourism's contribution to carbon dioxide emissions could grow by 150% in the next 30 years [5]. Arguably, if climate change is not properly addressed through adequate adaptive and or mitigation measures, it could lead to a shift in the attractiveness of destinations around the world [6,7,8,9,10,11,12]. While some locations could experience higher levels of attractiveness due to favourable temperatures, other destinations could become less appealing, causing a shift in visitation patterns [13]. The impacts of climate change on the tourism industry have been categorized into direct climatic impacts (e.g., redistribution of climatic assets among tourism regions), indirect environmental change impacts (e.g., water shortages, biodiversity loss), impacts of mitigation policies on tourist mobility (e.g., changes in tourism flows due to increased prices), and indirect societal changes (e.g., changes in economic growth) [5].

Destination Marketing Organizations (DMOs) are tasked with the role of marketing their region to travel consumers and travel trade intermediaries [14]. This study identifies the opportunities and challenges that climate change poses to Canada's tourism products in an urban as well as provincial setting. It then determines which DMOs have implemented initiatives to mitigate or adapt to impacts of climate change. Lastly, this study identifies barriers and incentives necessary to influence the tourism industry in

Canada's different regions to increase their level of participation in addressing effects of climate change. This paper will argue that although some DMOs have undertaken voluntary efforts to address climate change, leadership and education is absent and national legislation is needed if climate change is to be addressed in the tourism industry.

1.1. Climate Change and Tourism—A Canadian Context

Tourism is one of the leading growth sectors in Canada. Approximately 30 million non-resident travellers entered Canada in 2007—as well as approximately 34 million Canadians travelling within the country. Business and leisure travellers (both international and domestic) spent a total of $70.6 billion in Canada in 2007 alone [15]. Foreign travellers account for 23% of total Canadian tourism expenditures, making tourism an important export industry [15,16]. The impacts of climate change may positively affect Canada's key tourism product especially with references to the recreational and nature-based tourism industry in Canada [17]. Warm weather outdoor recreational activities such as summer festivals, campgrounds, golfing, and public beach access could experience benefits from the longer operating seasons [12,18,19]. Conversely, winter recreation activities may be negatively affected, which may result in a decline or elimination of sports such as skiing, snowmobiling, or tourism related events such as winter festivals [17,18]. For example, the Canadian Rocky Mountains experienced a 25% decrease in glacier coverage in the 20th century [17].

"Many of the species that Canada features in its tourism marketing (whales, otters, caribou, polar bears) are already endangered, and rapid changes in their habitat increase the risk of their extinction. Habitat changes such as the warming of Canada's boreal forest is encouraging the spread of new pests and diseases such as the pine beetle which reduce valuable timber inventories and blight the landscape and aesthetics sought by visitors".([9], p. 41)

As increasing studies about tourism and climate change outline [7,8,20,21,22,23], tourism will be affected and efforts are increasingly needed

1. Introduction

to curtail negative impacts. There are two measures within tourism that are used to address climate change: mitigation and adaptation. Mitigation usually refers to reducing greenhouse gas emissions so as to lessen the impact of tourism on climate change whereas adaptation refers to adapting or modifying actions or activities to adjust to a changing climate. Within a destination context, many tourism operations and organizations rely on marketing their offerings through a DMO, whether they are urban or rural based. Understanding what DMOs perceive climate change issues to be and the efforts they are taking to address such issues may provide some insight into climate change mitigation and adaption.

1.2. Destination Marketing Organizations

DMOs promote their regions with the intent of increasing tourism to the destination or improving its public image [24]. Most DMOs in Canada are private, not-for-profit industry associations with an independent board of directors. DMOs bridge various sectors within the tourism industry including: hotels, B&Bs, airlines and ground transit, attractions, tour operators, travel agents, convention facilitators, retail operators and restaurants. Funding for DMOs come from a combination of public funding and revenues generated through membership and service charges. Many Canadian municipalities have sought to increase their tourism revenue by establishing DMOs to market their region to travellers and travel trade intermediaries (e.g., travel agencies). In some municipalities, hotels can voluntarily charge a Destination Marketing Fee on hotel rooms, which is then channelled back to the DMO. Traditionally services provided by DMOs include promotional and advertising campaigns, publishing and distributing visitors' guides, online marketing, training, research and product development. One crucial market communication channel for DMOs has been the Internet, as information technology (IT) has transformed the tourism industry into a digital economy [25]. Although the integration of IT into the fabric of a DMOs overall marketing strategy is considered as an important key to success [26], the exponential explosion of information online has presented the DMOs with increasing challenges as well as opportunities in the information era. DMOs are challenged because of threats in their external environments such as the increased use of technology, changes in the industry's structure and market changes [26]. Additionally, many potential customers use alternative internet sources to find information about the destination, thereby challenging DMOs to provide relevant, added-value information which differentiates them from these alternative sources.

As there is little action on climate change within Canada by the government to mitigate or adapt at this time, DMOs could be uniquely situated to address the issues of climate change in the tourism industry. DMOs link the interests of the tourism industry to government policy-makers and consumer demand. DMOs are also equipped to identify shifts in the marketplace such as increased demand for climate-friendly tourism options, and provide guidance and education for industry to meet the demands of new climate change legislation. Due to this context, this research sought to identify the current level of awareness, and efforts to address climate change by DMOs and/or their members. A number of Canadian DMOs are located in urban centres as it is these regions that serve as transportation hubs for travellers en route to other parts of the surrounding area, including rural and protected areas. In Canada there are 10 provinces and 3 territories and the country offers a wide diversity of products ranging from ocean activities (whale watching, boating, surfing) on the east and west coasts to urban and natural attractions (wildlife viewing, provincial and national park activities, outdoor experiences, urban attractions) in the interior and arctic areas. Although there are other smaller DMOs that provide information about rural areas, the larger municipal DMOs offer the widest range of products that include activities beyond the urban experience. For example, Tourism Calgary located in western Canada offers information on activities and attractions in both the Banff and Jasper parks as well as tour experiences in the prairie areas. Tourism Halifax, located on the east coast, has members from many different areas of the Province of Nova Scotia including rafting companies and hiking tours that are located outside Halifax's urban area. The provinces also have DMOs which market the entire province or territory. Provinces and large municipalities were contacted for this study to gauge the widest area for study.

2. METHODOLOGY

This research sought to explore the current issues and opportunities for climate change and tourism from large municipalities and provincial/territorial DMOs perspectives. As DMOs often play an educational role as well as providing access to markets for its members, this research sought to understand the differing views across a country which has a large geographical area and is facing many different issues that are the result of climate change. Qualitative research was used in this study as it is based on the premise that social actors create emergent social processes and power structures which may refer to common patterns in social interactions [27]. This research is qualitative in nature, as there is no "formal" hypothesis to be tested [28]. This research is also inductive and interpretive, and in order to determine patterns, the following general questions were assessed including:

- challenges DMOs were facing as a result of climate change;
- mitigation / adaptation strategies;
- barriers or issues hindering the implementation of mitigation strategies;
- perceptions of demand by travellers for more sustainable forms of tourism; and
- tools and incentives needed to address climate change.

A full list of questions is outlined at the end of this paper.

This research compares the opinions of large municipal DMOs across Canada as well as territory and provincial DMOs. Although there are many levels of DMOs, large municipal DMOs were chosen in order to ensure that Canada's largest areas of Canada's tourism product were represented. Large municipal DMOs were also chosen for this study because much of Canada's tourism product is accessible through these municipal gateways including hotels, tour operators, attractions and events. As most visitors access Canada through municipal gateways, these DMOs often represent products that are located in national parks and other fragile areas as these members gain increased market access. Table 1 outlines the municipalities represented. A 70% response rate was obtained (19 of 27 DMOs who were contacted for this survey responded). Equal representation from across Canada was gained.

Table 1. Municipal DMOs Interviews and geographic location.

DMOs	
West Coast:	Vancouver
Prairies:	Saskatoon, Winnipeg
Rockies:	Calgary, Edmonton
Central:	Montreal, Toronto, Ottawa, Quebec City
Eastern:	Halifax
Northern/Arctic:	Whitehorse
Provinces/Territories:	Northwest Territories, Nunavut, Ontario, Manitoba, British Columbia, Nova Scotia, Prince Edward Island

Data were collected through semi-structured interviews. The measure was reliable as all interviewees were asked the same 17 questions with questions provided in advance to help avoid interviewer bias and interviews were conducted with CEO's to ensure data reliability. Informed and written consent was obtained from each respondent. Data collection took place in 2009. Due to the large geographic area, telephone interviews were conducted and were between one to two hours in length. Responses were then coded using 'latent coding' [29] and responses grouped together to form themes. It should be noted that there are some discrepancies between what is said by the interviewees and scientific facts, as respondents were not climate change experts, however their viewpoints are representative of industry and provide key insights into issues pertaining to mitigation and policy development in this area.

3. RESULTS

3.1. Impacts and Challenges of Climate Change

Respondents were first asked to rate the importance of climate change on the future viability of the tourism industry on a scale of 1 to 10, with 10 indicating extreme importance. This question was asked on a global context to gain an understanding of the interviewees' overall concern. Most respondents agreed that climate change was a very important factor in tourism's future viability (mean response: 7.8). Respondents were then prompted to discuss specific issues in relation to their geographical area.

Most of the respondents (located on the west and east coast and northern geographical areas) believed that climate change was extremely important (9 or 10/10) to the future viability of the tourism industry. Conversely, two respondents (located in the central Prairies) believed that climate change was of low importance (mean 5/10). These results may suggest a geographic divide in attitudes regarding climate change especially since, currently, most climate change impacts have been felt in coastal and arctic areas. Generally, respondents who identified climate change as extremely important also showed a heightened awareness of the impacts of climate change and identified environmental-sensitivity in their regions such as habitat and biodiversity loss

Respondents who did not consider climate change an important factor in the future viability of tourism believed existing research on climate change was not conclusive enough to change tourists' habits. One respondent expressed scepticism about existing science, "it's too early to tell the true effects of climate change on weather patterns and environment." Another respondent believed that climate change will not have a "dramatic impact on people visiting different parts of Canada and beyond" because people are "reluctant to give up leisure travel opportunities." This respondent represented a geographical area which has not felt many impacts from climatic change.

Respondents also identified a variety threats posed by climate change to the tourism industry. Responses were consistent with previous studies outlining impacts [13,30] such as unpredictable weather patterns and the impact on seasonal tourism, in particular on winter tourism. Changing weather patterns also posed a threat to infrastructure. One respondent noted that their municipality could not support an increase in summer tourism large enough to offset losses to winter tourism. Disease epidemics were a cited as a potential threat by one respondent (such as the impact of the SARS outbreak on Toronto's tourism industry in 2002). In some instances, respondents identified specific tourist attractions that were threatened by climate change. Arctic regions claimed that there are more overcast days which threaten tourism driven by the Aurora Borealis. Eastern Canadian areas saw increasing ice rain as a threat to winter festivals, and East Coast areas expressed concern about rising sea levels in coastal communities.

Several respondents felt the challenge posed by climate change on the tourism industry was a lack of awareness and action both within the industry and government and among consumers. Concerns included that there was still a lack of clarity on the issue of climate change, funding was not available to make necessary changes and there was a lack of knowledge of best practices to mitigate or adapt to impacts. One respondent felt that greatest threat was to ignore climate change until the impacts become unmanageable. In some cases, efforts to mitigate or adapt to climate change were seen as a threat to the industry. One municipality was concerned that businesses would cut travel to reduce their carbon footprint which could disrupt the convention and festivals/events sector. Respondents also identified potential regulation of the airline industry and the costs that would be incurred to meet those regulations as a major threat. At the time of the survey, a recession and economic crisis was weighing heavily upon the tourism industry. Several respondents saw the economic crisis as a major threat because businesses were struggling to make profits and therefore were less inclined to adopt mitigation strategies that might require an investment of capital.

3.2. Opportunities for Canadian Tourism Industry

Respondents were then asked to identify potential opportunities for the Canadian tourism industry with regard to climate change. The two most commonly cited opportunities were the potential advantages of warmer weather (as Canada has historically been a colder weather climate) and the potential for Canada to leverage itself as an environment-friendly tourism destination. Respondents felt that Canada's tourism industry was less susceptible to climate change than other nations, and Canada could become a more appealing destination as such. As temperatures become warmer, potential for new tourism opportunities could develop. Some felt that the northern region would benefit economically from the opening of the Northwest Passage. One respondent noted: "Many times we see people from the United States come... because there is sometimes no longer snow in the Northeast part of the United States." There were several other opportunities identified by respondents. One DMO suggested that Canada has a strong climate research infrastructure in place, and Canadian tourism could benefit from being on the forefront of scientific research. Another saw the issue of climate change as an opportunity to educate and motivate consumers and business to adopt more sustainable habits and practices. Interestingly DMOs did not discuss their role in helping their members.

3.3. Respondents' Perception of Consumer Attitudes

All respondents agreed that consumers are more aware of the impact of climate change than in the past. Increasingly, travellers are inquiring about environmental practices. However, most respondents felt consumers are not changing their behaviours as quickly their perceptions, and that consumers are not well-informed enough to make the right behavioural changes. This is consistent with other studies [20,31,32].

Respondents expressed concern about travellers' willingness to pay a premium for environmentally-friendly practices. One DMO noted a 2007 Lonely Planet study that concluded 93 percent of tourists would prefer to travel to destinations with environmentally friendly practices, however, another suggested there is still reluctance to assume the additional expense. Most respondents believe there is a market for environmentally-friendly tourism although it is still seen as a niche market that is susceptible to economic factors. The industry wants to see returns on their investment in green initiatives, but there is scepticism about the level of consumer demand. At the present, DMOs see the business incentive for green initiatives as a 'tie-breaker' that could provide competitive advantage when travellers must choose between destinations. Respondents did, however, note that the product which many visitors come to see are their members and therefore research is needed about consumer demand.

3.4. Initiatives Undertaken by DMOs

The interviews then investigated ways DMOs had taken action to mitigate or adapt to impacts on climate change. This question was phrased as both internal measures and also as actions taken to reach out to members. Most respondents had not taken any concrete actions as they noted that their role was to market a destination, not be a climate change expert. A few (three) respondents had taken no actions to reduce their impact while some respondents have instituted measures such as carbon offsetting for travel and member education programs (those located in the arctic or coastal areas).

Internally, activities that DMOs were introducing included setting up green teams and environmental policies to reduce impacts, however this was not necessarily in response to climate change pressures or impacts specifically. DMOs have also produced and distributed literature to members detailing best practices for reducing environmental impacts. When asked if the respondent organizations were sufficiently well briefed on the topic of climate change to develop internal policies and action, half of the respondents felt they were not sufficiently informed. Three respondents felt that it was not the priority of a DMO to develop environmental policies and that the task should be left to government regulatory agencies.

There were a few specific actions that were taking place. One DMO is currently undergoing an audit of members' practices to determine their next course of action and to assess who is doing what sustainably. Another DMO undertook a survey of members to determine which had instituted their own environmental policy—however, again this was not directly related to climate change. Roughly half had a written policy in place for sustainability or environmental management but none were specifically climate change policies. DMOs representing Canada's three largest municipalities (Toronto, Vancouver, and Montreal) have introduced carbon offsetting for employee travel. Several other respondents are currently investigating carbon offsetting as a future option. Only one respondent rejected carbon offsetting as an option: "I don't think we really believe in it. My personal belief is that it stops countries from cleaning up their act." Some DMOs have initiated research on the impacts caused by members and interviewees did outline efforts done by some of their members on mitigation and adaption, however, it was limited.

Some DMOs outlined member practices such as hotels or attractions obtaining LEED certification and energy reduction measures by accommodations and transport providers. These were few in nature.

3.5. Barriers to Implementing Initiatives

Respondents were then asked to identify barriers to implementing initiatives. The most commonly identified barriers were economics and a lack of policy or incentives for the DMOs to make efforts (half of respondents). Education and economics were seen as interrelated, and a combination of both was needed to overcome technical barriers. Half of respondents identified economic challenges as the dominant a barrier to implementing initiatives both internally and for their members. At the time of the survey the global financial crisis and recession were greatly impacting the tourism industry. Several respondents noted that the economic crisis could hamper progress toward achieving environmental goals. With money in short supply, businesses were focusing their resources on the best returns-on-investment. Several DMOs viewed green tourism as a niche market, and a less immediate concern. One respondent noted: "…climate change is what I term a 'luxury issue'. When we have time, we'll move forward and deal with it. "

One third of respondents also identified lack of education as a major barrier to implementing initiatives. Ways in which respondents defined education varied. Some respondents felt that there needed to be more education to the public, while others felt that industry needed more educational resources on the issue of climate change. Respondents wanted information on best practices from reliable sources such as government organizations or scientific bodies and all DMOs noted that they felt that there was a lack of action due to non-existent policies or leadership from government. They also wanted to know about effective strategies that have been implemented by other businesses and industries. Respondents wanted specific return on investment information for climate friendly practices and information on the business incentive for implementing initiatives. Specifically, they wanted to know what the market for green tourism is, and how to communicate effectively with them.

The technical barrier often cited was the lack of professional expertise on the issue among DMO personnel. DMOs are chiefly marketing organizations, and they often lack the skills set or resources to dedicate to environmental initiatives. This technical barrier could be overcome by an investment in personnel or through educational programs. Investing in personnel has its own difficulties. As membership organizations, DMOs are accountable to the demands of their members. One DMO noted "I can go to people in our office and say you should focus more on selling (us) as a green destination, but they are driven by what the clients are asking for. And, if the client is not asking for it, then they're not going to provide it." As with previous responses, many respondents discussed general issues of environmental management interchangeably with mitigation efforts while there was no real discussion of adaptation.

3.6. Strategies and Solutions for Adaptation and Mitigation of Climate Change Effects

DMOs were asked what actions would be most helpful to support tourism suppliers in Canada with respect to climate change. Again, coastal, mountain and northern DMOs were the most forthcoming with solutions as it is these regions of the country which have to date, experienced the greatest negative impacts. Regulatory incentives such as tax rebates for operations that achieve benchmarks, or penalties for organizations that fail to comply with standards were the most popular response. Education, communication, access to research and information sharing once again factored into responses, as did the need to establish a standardized certification body to set industry benchmarks. Most DMOs, regardless of location, said that industry is looking to the government for guidance on climate change initiatives. Respondents commonly expressed a desire for accurate information and standardization from the government. The respondents believed business will act more quickly where there is a financial or legal motivation. Respondents suggested most commonly that regulatory incentives should be introduced to the industry.

"Whether it is a positive tax treatment, or investment capital, or whatever the case may be, that gets business to change their practices. The other approach is penalize them, individuals or business for bad practices, but either way it comes down to monetary decisions. "Ultimately, you need regulation on [tourism suppliers]. You need people to be forced into doing it."

Respondents also wanted a standardized certifying body to police the industry and provide consumers with accurate information. Respondents point to several green certifications that are of varying quality and questionable value and currently none relates to Canada specifically. It was thought that an industry standard could more effectively hold businesses accountable for their claims and provide an effective marketing tool.

4. DISCUSSION

Although this study was on Canada, DMOs from around the world should take heed. As public-private partnerships, DMOs represent the varied interests of many tourist sectors within their geographic region, and therefore they are keenly aware of the impacts wrought by changes in the regional marketplace, and by policy changes on their client organizations. DMOs are also positioned well to assess the threats posed by climate change on their region as they represent a large number of varied tourism attractions and offerings to a diverse audience. This research found, however, that DMOs are not taking a leadership role in addressing impacts, or educating its members about climate change as they primarily see themselves as a marketing agency and are focusing on short term promotional issues rather than long term competitiveness.

This research supports previous research [31,33,34] in that there is a lack of knowledge by the consumer and industry. DMOs noted that although they see an increased interest by consumers in environmental travel, most respondents felt consumers are not changing their behaviours as quickly as their perceptions, and that consumers are not well-informed enough to demand more sustainable options. Most respondents see the link between their behaviour and climate change and some efforts have been made to limit environmental impacts although these are preliminary and not necessarily specifically focused on climate change. As DMOs' primary focus is on marketing and value for their membership fees, a short term return on investment was seen to be a key issue. Although much research [7,19,20] has demonstrated the longer term impacts of climate change (especially to mountain, arctic and coastal regions), DMOs and many of their members are not considering the longer term impacts. Clearly there is a need for more education for both the industry and the consumer. If both consumers and industry are informed on the issues of climate change, consumer demand for green initiatives will likely increase and the industry will be prepared with the necessary tools and knowledge to meet that demand. Furthermore, as regulation is instituted, industry will need to know how to meet regulatory requirements in a cost effective way. Despite this need, there has been little pressure placed on the industry by the government to implement climate change initiatives. The result, therefore, is a piecemeal approach to climate change mitigation and adaption. From this research, it was clear that it should be the role of federal government to lead the movement toward climate change action by developing a comprehensive climate change strategy related to tourism and assisting the industry in developing and implementing adaptation and mitigation strategies.

DMOs in different parts of Canada have different thoughts towards the severity of climate change. This is not surprising due to the vast geographical area and the many different climatic regions that the country encompasses. This research found that, although there are some opportunities for the Canadian industry to take advantage of the positive impending climate changes (increased seasonality, new areas opening up for nature based tourism and the consumers demand for destinations and companies which are seen to be taken a positive stance), most initiatives to counter or adapt to the negative aspects (movement to alternative forms of energy, education, planning, etc.) are voluntary and few in number. Although climate change is becoming an increasingly important issue to the majority of DMOs interviewed, there is little climate change action and interviewees in all geographic regions are looking to government for guidance. Some authors state that regulation [20,34] and policy initiatives are the answer to regulating emissions and adapting to climate change impacts and this research agrees with such findings. As per responses from DMOs, unless Canada accepts politically its responsibility for reaching Kyoto, regulation in the tourism sector is highly unlikely as tourism is intrinsically linked to other economic sectors. There is a need for partnerships between government and industry to foster knowledge sharing, however, government must take a leadership role to ensure regulations are set and the tourism industry reaches its needed targets for GHG reduction [9].

5. CONCLUSIONS

Changing weather patterns make seasons less predictable, threatening the sustainability and viability of tourism sectors across the nation. Tourism also contributes to climate change and environmental degradation, effectively undermining the environments that are its most vital resource. Consumer awareness on the issue of climate change has had a limited impact on the tourism industry, but is likely to increase even more as education continues to be a key focus of climate change mitigation practices. It is essential that the very body that attracts the consumer and promotes the product—the DMO-stays on top of the challenges which affect it. This research looked at DMOs views and actions about climate change as DMOs market and represent much Canada's tourism product. Because of this role, it was thought that they could be uniquely situated to act as educators to their client-members on the issue of climate change, and as information providers to planners for future threats to the industry. This was not the case. In Canada, although climate change is becoming an increasingly important issue to many DMOs and their members, there is little action for climate change mitigation or adaptation.

As DMOs situate themselves between public and the private interests, having both concern for the prosperity and sustainability of their industry and the region as a whole, it was thought that they could be instrumental in communicating the need to mitigate and adapt the effects of climate change to a broad group of stakeholders. The respondents, however, felt they did not have the resources to provide adequate training and planning for climate change mitigation. A critical issue for mitigating the effects of climate change is that a holistic and communicative approach that balances tourism development with other activities is necessary as it is difficult to regulate climate change and tourism as a separate function from other issues. This research concludes that the following action be undertaken to further this agenda:

- Improve DMO education and awareness of climate change and provide them with the tools and best practices needed to measure impacts
- Establish incentives for the industry to achieve environmental benchmarks and to adopt sustainability measures
- Enforce accountable industry-wide standards to reduce environmental impacts caused by industry
- Educate consumers of the impacts of their travels on their host destinations to foster demand
- Devise regional plans for the adaptation of the tourism industry to climate change impacts

The responses provided by the DMOs for this research illustrate a clear need for more communicative and policy-based action with regard to mitigating the effects of climate change. The DMOs from this study note that they and the industry are looking to the government for guidance. Currently in

Canada, policy does not seem to be forthcoming. If industry is going to shift, there must be more regulatory benchmarks to get the industry to understand their current situation and to respond. Voluntary benchmarks have been set by some industry leaders, but economic factors and the competitive marketplace constrain wider action. For actions to occur, an overarching jurisdictional body that enforces adherence to benchmarks industry-wide is preferable to voluntary benchmarks. At the time of writing, a nationwide certification scheme was being proposed and piloted in Canada, however, voluntary, often market-driven, benchmarks may appeal to the eco-conscious traveler, but there is concern about the long-term viability of this market.

DMOs are recognizing an increase in consumer demand for operators who offer environmentally-friendly practices, but these consumers are still considered a niche market, being too small a proportion of the overall marketplace to provide a sustainable industry. As environmental education increases, more consumers are likely to demand that tourism operators adopt responsible methods which may have a positive effect on mitigating climate change. DMOs see the potential for eco-friendly travel as a short term opportunity for market differentiation and this may offer some incentive.

DMOs could function as long-term planning entities within regions. Businesses within the tourism industry are seen as being oriented toward short-term sustainability and this may have been further exacerbated by the recent global financial crisis. Though DMOs are chiefly marketing organizations, they have the potential to recognize and respond to the impacts of climate change in their regions. In some regions, the effects of climate change have already been experienced. In other regions, the effects are eminent. DMOs face increasing competition from other community-based or industry organizations that provide marketing services. For example, global events, individual websites or social media venues are often perceived as more relevant to inform the individual or organization about tourism options in a destination. As more people 'experience' a place, more focus will be put upon the attractiveness of a destination. As DMOs are feeling the squeeze from other forms of technology being used as a promotional tool and other impacts, education and leadership regarding climate change could become an added value proposition that they could offer their members. The DMO must be aware of how these issues affect their region, and prepare their members to adapt.

While this case study was Canadian, implications for DMOs in other regions can be applied. It is clear that DMOs in northern regions will feel different affects from those in the South, and that destinations in the deserts and prairies will have different impacts from those on the coasts, arctic areas and in mountainous regions. For DMOs to serve as effective educators, they need access to adaptation and mitigation strategies and incentives that are considerate of their regions, while also having awareness about climate change on a global perspective. This research found that DMOs as marketing and educational bodies do little to work with their client members about climate change impacts.

It is important to note that the very nature of climate change is linked to a bigger issue—the dependency of tourism on a declining supply of natural resources. Without longer term strategic planning, the tourism industry in Canada and destinations globally, will not be equipped to deal with a changing future. The long-term, global threat of climate change affects all industries including tourism but until its effects are directly felt in terms of human life or prosperity, little will likely be done. Perhaps only when higher operating costs due to the cost of carbon, adapting to negative consequences such as snow making or flood defence systems or rising sea levels will changes be made. When destinations see reduced consumer demand as behaviour shifts towards sustainable tourism options, and brand images are negatively affected when evaluated on their sustainability credentials, will destinations react. Unfortunately, the tourism industry's lack of proactive measures may result in all the above—and these consequences will be hard to reverse. As public-private partnerships, DMOs could be uniquely positioned to undertake a role as educators, unfortunately they claim they are not equipped with the technology, finances or the knowledge to address these issues at the present time. If tourism is to move forward in this changing environment, something must be adjusted. Tourism is faced with two options. First, more effort is needed to provide industry with the necessary tools to transform their product into climate-friendly attractions—perhaps through auditing and/or improving environmental practices. Second, DMOs themselves need to realize that their very competitive position depends on adapting and changing for the future or they may soon become obsolete.

This study looked specifically at the Canadian industry but the issue of climate change is global in nature. Further research to assess the ways other municipal marketing organizations have responded to the issue of climate change could provide insight into effective methods employed elsewhere and spur further awareness of the issue to the wider field of marketing organizations.

REFERENCES AND NOTES

1. WEF. Towards a Low Carbon Travel & Tourism Sector, Available online: http://www.weforum.org/pdf/ip/att/ ClimateChangeReport Copenhagen.pdf (accessed on 19 September 2009).

2. UNEP/UNWTO. Climate Change and Tourism: Responding to Global Challenges, Available online: http://www.unwto.org/sdt/news/en/ pdf/climate2008.pdf (accessed on 10 August 2008).

3. IPCC. Annex 1: Glossary, Available online: http://www.ipcc.ch/pdf/glossary/ar4-wg3.pdf (accessed on 16 September 2009).

4. Climate Change 2007: Mitigation of Climate Change; Contribution of Working Group III to the Fourth Assessment Report of the Intergovernmental Panel on Climate Change; Cambridge University Press: Cambridge, UK, 2007.

5. UNWTO. UNWTO World Tourism Barometer, 2008. Available online: http://www.world-tourism.org/facts/wtb.html (accessed on 17 September 2009).

6. Dodds, R.; Kelman, I. How climate change is considered in sustainable tourism policies: A case of the Mediterranean Islands of Malta and Mallorca. Tourism Rev. Int. **2008**, 12, 57–70.

7. Gössling, S. Global environmental consequences of tourism. Global Environ. Change **2002**, 12, 283–302.

8. Gössling, S.; Hall, C.M. Uncertainties in predicting tourist flows under scenarios of climate change. Climatic Change **2006**, 79, 163–173.

9. Dodds, R.; Graci, S. Canada's tourism industry—Mitigating the effects of climate change: A lot of concern but little action. Tourism Hospit. Plann. Dev. **2009**, 6, 39–51.

10. Peeters, P. Mitigating tourism's contribution to climate change—An introduction. In Tourism and Climate Change Mitigation Methods, Greenhouse Gas Reductions and Policies; Peeters, P.M., Ed.; Stichting NHTV Breda: Breda, The Netherlands, 2007; pp. 11–26.

11. Perry, A. Impacts of climate change on tourism in the Mediterranean: Adaptive responses. In Climate Change in the Mediterranean: Socio-Economic Perspectives of Impacts, Vulnerability and Adaptation; Giupponi, C., Shechter, M., Eds.; Edward Elgar Publishing: Cheltenham, UK, 2003; pp. 279–289.

12. Scott, D.; McBoyle, G.; Schwartzentruber, M. Climate change and the distribution of climatic resources for tourism in North America. Climate Res. **2004**, 27, 105–117.

13. Amelung, B.; Nicholls, S.; Viner, D. Implications of global climate change for tourism flows and seasonality. J. Travel Res. **2007**, 45, 285–296.

14. Choi, S.; Lehto, X.L.; Oleary, J. What does the consumer want from a DMO website? A study of US and Canadian tourists' perspectives. Int. J. Tourism Res. **2007**, 9, 59–72.

15. TIAC/AITC (Tourism Industry Association of Canada). Tourism Facts, 2009. Available online: http://www.tiac.travel (accessed on 15 December 2009).

16. CTC. Facts and Figures Year Review, Available online: http://www.corporate.canada.travel/docs/research_and_statistics/stats_and_figures/tourism_year-in-review_2006_web_eng.pdf (accessed on 26 January 2008).

17. Scott, D.; Jones, B. Climate Change & Nature-Based Tourism: Implications for Park Visitation in Canada; Department of Geography, University of Waterloo: Waterloo, Canada, 2006.

18. Scott, D.; Jones, B.; Abi Khaled, H. The Vulnerability of Tourism in the National Capital Region to Climate Change; Technical Report to the Government of Canada's Climate Change Action Fund; University of Waterloo: Waterloo, Canada, 2005.

19. Scott, D.; Jones, B.; Lemieux, C.; McBoyle, G.; Mills, B.; Svenson, S.; Wall, G. The Vulnerability of Winter Recreation to Climate Change in Ontario's Lakelands Tourism Region; University of Waterloo: Waterloo, Canada, 2002; Department of Geography Publication Series Occasional Paper 18.

20. Becken, S.; Hay, J.E. Tourism and Climate Change: Risks and Opportunities; Channel View Publications: Bristol, UK, 2007. [Google Scholar]

21. Tourism, Recreation and Climate Change; Hall, C.M., Higman, J., Eds.; Channel View Publications: Bristol, UK, 2005.

22. Nicholls, S. Climate change, tourism and outdoor recreation in Europe. Manag. Leisure **2006**, 11, 151–163.

23. Patterson, T.; Bastianoni, S.; Simpson, M. Tourism and climate change: Two-way street, or vicious/virtuous circle? J. Sustain. Tourism **2006**, 14, 339–348.

24. Morrison, A.M. Hospitality and Travel Marketing, 3rd ed.; Delmar: Albany, NY, USA, 1998.

25. Buhalis, D. Tourism in an era of information technology. In Tourism in the Twenty-First Century; Faulkner, B., Moscardo, G., Laws, E., Eds.; Continuum: New York, NY, USA, 2000; pp. 163–180.

26. Gretzel, U.; Yuan, Y.; Fesenmaier, D. Preparing for the new economy: Advertising strategies and change in destination marketing. J. Travel Res. **2000**, 39, 146–156.

27. Warren, C.A.B.; Karner, T.X. Discovering Qualitative Methods; Oxford University Press: New York, NY, USA, 2010.

28. Patton, M. Qualitative Evaluation and Research Methods, 2nd ed.; Sage Publications: Thousand Oaks, CA, USA, 1990.

29. Neuman, W.L.; Robson, K. Basics of Social Research; Pearson: Toronto, Canada, 2009.

30. Wolfsegger, C.; Gössling, S.; Scott, D. Climate change risk appraisal in the Austrian ski industry. Tourism Rev. Int. **2008**, 12, 13–23.

31. Dodds, R.; Leung, M.; Smith, W. Assessing awareness of carbon offsetting by travellers and travel agents. Anatolia **2008**, 19, 135–147.

32. Scott, D.; Jones, B.; Konopek, J. Exploring potential visitor response to climate-induced environmental changes in Canada's Rocky Mountain National Parks. Tourism Rev. Int. **2008**, 12, 43–56.

33. Becken, S. How tourists and tourism experts perceive climate change and carbon-offsetting schemes. J. Sustain. Tourism **2004**, 12, 332–345.

34. Gössling, S.; Broderick, J.; Upham, P.; Ceron, J.P.; Peeters, P.; Strasdas, W. Voluntary carbon offsetting schemes for aviation: Efficiency, credibility and sustainable tourism. J. Sustain. Tourism **2007**, 1, 223–248.

CHAPTER 7

Destination Image Perception of Tourists to Guangzhou—Based on Content Analysis of Online Travels

Xiaohong Li

School of Management, Jinan University, Guangzhou, China

ABSTRACT

It is said that Guangzhou is an indefinable city. This paper analyzed the online travels from www.mafengwo.cn to study the destination image perception of tourists to Guangzhou. Content analysis was used to investigate the destination image perception of tourists from two aspects, namely cognitive image and affective image. Then by comparing the image of Guangzhou positioning in the literature with the online travels image, the paper concluded that Lingnan, Ram City and delicious food were more important in tourists' perception. The results can provide reference when the government promotes the Guangzhou destination image.

KEYWORDS

Guangzhou, Destination Image, Online Travels, Content Analysis

1. INTRODUCTION

It is said that Guangzhou is an indefinable city. There was a time that Guangzhou government used the image of Flower City, Business City and Lingnan Cultural City to advertise Guangzhou, but failed to get good results. In 2015, an advertising video called "Guangzhou was born because of you" was broadcast, it had caused extensive dispute. It was because that the video just displayed the modern city landscape, but neglected the characteristics of Guangzhou, such as Lingnan culture landscape. At the same time, from April 15, 2015, Southern Metropolis Daily made series report: How to develop tourism, offer advice and suggestions to make Guangzhou be a tourism city. As a result, different specialists from various perspectives expressed their opinions. A number of scholars studied the destination image of Guangzhou, however, most of them took qualitative research according the history of Guangzhou. Some of them came up with the strategies to promote Guangzhou image. While in the quantitative research, some scholars took empirical study to explore food cultural landscape of Guangzhou, and others studied the tourists' cognitive difference system. However, there is no qualitative research from tourists' perspective to study the destination image perception to Guangzhou.

In the era of internet developing rapidly, a large number of tourists publish their travel experience using words to record their tourism for other readers. These travels are not only the writers' experience, but also good study materials. Meanwhile these materials are free and easy to receive. In China, the places using these travels and blogs to study the destination image are Tibet, Beijing, Taiwan, and so on. Some of them using content analysis method just analyzed the high-frequency of the words, but haven't analyzed the sentences as well as the described paragraphs while some of them just analyzed the sentences and described paragraphs written by different writers. Some papers analyzed small amount of travels individually. Therefore, we think we can apply content analysis method to analyze the travels combines the high-frequency words with sentences and described paragraphs and use these materials to study Guangzhou destination image. As a result, this paper analyzed the travels about Guangzhou to investigate tourists' destination image perception. The results of tourists' destination image perception can provide reference when the government promotes the Guangzhou destination image.

2. LITERATURE REVIEW

2.1. Tourism Destination Image

Tourism destination image has been studied in 1970s [1], the concept was given by Hunt in the early years, and he considered tourism destination image was someone's impression to his/her non-resident place, which was a pure subjective conception [2]. Crompton and Kotler (1979) pointed out that tourism destination image was a comprehensive of belief, opinion and impression that tourists to a place or destination [3]. Echtner and Ritchie summarized tourism destination image was a tourist's opinion to the whole destination and the whole characteristics, and based on it, they proposed a destination image framework, i.e. holistic-attribute, functional-psychological, common-unique [4]. The definitions of tourism destination image mentioned above are from the demand of tourists' perspective, they stressed from cognitive psychology and the perception of individual and community. On the other hand, some authors define tourism destination image from supply perspective. They distinguished two kinds of image, which were projected image and received image (Koteler and Barich, 1991) [5]. Projected image was regarded as people's opinion and impression to the destinations they paid close attention to, actually, it could be seen as a duplication of symbol and meaning, which was constructed and spread in society (Nuria and Jose, 2005) [6]. Received image was formed by projected information, tourists' desire, motivation and previous knowledge, experience, preference and other personality (Bramwell and Rawdig, 1996) [7]. Baloglu and McCleary tended to believe image was constituted by tourists' rational and emotional express, and they connected each other closely. On the one hand, perceptual/cognitive appraisal involved individual's brief and acquaintance, while affective appraisal involved the emotion to the destination or attaching to it. Academic community gets consensus on two points, namely cognitive appraisal is the fundament of affective appraisal, and affective appraisal derives from the cognition of objects [8]. Image of a destination is the result of cognitive appraisal and affective appraisal. Grosspietsch (2006) distinguished two kinds of tourism destination image, they were tourists' perceived image and destinations projected image, and tourists' perceived image were cognition and impression of potential and actual tourists to destinations. The latter one were tourism operators tended to set up the image in tourists' mind [9]. Some scholars divided tourism destination image according to the visiting time, which were original image and re-evaluation image (Selby and Nigel, 1996) [10].

From the definition of tourism destination image mentioned above, the general characters of them are listed as follows: Firstly, they stresses the perception of individual. Secondly, it is individual's cognition to one particular tourism destination. Finally, the perception isn't always the same as the reality of the destination, which exists subjectivity.

In China, the study of tourism destination image began in 1980s, and it became study hotspot in 1990s. Qiu's article, Analysis the Tourism Image of China, was considered the earliest paper to study destination image in China [11]. Li (1999) has written a book called Tourism Destination Image Planning: Theory and Practice is the first book to discuss destination image systematically [12]. Scholars' definition of tourism destination image not only focused on tourist's individual opinion, but also according to the objective situation of the destinations. By literature research, the study of tourism destination image in China were mission and practice oriented, scholars tended to explore image from destination perspective. However, some scholars focused on people's perspective, they mainly discussed the influence of perception image, the formation of perception, marketing management and empirical studies of perception image.

Song (2000) came up with the idea that tourism destination image had perception and non-perception two parts, and then he analyzed the influence on the perception and non-perception to tourists [13]. Zhang, Lu

2. Literature Review

and Zhang (2006) found that tourism destination image followed the law of diminishing space [14] . Liu, Wu and Xiao (2006) discovered that the order of different tourist destinations would affect tourists' perception [15] . Zhou (2007) proved that the perception of destination image was affected by the watching way, cultural environment, medium and demographic feature [16] . Liu et al. (2015) studied the influencing factors of destination image, from high to low, they were tourism facilities, atmosphere, environment, service price and service, and they took empirical research by the domestic tourists traveling in Beijing during National Day [17] . Scholars in China are apt to consider it is doubtful of explaining tourism destination image from subjective perspective, because people's perception to one destination also need to base on objective situation [18] .

2.2. Content Analysis in Tourism Study

Content analysis method arises at the beginning of the 20th century, which was first used in the press. American famous communication expert Berenson (1952) defined content analysis method as a research method which was objective, systematical and quantificational described distinct communicate content. After that, it was widely used in journalism, communication, library-information, psychology and other social sciences [19] .

Content analysis can be divided into three types, they are interpretation content analysis, experiment content analysis and computer-assisted content analysis [20] . Interpretation content analysis is a method that researchers read, understand and explain the content the writers tended to convey. Experiment content analysis is a method combines qualitative and quantitative researches. Qualitative content analysis is like cloze tests which describes and ratiocinates the connection and structure of each concept, while quantitative content analysis mainly analyses the words frequency. Recently, the development of computer makes the computer-assisted content analysis has developed fast.

When searching for literature, we found that content analysis used in tourism study mainly was tourism destination image, tourism destination marketing, tourists' experience, tourists' motivation, tourism concept and so on. Content analysis using in tourism destination image focused on the perception of tourism to one destination (Stepchenkova and Morrison, 2006) [21] , the comparison between different mediums of one destination (Stepchenkova and Zhan, 2013) [22] and the formation of tourism destination image (Llodrà-Riera et al., 2015) [23] . The content of content analysis was tourism official websites, user-generated content websites, newspaper, advertisement, blogs and government official documents.

In China, content analysis using in tourism study was after 2000, and the quantity was relatively few. The main study was about the definition of concept, destination image, tourists' motivation, the satisfaction of tourists and tourism marketing (He, 2012) [24] . The object of content analysis was online travels, online communities, sina microblogs, photos, destination official websites and blogs. Most of the research analysed the texts, and based on the internet, by using computer-assisted technology, such as Nvivo, Spss and Rost.

3. METHODOLOGY

3.1. Select Online Travels

We chose www.mafengwo.cn to be the travels analysis website, the reasons are listed as follows:

Firstly, during our research period, www.mafengwo.cn was often ranking first in Alexa among the specialized travels websites in China. Secondly, www.mafengwo.cn was a website providing user-generated content for the users, and the website has accumulated a large number of users, the quality of travels they published about Guangzhou were up to 5000. Finally, comparing with Ctrip and other travels websites, the quality of travels about Guangzhou in www.mafengwo.cn was highest. As a result, we chose www.mafengwo.cn to be the travels analysis website.

3. Methodology

We sorted the travels about Guangzhou according to the newest time (the expiration date was April 11, 2015), we got 891 travels, then we numbered and analyzed them. The release time of them were from August 31, 2014 to April 11, 2015. After selecting, we finally got 241 travels. The rules we selected the travels were listed as follows: Firstly, got rid of travels which were all photos without words. Secondly, got rid of travels without travelling places or features of the places. Thirdly, got rid of unfinished travels. And then got rid of travels which were not narration. Finally, got rid of travels besides Guangzhou including other cities.

3.2. Process the Online Travels

The rules we process these 241 travels were as followed: Firstly, deleted the pictures and videos in the travels, we just analyzed the words. Secondly, in order to make sure the expression and be convenient to count, we substituted Xiaomanyao for Cantonese Tower, substituted Jiulong Lake Park for European Small Town and so on. Thirdly, in order to guarantee the statistic scientificity, we substituted the words which could be count repeated, such as we substituted Pearl Night for Pearl River Night Cruise, preventing the double count of Peal River. Afte processing of the 241 travels, we got about 750 thousand words, the average words of each travels were about 3100.

3.3. Research Software

This paper used Rost CM6.0 to analyze the travels content. Rost was designed by Dr. Shen and his Rost Virtual Learning Team, who worked in Wuhan University. This software is specialized in Chinese content analysis, which can capture the website in real time and convert to txt to save. What's more, it can count the words frequency, analyze social network, do semantic analysis and emotion analysis. Rost CM6.0 is popular, the users are all over the world, including Cambridge University, Loughborough University, Texas A&M University, Peking University, Sun Yat-sen University et al.

3.4. Content Analysis Category

The classification of perception tourism destination image is based on Baloglu and Mc Cleary (1999), which are cognitive image and affective image. The cognitive image including five factors, which Beerli and Martin summarized in 2004, namely natural and cultural resource, tourism and leisure facility, atmosphere, social situation and environment and beach [25].

Russel and his colleagues (1980) pointed out that affective quality could be measured according to a two-di- mensional bipolar space model, which could be defined by eight variables (Figure 1): pleasant, exciting, arousing, distressing, unpleasant, gloomy, sleepy and relaxing, but only two of the scales (pleasant-unpleasant and arousing-sleepy) were proved theoretically needed to represent affective space and images [26].

The content analysis category is three parts, the first part was basic information of the travels writers, including the amount of Guangzhou and non-Guangzhou tourists, male and female tourists. The second part was the perception tourism destination image, we based on Baloglu and McCleary (1999)'s classification, which destination image was divided into cognitive image and affective image, and cognitive image was divided into five factors according to Beerli and Martin, but in accordance with the content of travels of Guangzhou, we adjusted the five factors, so we just analyzed natural and cultural resources, tourism facilities and services, social environment and atmosphere. The affective image was based on Russel and his colleagues' (1980) two-dimensional bipolar space circumplex model. And the last part was comparing the image of Guangzhou positioning in the literature with the online travels image.

4. RESEARCH ANALYSIS

4.1. The Basic Information of Writers of Selected Travels

We have chosen 241 travels, in which most of the writers were non-Guangzhou, and the travels of non-Guang- zhou to Guangzhou ratio was

1.87 to 1. Moreover, the number of words of non-Guangzhou was almost twice the Guangzhou in total. The mainly reason was that to Guangzhou travelers, it was daily leisure activities, so the stay time was short, and the scenic spot was often one or two. As a result, the number of words was relatively fewer. According to writers' gender, the number of female to male ratio was 2.39 to 1, almost the same as 891 travels. Generally speaking, comparing to males, females wrote more words, their travels were longer than males'. It was obvious that female tourists tended to share their tourism experience by travels and convey more content. After reading 241 travels, we found that most of the writers were choosing independent travel, instead of package tour, and the stay time was 3 - 5 days with their families or friends. The basic information about the travels writers sees Table 1.

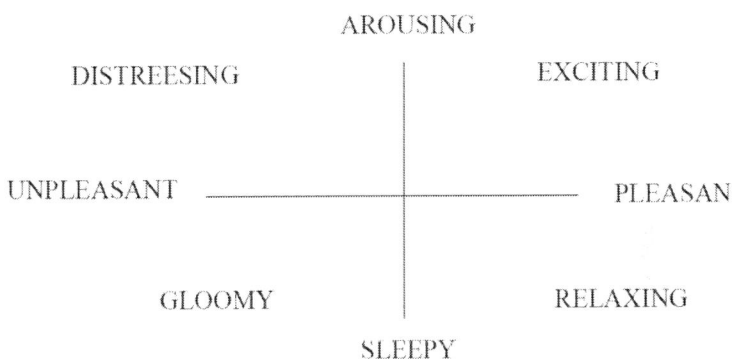

Source: Russel, J. A., and G. Pratt (1980) a Description of Affective Quality Attributed to Environment. *Journal of Personality and Social Psychology*, 38, 311-322.

Figure 1. Two-dimensional representation of a circumplex model of the affective quality (appraisal) attributed to places/environments.

4.2. The Cognitive Image of Selected Travels

4.2.1. Natural and Cultural Resources

We have extracted 50 first high-frequency words, and found that cultural resources were more than natural resources, which coincided with the situation of Guangzhou tourism resources. Chimelong, Cantonese Tower and Shameen were mentioned a lot by the writers in travels. Chimelong became famous not only it was most welcomed artificial theme park, but also two TV shows called Dad Where Shall We Go and Wonderful Friends were taken films here by Hunan Satellite. With the broadcast of these shows, they expanded the popularity of Chimelong, the phrases Dad where shall we go have been mentioned 40 times, and one travel note wrote, "I have heard that Chimelong is an interesting place, after watching the show, her dad made a decision to go for a real Dad Where Shall We Go. YCQ is very excited when hearing we would go to the show place" (No. 0724). Cantonese tower is a landmark building in Guangzhou, which represents new appearance of Guangzhou. Shameen building group is western-style houses, which also attracts tourists' attention. According to The Classification, Investigation and Evaluation of Tourism Resource (GB/T18972-2003), we classified the 50 high-frequency words, and saw Table 2.

According to Table 2, we found that tourists preferred to visit cultural resources in Guangzhou, and mainly visited the synthetic human cultural tourism sites, the second one was living places and communities. In synthetic human cultural tourism sites, they preferred the recreation and amusement resorts, and Chimelong Holiday Resort was most popular. Garden recreational areas were also mentioned a lot. Universities became a tourism resource, such as South China Agricultural University (SCAU). In March, the cherry blossom festival had attracted many people, and one travel note had mentioned that, "in the middle of March, it's the best time to appreciate the cherry. With the popularity of SCAU in Guangzhou, the flow of people is very large, but if you want to appreciate the flowers, you'd better not to choose the weekend" (No. 0068). The proportion of park was large, Yuexiu Park was famous, and when mentioning Yuexiu Park, Five-Rams Stone Sculpture attracted people's attention, because it was landmark building there. The frequency of Five-Rams Stone Sculpture was 126, which illustrated the image of ram city went deep in people's mind. In living places and communities, streets and buildings with Lingnan style got much concern, such as Shangxiajiu Pedestrian Street, Beijing Road, and Xiguan Mansion.

4. Research Analysis

Table 1. The basic information of 241 travels.

Writers	Classification	Numbers	Words	Average words of per travels
According to the resident place	Guangzhou	84	158,777	1890
	Non-Guangzhou	157	593,320	3779
According to the gender	Male	71	204,751	2884
	Female	170	547,112	3218
Total		241	752,097	3121

Table 2. The classification of 50 high-frequency words.

Main class	Sub class	Fundamental class	Concrete content	Frequency	Total
Natural resources	A. Land scenery	AA. Synthetic natural tourism areas	AAA. Mountain tourism areas	Baiyun Mountain, Hulu Mountain, and Baishuizai Scenic Spots	372	372
	B. Water area landscape	BA. Stream segments	BAA. Sightseeing and recreational segments	Pearl River	400	400
Cultural resources	F. Architecture and facilities	FA. Synthetic human cultural tourism sites	FAA. Test sites for teaching and scientific research	The Huangpu Military Academy, Sun Yat-sen University, South China Agricultural University, and Guangzhou Higher Education Mega Center	517	
			FAB. The recreation and amusement resorts	Chimelong Holiday Resort and Jiulong Lake Resort	1703	
			FAC. Religion and sacrificial places	Shishi Sacred Heart Cathedral and Guangxiao Temple	549	
			FAD. Garden recreational areas	Baomo Garden, Lingnan Impression, Nanyue Garden, Yunxi Ecological Park, Shimen National Forest Park and Yuexiu Park	1224	5938
			FAE. Places of cultural activities	Guangdong Provincial Museum, Guangzhou Museum, Fangsuo Library and Guangzhou Library	297	
			FAG. Places of social and trade activities	Taikoo Hui and Tee Mall	135	
			FAH. Exhibited places of animals and plants	South China Botanical Garden and Millions of Sunflowers Garden	326	
			FAK. Scenery enjoy spots	Cantonese Tower	1187	
		FB. Single places for cultural or sports activities	FBC. Demo rooms	Haixinsha Island	209	209
		FC. Landscape and appertaining architectures	FCI. Square	Huacheng Square and Haizhu Square	255	255
		FD. Living Places and Communities	FDB. Characteristic streets	Shameen, Shangxiajiu, Beijing Road, Dongshan Western-style Houses, Xiguan Mansion and Litch Bay	2869	4561
			FDC. Characteristic communities	Redtory, Xiaozhou Village, Shawan Ancient Town and Xitou Village	757	
			FDD. Celebrity residence and historic commemorative buildings	Liugengtang, Lin Zexu Memorial Hall, Sun Yat-sen Memorial Hall and Yuyin Garden	265	
			FDE. Academies	Chen Clan Temple	670	
		FE. Burial places	FEB. Tombs(group)	Tomb of the Nanyue King	163	163
		FF. Transportation buildings	FFC. Ferry and dock	Tianzi Marina, the Ancient Huangpu Harbor and Taigu Marina	109	109
		FG. Sections of dyke	FGD. Sections of dyke	The 19th Chung	36	36

4.2.2. Tourism Facilities and Service

1) **Catering Analysis**

Among the catering words, morning tea (zaocha), dessert and Taotaoju (a restaurant) were the first-three- mentioned words. Morning tea was a special breakfast activities in Lingnan area, which has become one of the activities tourists should experience. One travel note said, "In the early morning, we begin our morning tea, it is an activities must take in Guangzhou, to Cantonese, ordering a bottle of tea, watching newspaper and chatting with each other are everywhere in restaurants" (No. 0572). When talking about desserts, Nanxin Milk Dessert was very famous. Taotaoju was one of the time-honored brand shops in Guangzhou, so it also became a special catering to tourists. We analyzed the 50 high-frequency catering words and sorted them out (see Table 3).

According to the 50 high-frequency catering words, we found that tourists mainly chose the time-honored brand shops, and they mainly chose the famous food, for example, the fish skin and Tingzai Porridge of Chentianji, the double-skin milk and beef three star of Nanxin Milk Dessert. The word delicious food has been mentioned up to 302 times and foodie up to 94, that's to say, several number of tourists considered themselves as foodies, the trip to Guangzhou was a trip of delicious food. Some title of the travels have expressed their wish for delicious food in Guangzhou, such as Trip of Delicious Food in Guangzhou in June 1 (No. 0572), Trip of Delicious Food in Guangzhou, I am a Foodie, just Go (No. 0393). There were several time-honored brand in Baohua Road, such as Baohua Noodle House, Chentianji, Shunji, Lin Heung Lau, Nanxin Milk Dessert and so on, therefore, it attracted lots of tourists.

2) **Accommodation Quality**

Generally speaking, tourists gave positive evaluation to accommodation in Guangzhou, based on the content of the travels. Hotel has been mention for 967 times, but not all the travel notes have evaluation to hotel, by analyzing the content, we found that Chimelong Holiday Resort was the most mentioned scenic spots, so Chimelong Hotel was also mentioned a lot. A part of families visiting Chimelong Holiday Resort chose to live in Chimelong Hotel, one travel note said, "we lived in Chimelong Hotel, comparing to other hotels, it was expensive, when we booked this hotel, it was about 800yuan per night (not including breakfast and tickets), but the environment was good and safe, moreover, it was very convenient to Chimelong Park" (No. 0556). However, most of the tourists considered the fare of Chimelong Hotel was high, one travel note wrote that, they couldn't afford the room fare of Chimelong Hotel (No. 0194). Another travel note mentioned, "it was said that the environment of Chimelong Hotel was excellent, but we couldn't accept the price, searching for the hotels around, we finally chose the Atlan Hotel" (No. 0152). Then we found that seldom of them chose hostels, but the young tourists would choose the economy hotels because of the convenient transportation and low price. The evaluation of economic hotels was different from different hotels, but most of them thought it was lower price and convenient. One travel note wrote that there were many economic hotels here, such as 7 days and home inn, average price of them was 300 yuan, which he could accept, and it was convenient to go to subway, shopping and eating (No. 0153).

3) Transportation

From Table 4, we found that the main means of transportation tourists taken to Guangzhou were trains and air planes, and the in city transportation was metro, one travel note mentioned that, the coverage of metro in Guangzhou was very large, which was not less than Shanghai and Beijing. As a result, if you need to plan a long trip in Guangzhou, you can take metros as a main transportation (No. 0563). Besides, buses were also the main transportation in Guangzhou, "bus in Guangzhou is also very convenient, we have suffered heavy rain and we couldn't take a taxi but take a bus" (No. 0149), one travel note said. However, most evaluation of taxi was negative, they thought taxis in Guangzhou needed to wait for a long time, sometimes you even needed to rush to get a taxi, and the drivers had the suspicion to cheat the customers. One travel note said, "we have made a decision to spend much money to take a taxi to go back hotel, but after arriving, we haven't seen a taxi" (No. 0726), another travel note said, "we have been waiting for about half an hour, but haven't see a taxi, it's very fewer, moreover we can't get a taxi because of too much people waiting here" (No. 0530).Tramcars served as a means of transport with sightseeing and communication function were run in the end of 2014, which got much attention. To tourists, it was a sightseeing transportation. "When the tramcar gets in the station, it looks like a famous star, people put up their phones and cameras to take photos for the tramcar (No. 0415),"one travel note said.

Table 3. The classification of the 50 High-frequency catering words.

Classification	Concrete content	Frequency
Time-honored brand	Taotaoju, Yinji, Nanxin Mile Dessert, Baohua Noodle House, Lin Heung Lau, Shunji, Guangzhou Restaurant, Chentianji, Panxi Restaurant, Bingsheng Seafood Restaurant, Huishijia, Dayang, Diandoude, Lan Kwai Fong, Wuzhanji, Xiangqun, Renxin and Baihua	1688
Food	Dessert, double-skin milk, snack, fish skin, beef offal, beef, durian, honey, shrimp wonton, zingiber unixed milk, herbal tea, sushi, beef three star, radish, shrimp dumpling, Tingzai Porridge, dim sum, seafood, sour pig's knuckles, rice roll, milk tea and ice-cream	1716
Other	Morning tea, restaurant, popular, Baohua Road, tea canteen, barbecue, Starbuck, foodie, food and time-honored brand	1147

Table 4. Transportation words mentioned by travels.

Means of transportation	Frequency
Metro	972
Train	326
Bus	176
Air plane	145
Taxi	18
Ferry	19
Tramcar	14

4.2.3. Leisure and Recreational Activities

Seldom of tourists talked about the leisure and recreational activities in Guangzhou, and their recreational activities were shopping, taking tramcars, taking Cantonese tower sightseeing elevator and so on. Shangxiajiu Pedestrian Street and Beijing Road were good places for shopping, goods were cheap, "Shopping in Shangxiajiu is very convenient (No. 0135)" one travel note said. Secondly, some malls, such as Teemall, One Link Walk and Grand Buy, were also good places for shopping. Tourists were generally positive when talking about shopping, "There are lots of brands in Teemall and One Link Walk, shopping here is your best choice (No. 0420)"one travel note said. Shopping in night market was a characteristic and recreational activities in Guangzhou, "In the evening, the light on the road and the neon light of shops are light up, which brings you get into a splendid night scenery, it is so charming (No. 0420)" one travel note said.

4.2.4. Social Environment

The analysis of social environment was mainly from three aspects, namely the friendliness of local residents, the appearance of the city and the obstacle of localism. When talking about the friendliness of local people, some of them said, "The Cantonese entertain friends very well. (No. 0531)" one travel note said. Another travel note said, "Guangzhou is a good place, people here are hospitable (No. 0792)" Tourists thought Guangzhou was a large and prosperous city, "You can find out all the features of international metropolitan, which are tall buildings, wide and neat streets and people everywhere (No. 0665)" one travel note wrote. "Look at the shops on the both side of the pedestrian streets, think about the flourishing from millennium to now, such a feeling of traversing occurred me"(No. 0320) a travel note said. When talking about the obstacles of localism, the main localism of Guangzhou was Cantonese, which became a feature as well as an obstacle. One tourist said, "Buses here are broadcast mandarin and Cantonese, when it broadcast Cantonese, I have no idea what it is talking about, but this makes me feel I have been to south China, and realized Lingnan culture (No. 0028)."

4.3. The Affective Image of Selected Travels

From Table 5, we knew that most of the tourists gave positive evaluation to their trip. To tourists, the trip to Guangzhou was pleasant and arousing, scholars have pointed out that the essence of tourism was enjoyable experience [27] or poetic dwelling [28], as a result, the purpose of tourism was for happiness. "The time of happiness is always short (No. 0028)" one travel note said. Another said, "The staff along the way waved for us, what a happy journey" (No. 0187). However, some evaluations of Guangzhou were unpleasant and sleepy, the main reasons were that tourists were looking forward to the attractions and catering, but the experience was not perceived as high as expectation, and therefore they were disappointed. "The location is not convenient, it is far away from the metro station... it is the most disappointing station of the trip (No. 0324)"one tourist said. "The taste is just so so, it is not tasty as expectation, a little bit disappointed" (No. 0371) "After entering the interstellar battle, I feel so bored, it is just travel to 3D space, I almost sleep (No. 0656) another two travels wrote down.

Table 5. Analysis of emotional words of travels.

Dimensionality	Concrete content	Frequency
Pleasant	Happy, pleasant, interested, enchanted, satisfied, joyful...	789
Unpleasant	Unhappy, unpleasant, disappointed, frustrated...	354
Arousing	Aroused, excited, surprised, amazed, surprised...	263
Sleepy	Bored, sleepy, drowsy...	47

As a result, most of the affective appraisal which tourists to Guangzhou is positive, while the negative appraisal is because of the gap of expectation and perception to attractions and catering.

4.4. Analysis of Guangzhou Destination Image

We searched for the literature about Guangzhou destination image in CNKI, and got 3 papers about Guangzhou image positioning (saw Table 6). The four authors positioned Guangzhou destination image from different aspects, and we combine the position with the travels content, than we analyzed the frequency of Guangzhou destination image words (saw Table 7).

Combining Table 6 and Table 7, we found that tourists mentioned delicious food, Lingnan and Ram City a lot in their travels, Lingnan architectures and delicious food are two characteristic of Guangzhou, and the image of ram city impressed tourist most. While history and culture are seldom mentioned by tourists, which illustrated that tourists paid more attention to modern tourist attractions. Therefore, delicious food, Lingnan and Ran city are the prominent perceived by the tourists in their travels.

5. CONCLUSIONS

By analyzing the travels of Guangzhou in www.mafengwo.cn, we explored the tourism destination image perceived by the tourists. We analyzed the tourism destination image from two aspects, namely cognitive image and affective image. The conclusions we got were listed as follows.

Firstly, the writers of the travels were female and non-Guangzhou, they usually traveled with friends and relatives for 3 - 5 days and chose independent travel.

Secondly, in the cognitive image analysis, they preferred recreation and amusement resorts of synthetic human cultural tourism sites. Moreover, universities also became tourist attractions. In living places and communities, tourists paid attention to streets with Lingnan style. Morning tea, dessert and Taotaoju were mentioned first three words, and tourists would go to time-honored brand shops and ordered the recommended food. When talking about the accommodation in Guangzhou, most of them gave positive evaluation to accommodation, tourists who visited Chimelong Holiday Resort would choose Chimelong Hotel, but they considered the fare of Chimelong Hotel was high. Seldom of tourists would choose hostels, the young preferred economic hotels because of the convenient transport and low price. The main means tourists chose to get to Guangzhou were trains and air planes, metros were intracity transport, however, the evaluation of taxi was negative. Most evaluation of leisure and recreational activities were positive, they considered Shangxiajiu and Beijing road were good places for shopping, goods here were cheap, and the malls were also suitable for shopping. Shopping in night market, taking tramcar were the leisure and recreational activities as well. When talking about the social environment, tourists thought local residents were friendly and enthusiastic, and Guangzhou was a large and prosperous city. Cantonese was characteristic of Guangzhou as well as the communication barrier.

Thirdly, most affective evaluation of Guangzhou was positive and the negative evaluation was because of the gap of expectation and perception on tourism attractions and food.

Table 6. Literature of the positioning of Guangzhou destination image.

Author	Year	The positioning of Guangzhou image
Yu	1998	Ecology city, the door of south China and tourism center of Lingnan landscape culture
Huang and Zhang	2007	International metropolis and famous cultural city
Hu	2009	Lingnan and millennium city, food paradise and popularity

Table 7. The frequency of Guangzhou destination image.

Guangzhou Image	Frequency
Delicious Food	302
Lingnan	236
Ram City	126
Flower City	40
History and Culture city	21
Sui City	9

Finally, by analyzing the positioning image of Guangzhou suggested by the scholars and the frequency of Guangzhou destination image of the travels, we knew that tourists agreed that Guangzhou was a metropolis with the feature of Lingnan culture, and Ram City impressed tourists most, besides Guangzhou was a paradise of delicious food.

ACKNOWLEDGEMENTS

This study was supported by Jinan University's Scientific Research Creativeness Cultivation Project for Out- standing Undergraduates Recommended for Postgraduate Study in 2014.

REFERENCES

1. Lei, Y. (2013) The Process and Comparison of Destination Image in China and Abroad in Latest Five Years. Journal of Shijiazhuang University, 6, 67-71.
2. Hunt, J.D. (1975) Image as a Factor in Tourism Development. Journal of Travel Research, 13, 1-7.
3. Crompton, J.L. (1979) An Assessment of the Image of Mexico as a Vacation Destination and the Influence of Geographical Location upon the Image. Journal of Travel Research, 18, 18-23.

References

4. Echtner, C. and Ritchie, J.R.B. (1993) The Measurement of Destination Image: An Empirical Assessment. Journal of Travel Research, 24, 3-13.
5. Kotler, P. and Barich, H. (1991) A Framework for Marketing Image Marketing Image Management. Sloan Management Review, 32, 94-104.
6. Nuria, G.E. and Jose, A.D.B. (2005) The Social Construction of the Image of Girona: A Methodological Approach. Tourism Management, 35, 402-426.
7. Bramwell, B. and Rawding, L. (1996) Tourism Marketing Images of Industrial Cities. Annals of Tourism Research, 23, 201-221.
8. Baloglu, S. and McCleary, K.W. (1999) A Model of Destination Image Formation. Annals of Tourism Research, 26, 868-897.
9. Grosspietsch, M. (2006) Perceived and Projected Images of Rwanda: Visitor and International Tour Operator Perspectives. Tourism Management, 27, 225-234.
10. Selby, M. and Nigel, J.M. (1996) Reconstruing Place Image: A Case Study of Its Role in Destination Market Research. Tourism Management, 17, 287-294.
11. Qiu, Y.M. (1986) Analysis the Tourism Image of China. Economy Problem, 8, 56-57.
12. Li, L.L. (1999) Tourism Destination Image Planning: Theory and Practice. Guangdong Tourism Publishing Company, Guangzhou.
13. Song, Z.H. (2000) Study on Tourism Destination Image from Tourists' Aspect. Tourism Tribune, 15, 63-67.
14. Zhang, H.M, Lu, L. and Zhang, J.H. (2006) The Influence of an Analysis of the Perceived Distance on Tourism Destination Image. Human Geography, 5, 25-30.
15. Liu, R.W., Wu, D.T., Xiao, X. and Lei, Y.Z. (2006) The Research on the Effect That Time Sequence of Tourists Apperceiving Destination Image Has on Destination Image Apperceiving of Tourists. Economic Geography, 1, 145-150.
16. Zhou, J. (2007) Tourism Perceiving Image and Creation Tourist Destination's Image. Journal of Gulin Institute of Tourism, 3, 353-356.
17. Liu, Z.X, Ma, Y.F., Li, S., Niu, Y.L. and Wei, T. (2015) Evaluation on the Factors Influencing the Tourism Image of Beijing Based on Tourists'

Cognition and Perception. Journal of Arid Land Resources and Environment, 29, 203-208.
18. Wang, L., Liu, H.T. and Zhao, X.P. (1999) Research on the Definition of Tourism Destination Image. Journal of Xi'an Jiaotong University (Social Sciences), 19, 25-27.
19. Bo, W. (1997) Study on the Method of Content Analysis. Chinese Journal of Journalism & Communication, 4, 56-60, 69.
20. Qiu, J.P. and Zou, F. (2004) Research about Content Analysis Method. Journal of Library Science in China, 2, 14-19.
21. Stepchenkova, S. and Morrison, A.M. (2006) The Destination Image of Russia: From the Online Induced Perspective. Tourism Management, 27, 943-956.
22. Stepchenkova, S. and Zhan, F.Z. (2013) Visual Destination Image of Peru: Comparative Content Analysis of DMO and User-Generated Photography. Tourism Management, 36, 590-601.
23. Riera, I.L., Ruiz, M.P.M., Zarco, A.I.J. and Yusta, A.I. (2015) A Multidimensional Analysis of the Information Sources Construct and Its Relevance for Destination Image Formation. Tourism Management, 48, 319-328.
24. He, Y. and Yang, X.X. (2012) Literature Review and Application of Content Analysis Method in Tourism Research in China. Journal of Southwest Agricultural University (Social Science Edition), 10, 6-12.
25. Beerli, A. and Martin, J.D. (2004) Factors Influencing Destination Image. Annals of Tourism Research, 31, 657-681.
26. Russel, J.A. and Pratt, G. (1980) A Description of Affective Quality Attributed to Environment. Journal of Personality and Social Psychology, 38, 311-322.
27. Yang, Z.Z. (2014) On the Essence of Tourism. Tourism Tribune, 29, 13-21.
28. Xie, Y.J. (1998) On the Nature and Features of Tourism. Tourism Tribune, 13, 41-44, 63.

CHAPTER 8

An Evaluation of Destination Management Systems in Madagascar with Aspect of Tourism Sector

Daré Aurélien, Rakotonirina Jeremy Desiré

School of Economics and Management, China University of Geosciences, Wuhan, China

ABSTRACT

This paper describes the Destination Management Systems (DMS). Madagascar is a developing country in Africa and has a big population. Madagascar is one of the beautiful countries with its attractive beautiful scenery. These sceneries, lakes and beaches of Madagascar always attract travelers from all over the world to come to Madagascar to enjoy the nature, so a portion of this paper discusses about the tourism sector of Madagascar. No doubt Madagascar is gaining some money from tourism sector, but it can be increased if there can be a strong DMS for the traveler. The overall Destination Management Systems (DMS) and tourism sector of Madagascar have been discussed in this paper. As per conclusion, we can say DMS could increase visitor traffic, attract the right market segment with the provision of an accurate and up to date comprehensive electronic database. DMS also supports the wide distribution of destination information online. DMS can also help to create more efficient internal and external networks, which can have long-term positive effects on the local economy in achieving competitiveness.

KEYWORDS

Madagascar, Tourism, Destination Management Systems (DMS), Economy

1. INTRODUCTION

The Republic of Madagascar is a country located in Eastern Africa, consisting of the world's fourth largest island and some smaller islands in the Indian Ocean. Madagascar has a coastal plain, high plateau and mountains. Major rivers include the Betsiboka, Onilahy, Mangoky and Tsiribihina. The east coast of Madagascar has lowlands leading to steep bluffs and central highlands. The Tsaratanana Massif in the north has volcanic mountains. The west coast has many protected harbors and broad plains, while the southwest is a plateau and desert region. The largest city and capital is Antananarivo. Other important cities are Antsirabe, Mahajanga and Toamasina. The highest peak is the Maromokotro, reaching 2876 meters above sea level. Madagascar, the 4th largest island in the world with a surface area of 587,014 km, is known in the world due to the quality and diversity of its natural wealth of flora and fauna. The country also enjoys great geomorphologic diversity, such as the Tsingy de Bemaraha in the north, a UNESCO World Heritage site or the Massif in the south (Figure 1). The Republic of Madagascar is an island with a total area of 587,040 sq. km (land: 581,540 sq. km, water: 5500 sq. km) located in the southern parts of Africa in the Indian Ocean, East of Mozambique. The tourism sector is a dominant source of foreign exchange earnings. The sector accounts for one fourth of earnings, originating from services and is growing at 20% per year. The country is endowed with superb tourist sceneries (Diego Suarez's region—Antsiranana—in the north of the country for instance).

1. Introduction

Figure 1. Madagascar globally location and political map.

A Destination Management System (DMS) is a term for a specialized services company possessing wide- ranging local knowledge, expertise and resources, specifying in the design and application of events, activities, tours, transportation and program logistics. A DMS provides a ground service founded on local knowledge of their given termini. These facilities can be in the shape of transportation, hotel accommodation, restaurants, activities, excursions, conference venues, themed events, gala dinners and logistics, meetings, incentive schemes as well as helping with overcoming language barriers. These functions usually performed by travel agents and tour operators.

The types of DMO can be:

- National Tourism Authorities (NTAs) or Organisations (NTOs), responsible for management and marketing of tourism at a national level.
- Regional, provincial or state DMOs, responsible for the management and/or marketing of tourism in a geographic region defined for that purpose.
- Local DMOs, responsible for the management and/or marketing of tourism based on a smaller geographic area or city/town.
- Product Based—bringing together stakeholders related to a specific type of tourism product (i.e. Bird watch- ing, adventure, etc.).

Destination Management Organization will be a leading and most significant body to manage a Destination developing into a "total destination management system". The DMS will become the final source of strategic aptitude and grow as a communication center for a destination. The DMSs community will be required after by Govt. organizations and large establishments for information and leadership [1] . In general, the modest performance of administrations is cleared from the input and output side. The input quantity is based on physical and human capital award and research and growth expenses. The output side covers profitability, market share, productivity, growth and so on [2] . The output side is market share both in the number of entrances and the amount of tourism revenues, productivity and so on. As [3] Pearce (1997) implies, a competitive investigation refers to comparative studies. Therefore, destination attractiveness can be calculated both quantitatively and qualitatively. Quantitative performance of a destination can be measured by looking at

1. Introduction

numbers such as annual numbers of tourist arrivals, amount of annual tourism receipts, level of expenditure per tourist, length of overnight stays. Destination Marketing System (DMS) can act as organizers to accomplish the strategic objectives of the purpose. [4] Buhalis (2000) declared about these as attract the long-term feasibility of the local population, provision of visitor satisfaction and maximizing profits for Small Medium-Sized Tourism Enterprises. In the present time the demand of new techniques of information has been increased. With new information technologies obligating heightened over the last decade it is of paramount importance DMS in world and some developed countries like Madagascar re-engineer their business process, develop new business models and take advantage of these new tools [5] . Destination Management Systems (DMS) can be described as the IT structure of the Destination Management Organization DMO [6] . DMS should be able to act as an allowing machinery to assimilate the different services and products from the tourism industry. In the field of information and communications technologies (ICTs), Madagascar is presently performing a keen interest, with in specific a firm willingness on the part of the government to draw up a determined national policy with the goal of mixing the country in the global information society by 2010. It is in this environment that we are exploring the future of e-tourism in Madagascar and, in specific, the possibility of a destination management system for Madagascar being set up.

DMS should not only help to hand pre-trip, post arrival information requests, but also assimilate an availability and booking service too [7] [8] . DMS could increase visitor traffic, attract the right market segment with the provision of an accurate and up to date comprehensive electronic database [9] . DMS can also help to create more efficient internal and external networks, which can have long-term positive effects on the local economy in achieving competitive advantage. DMS also support the wide distribution of destination information online. WTO (2001) has showed that a number of new electronic circulation channels are developing through online travel agencies, search directories and destination portals. Therefore an important technical thought in the design of DMS is the development of open systems so that links can be developed with alternative circulation channels [9] .

Alternative circulation channels are mandatory to support DMS to interface and distribute information through websites, TICs, call centers, kiosks and traditional marketing channels [10]. Therefore DMS need to realize it to act as an enabler in sustaining modest improvement [11] [12].

2. TOURISM SECTOR OF MADAGASCAR

Tourism is a multidimensional movement that includes not only drives the people from one place to another, but also undertakes both directly and indirectly interweaved towards the easing of this process [13] - [15]. The tourism industry is consequently a compound phenomenon which includes deal processes that are determined by global effects of multinational tourism and travel corporations, geo-political and other broader forces of economic change [16]. As such, tourism produces vast economic movement worldwide and has grown to be one of the world's largest businesses in terms of volume and revenue created. The development of the tourism industry has been remarkable, with international tourist arrivals growing from 25 million in 1950, to 277 million in 1980, to 438 million in 1990, to 681 million in 2000, and currently standing at 880 million [17]. The tourism industry can be a major source of government income in Eastern Africa especially for those countries whose GDP role by the industry is expressively high.

The demand of Information & Communication Technologies (ICTs) is raising the performance of tourism firms, but also of tourism destinations at a macro-economic level is widely encouraged [18] - [20]. The entirety of the ICTs advanced by Destination Management Organisations (DMO) for marketing their destinations represents the Destination Management System (DMS) of the destination. In general, DMS are considered as inter-organizational ICT directing to link the physically disconnected tourism supply with the tourism demand. For example, [21] defined the DMS as an inter-organizational system that contacts tourist products, suppliers and offers, with consumers and mediators in order to enable easy access to complete and up-to-date destination information and to allow reservations and purchases. A more full DMS definition was newly established from a Delphi study that took into concern the responses of several tourism stakeholders [22] : "...DMS are systems that consolidate and distribute a comprehensive range of tourism products through a variety of channels and platforms, generally catering for a specific region, and supporting the activities of a DMO within that region. DMS attempt to utilise a customer centric approach to manage and market the destination as a holistic entity, typically providing strong destination related information, real-time reservations, destination management tools and paying particular attention to supporting small & independent tourism suppliers".

So it's clear from the above mentioned statement that the major role of DMS is to act as electronic in-between providing functionalities connected to e-distribution, e-marketing and e-sales for the whole destination and its tourism suppliers. Indeed, the impact and the necessity of DMS for the survival and the competitiveness of small and medium tourism enterprises (SMTEs) is considered so indispensable, specifically when considering that although SMTEs represent the majority of firms at many destinations, they lack the technological, managerial and financial resources for exploiting ICTs for e-commerce and e-marketing purposes. Therefore, SMTEs heavily depend on DMS for having an e-presence and an alternative e-distribution channel that may in turn reduce their dependencies on tour operators and other intermediaries [19] [22] [23] . Moreover, it should also be highlighted that DMS should serve a much wider role that first aims to support a sustainable development and management of destinations by supporting the marketing and development of sustainable forms of tourism. In Figure 2, we can see the upstream and downstream values of overall tourism sector.

3. THE TOURISM VALUE CHAIN

Tourism can be called as a leading economic activity in many Small Island Developing States (SIDS) and a key component of their progress strategies [25] . The Indian Ocean Islands, of which Madagascar, Mauritius, Reunion,

Maldives and Seychelles remain the most important tourist destinations, depend on to varying extent on tourism for economic development. The involvement of the travel and tourism industry (TTI) to these island economies cannot ignore. Table 1 provides a summary of the main features of the TTI for the above-mentioned islands in 2010 with respect to the input of tourism to Gross Domestic Product (GDP), direct employment, earnings from tourism, and investment in tourism. Obviously, these data recommend that Maldives and Seychelles are the most reliant on the tourism industry for economic development and Madagascar was gaining 12.7% from the tourism sector to the GDP of the country in 2010. However, the highest incomes and investment in tourism are for Mauritius and Madagascar individually, suggesting that the attractiveness of these islands varies. In fact, Mauritius and Barbados are the only two islands to rank in the top 50 countries, 40th and 30th respectively, for overall destination competitiveness on the Travel and Tourism Competitiveness Index [26].

Travelers from all over the world have always been attracted to Madagascar by its abundant of natural resources (fauna, flora, geomorphology, and climate) and famous and well known cultural resources. At the present time Tourism in Madagascar has been increased over the last twenty years with an annual growth rate which can see in Figure 3. Madagascar is progressively positioning itself as a prime eco-tourism destination. Madagascar is not only gaining revenue from this sector but also a number of Madagascar citizens are belong to this sector for their income, so it also one of the main source of people. In Table 1, we can see the number of people belong to this profession.

In some countries where tourism plays a very significant role ensuring economic growth, it is collective to find cases in which the pressure to grow the sector justified an extremely fast growth and absence of destination planning. So, especially in destinations which have attentive on the so-called Sun and Sea Tourism, physical problems arose, originating from extreme attentions at different, though connected levels.

3. The Tourism Value Chain

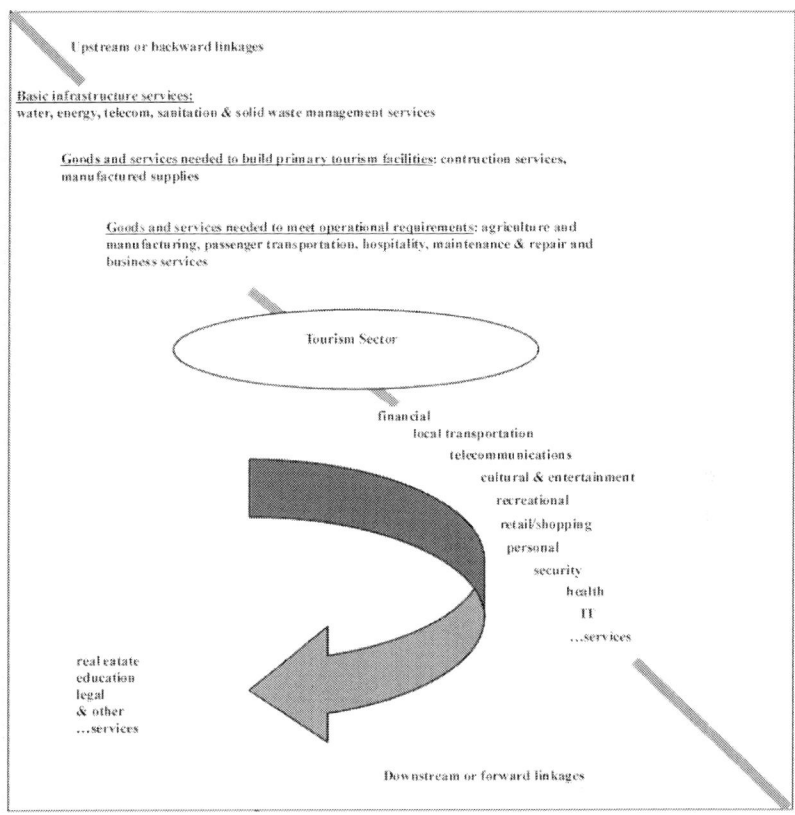

Figure 2. UNCTAD (2007) [24].

Table 1. Tourism employment in select Eastern Africa countries within 2010.

	Direct and Indirect Employment through Tourism	
	No. Employed	% Working Population
Seychelles	22,000	56%
Madagascar	455,000	10.1%
Kenya	439,000	7.3%
Ethiopia	Over 1.6 million	6.9%
Rwanda	122,000	6.6%
Tanzania	624,000	6.3%
Uganda	381,000	5.9%
Comoros	10,000	5.4%
DRC	472,000	3.1%
Burundi	53,000	3%

Source: WTTC, 2010 [37].

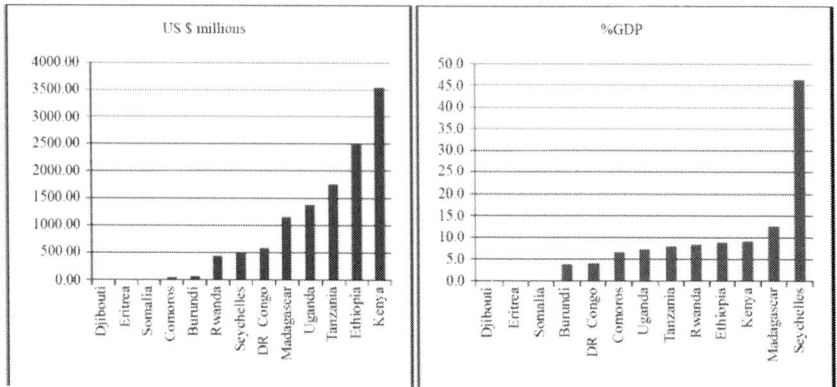

Figure 3. Travel and tourism GDP contribution in Eastern Africa, 2010. Sources: WTTC, 2010 [27].

4. SOME EFFECTS OF TOURISM DEVELOPMENT IN EASTERN AFRICA

Possession and constant growth of relevant tourism abilities and knowledge is essential in the creation of human capital, which is an acute resource for the development of a modest sustainable destination [28]. In the Eastern Africa region, the absence of relevant expertise and knowledge has been a problem for effective tourism development [29]. There have been concentrated efforts, particularly by the more mature tourism destinations, to address this challenge.

However, it is equally important to note that the level of local investment in the tourism industry and the general contribution of the industry in addressing the various developmental challenges have been insignificant in the region [30]. This is could be credited to the fact that these tourism training creativities have been in retort to specific skills lacks within the industry, which is conquered by large initiatives, instead of laying stress on basic skills and knowledge to facilitate, for instance, small enterprise development [30]. The rigidity of the training programmes in Eastern Africa especially in Madagascar has caused in very few people being equipped with the suitable skills and knowledge necessary to raise local efficiency and produce wealth through tourism development [28] . One more major challenge facing tourism development in the Eastern Africa region, especially in countries where tourism subsidizes meaningfully to GDP, is that benefits accumulating from industry do not trickle down to local societies, even in those areas where tourist activity is leading. This could be credited to lack of sufficient social capital.

Literature have shown that the more the local public participation in tourism development through the various community-based initiatives, the higher the welfares that accumulate to them [31] . Community-based initiatives, consequently, have the possible to benefit communities and offer a progressive route through which support networks could be well-known [32] . SWOT Analysis of Madagascar's Tourism Sector can be seen in Table 2. This analysis has been done by United Nations Economic Commission for Africa UNECASRO-EA, 2011.

5. STRATEGIES FOR IMPROVING TOURISM LINKAGES IN MADAGASCAR

Due to the fact that tourism has been recognized as one of the key sector to drive economic growth and that the usual model has not brought about expressive socio-economic growth, a number of approaches have been planned to enhance tourism economic linkages. [34] for illustration, suggests several measures that evolving countries can pursue to improve tourism economic linkages including: a) boosting procurement from local enterprise; b) partnerships with neighbours; c) supporting local cultural and heritage product development coupled with enhancing the capacity of local entrepreneurs to deliver these products and d) encouraging tourist expenditure within the local economy. Attaining this goal will need suitable policy interferences that encourage local community involvement in tourism, such as capacity-building for local businesspersons to permit them meet industry standards, and those policies that encourage stockholders to partner efficiently with local communities. Additional, it is significant to note that important improvements to the tourist industry can be realized through regional addition, thus cumulative the abilities of obtaining inputs within the area (e.g., through Madagascar manufactures). In adding, new notions of tourism progress that supposedly improve linkages including ecotourism and community-based tourism will have to be trained to ensure a balanced method in meeting respective country significances under broad sustainability measurements of economic, environmental and socio-cultural concerns.

6. CONCLUSIONS AND RECOMMENDATIONS

As we can say Madagascar is a paradise on earth due to its beauty and abundant attractive scenery. The government has a big portion of GDP which gets from this sector, but as per my conclusion, it can increase more if we try to develop the strong DMS in the country along with strong ICT system. The research proves that DMO within Madagascar does understand the importance of realizing DMS and those they can deliver the destination with modest advantage and long-term welfares. An important number of DMS are not subcontracting DMS regardless of the advantages can be gained. The DMS still prefer to purchase the system and few proceed to build their own DMS, lack of funding and scientific expertise are seen as two of the major obstacles. DMS not only help the traveler, but also increase the economy of any country, a country like Madagascar, which is a developing country, not only need to create these types of system itself, but the government should also encourage DMS system in the country for the foreigner investors. The vendors of DMO are also responsible to create a confirmatory environment for the tourists, develop the market demand and driven tourism packages, expand the service quality and product development of local suppliers and improve the environmental performance of private businesses. To support sustainable tourism development, promote responsible travel and give incentives to travelers to understand and protect the local culture and environment they are visiting. Government of Madagascar should take some serious and suitable steps for the betterment of this sector because SWOT analyses that the weakness and threats of this sector are high. Apart from it, a large number of Madagascar people attach with this profession, so by the betterment of this sector, government can promote the economy of the country as well as the life of people. A sustainable and developed DMS are in need of time for this sector within country.

Table 2. SWOT analysis of Madagascar's tourism sector.

SWOT analysis of Madagascar's Tourism sector	
KEY STRENGTHS • One of the world's top "biodiversity hotspots" • Rich natural heritage—more than 80% of Madagascar's flora and fauna are found nowhere else in the world and some taxonomic groups, including reptiles and amphibians, are over 95% endemic • Coastal attractions	**KEY WEAKNESSES** • Lack of supportive policies and regulations • Investment climate is not conducive to tourism development • Weak country credit rating • Unfavorable labour relations in the hospitality sector • Poor airline connectivity • Inadequately skilled labour-force • Inadequate funding for tourism initiatives • Stiff competition from neighbouring tourism destinations such as Seychelles and Mauritius
KEY OPPORTUNITIES • Investment finance through the "Funds For the Promotion of Private Business" • Tourism product diversification e.g. adventure, special interest etc. • Establishment of ecotourism investment zones • Emerging markets in Africa and Asia	**KEY THREATS** • Lack of an "open skies" policy • Perceptions of poor governance and political instability • degree of environmental and forest degradation • Climate change • Disaster vulnerability e.g. drought, cyclones, flooding, etc.

Source: (UNECASRO-EA, 2011) [33].

REFERENCES

1. Varghese, B. (2013) Intervention of Destination Management Organization's in Tourist Destinations for Branding, Image Building and Competitiveness—A Conducive Model for Karnataka. International Journal of Investment and Management, 2, 50-56.

2. Jacobson, D. and Andreosso-O'Callaghan, B. (1996) Industrial Economics and Organization. McGraw-Hill, Maidenhead.

3. Pearce, D.G. (1997) Competitive Destination Analysis in Southeast Asia. Journal of Travel Research, 35, 16-24.

4. Buhalis, D. (2000) Marketing the Competitive Destination of the Future. Tourism Management, 21, 97-116.

5. Gretzel, U., Yuan, Y.L. and Fesenmaier, D.R. (2000) Preparing for the New Economy: Advertising Strategies and Change in Destination Marketing Organizations. Journal of Travel Research, 39, 146-156.

6. Sheldon, P. (1997) Tourism Information Technology. CAB International, New York.

7. Buhalis, D. (1997) Information Technology as a Strategic Tool for Economic, Social, Cultural and Environmental Benefits Enhancement of Tourism at Destination Regions. Progress in Tourism and Hospitality Research, 3, 71-93.

8. Frew, A.J. and O'Connor, P. (1998) A Comparative Examination of the Implementation of Destination Marketing System Strategies: Scotland and Ireland. In: Buhalis, D. and Schertler, W., Eds., Information and Communication Technologies in Tourism1998. Springer, Wien-New York, 258-267.

9. Sheldon, P.J. (1993) Destination Information Systems. Annals of Tourism Research, 20, 633-649.

10. O'Connor, P. (2002) The Changing Face of Destination Management Systems. Travel and Tourism Analyst, No. 2, 1-25.

11. Lewis, R.D. (2002) Modelling Tourism Impacts Using IT Based DMS. In: Fesenmaier, D., Klein, S. and Buhalis, D., Eds., Information and Communication Technologies in Tourism, Springer, Wien-New York, 97-104.

12. Gretzel, U., Yuan, Y.L. and Fesenmaier, D.R. (2000) Preparing for the New Economy: Marketing Strategies and Change in Destination Marketing Organizations. Journal of Travel Research, 39, 146-156.

13. Smith, S. (1995) Tourism Analysis: A Handbook. 2nd Edition, Longman, London.

14. Cooper, C., Fletcher, J., Gilbert, D. and Wanhill, S. (1998) Tourism: Principles and Practice. Pitman, London.

15. Sharpley, R. and Telfer, D.J. (2002) Tourism and Development: Concepts and Issues. Channel View Publications, Clevedon.

References

16. Milne, S. and Ateljevic, I. (2001) Tourism, Economic Development and the Global-Local Nexus: Theory of Embracing Complexity. Tourism Geographies, 3, 369-393.

17. UNWTO (2010) Manual on Tourism and Poverty Alleviation—Practical Steps for Destinations. Madrid. http://pub.unwto.org/WebRoot/Store/Shops/Infoshop/4BEB/F726/BC4B/3784/9088/C0A8/0164/FDA7/100513_manual_pov_all_excerpt.pdf

18. WTO-World Tourism Organisation (2007) A Practical Guide to Tourism Destinationv Management. WTO Business Council, Madrid.

19. Sigala, M. and Marinidis, D. (2009) Exploring the Transformation of Tourism Firms' Operations and Business Models through the Use of Web Map Services. European and Mediterranean Conference on Information Systems 2009, Izmir, 13-14 July 2009, 1-13.

20. Wang, Y. and Fesenmaier, D. (2006) Identifying the Success Factors of Web-Based Marketing Strategy: An Investigation of Convention and Visitors Bureaus in the United States. Journal of Travel Research, 44, 239-249.

21. Chen, H.M. and Sheldon, P. (1997) Destination Information Systems: Design Issues and Directions. Journal of Management Information Systems, 14, 151-176.

22. Frew, A.J. and Horan, P. (2007) Destination Website Effectiveness: A Delphi Study-Based eMetric Approach. HITA Conference, Orlando, 49-80.

23. Frew, A.J. and O'Connor, P. (1999) DMS: Refining & Extending an Assessment Framework. In: Buhalis, D. and Schertler, W., Eds., ICT in Tourism, Springer, Wien, 398-407.

24. UNCTAD (2007) FDI in Tourism: The Development Dimension. United Nations.

25. De Villiers, D. (2005) Small Islands Big Stakes. Opening Speech, The United Nations Conference on Small Islands, Port-Louis, 10-14 January 2005.

26. World Economic Forum (2009) The Travel and Tourism Competitiveness Index 2009: Measuring Sectoral Drivers in a Downturn. World Economic Forum, Cologny.

27. WTTC (2010) Travel and Tourism Economic Impacts: Country Reports 2010. London.

28. Lazear, E.P. (2004) Balanced Skills and Entrepreneurship. American Economic Review, 94, 208-211.

29. Victurine, R. (2000) Building Tourism Excellence at the Community Level: Capacity Building for Community-Based Entrepreneurs in Uganda. Journal of Travel Research, 38, 221-229.

30. Doswell, R. (2000) African Tourism Training and Education: Hits and Misses. In: Dieke, P.U.C., Ed., The Political Economy of Tourism Development in Africa, Cognizant Communication Corporation, USA, 247-259.

31. Mitchell, R.E. and Reid, D.G. (2001) Community Integration: Island Tourism in Peru. Annals of Tourism Research, 28, 113-139.

32. Manyara, G. and Jones, E. (2007) Community-Based Tourism Enterprises Development in Kenya: An Exploration of Their Potential as Avenues of Poverty Alleviation. Journal of Sustainable Tourism, 15, 628-644.

33. UNECASRO-EA (2011) Towards a Sustainable Tourism Industry in Eastern Africa: A Study on the Challenges and Opportunities for Tourism Development. United Nations Economic Commission for Africa, Addis Ababa.

34. ODI (2006) Tourism Business and the Local Economy: Increasing Impact through a Linkages Approach. ODI Briefing Paper, March 2006. http://www.odi.org.uk/resources/download/1944.pdf

CHAPTER 9

Biosphere Reserve as a Learning Tourism Destination: Approaches from Tasik Chini

A. Habibah[1,2], I. Mushrifah[1], J. Hamzah[1,2], A. Buang[1], M. E. Toriman[2,3], S. R. S. Abdullah[1,4], K. Z. Nur Amirah[1], Z. Nur Farahin[1], A. C. Er[2]

[1]*Pusat Penyelidikan Tasik Chini, Faculty of Science and Technology, Universiti Kebangsaan Malaysia, Bangi, Malaysia*
[2]*School of Social, Development and Environmental Studies, Faculty of Social Sciences and Humanities, Universiti Kebangsaan Malaysia, Bangi, Malaysia*
[3]*Universiti Sultan Zainal Abidin, Terengganu, Malaysia*
[4]*Faculty of Engineering and Built Environment, Universiti Kebangsaan Malaysia, Bangi, Malaysia*

ABSTRACT

Biosphere Reserves (BR) are special areas or regions highly recognized for their conservation, logistic functions and sustainable development initiatives. However, not much work has explored into the BRs' roles or functions as tourism learning destination, especially during the early years of their recognition as BR. This article aims to identify the mechanism utilised in the learning tourism function at Tasik Chini Biosphere Reserve since its inception in 2009 to the present year of 2013. The results reveal that learning of science and culture of the locals are the two-tier perspectives utlised in conceptualizing a tourism learning destination. Activities introduced in the specific themes of The Sustainability of Tropical Heritage fulfil the fundamental need of deep learning of scientific research and learning of the BR's ecosystem, while the Ecosystem Health fulfils both deep and surface learning of the young visitors. The cultural knowledge of the community, on the other hand, offers a unique and authentic experience to the learners or visitors. As a learning tourism destination, the learning community, nevertheless, expects that the standard of tourism services should not be marginalised and must meet the high standard of tourism services. It is imperative that the science of Biosphere Reserve and the local culture are linked to set a holistic foundation in the creation of the learning programmes at the Tasik Chini Biosphere Reserve.

KEYWORDS

Learning Tourism Destination; Ecotourism; Tasik Chini Biosphere Reserve; Deep Learning

1. INTRODUCTION

Biosphere Reserves (BR) are highly acknowledged as areas or regions for undertaking learning, reconciling environmental issues and integrating approaches for sustainable development. This is especially apparent in the three complementary and mutual functions of biodiversity conservation; sustainable economic and human development; and logistics support for research, monitoring, education and information exchange [1-4]. During the United Nations Decade of Education for Sustainable Development (UNDESD, 2004-2013), another special responsibility given to the BR is to function as a learning laboratory or learning site, where evidence-based knowledge, iterative and practical principles are utilised to ensure sustainable development [1,5]. This mission and aspiration further demand salient approaches as learners now are not limited to only small groups of locals; instead, it is intended to cater to a wider international community [6,7].

1. Introduction

Whilst extensive documentations on successful learning functions or sites provide references to the policy makers and practitioners in the recent years [8-13], what constitutes a good model of a learning destination for a newly endorsed site still remains limited. There is no agreed and well defined learning destination that offers a holistic orientation and specific-site knowledge. To date, social learning [14], ecosystem approach [9,15] and systems thinking approach [5,16,17], learning tourism destination

(LTD) or TLA (Tourism learning area) [18-20], CARE approach (complexity, aesthetic, responsibility and ethics), edutourism [21] and experiential learning [22] have been introduced to foster the learning of sustainable development. Of specific roles, the Yukon entrepreneurs [23] have created an interactive tool in product development, and further suggest a change from "product" to "programme" and from "tourist" to "participants" in order to enhance the learning experiences, especially for the special niche market, including the educational tourists.

Although the above-mentioned approaches provide alternatives in realizing the learning objectives and functions of the BRs, some limitations on implementations and acceptance of the stakeholders are still apparent because of the diversity of socio-cultural, physical, economic and environmental background of both visitors and educators [24-27]. The existing literature has yet to provide a holistic mechanism that elucidates the roles of the stakeholders in initiating, inventing and developing the Biosphere Reserve experiences as the foundation of tourism learning destination.

Tasik Chini has been accorded as the first Biosphere Reserve in Malaysia in 2009. Rich in diversity of flora, fauna and culture of aboriginal tribe, Tasik Chini is a sensitive area of class 1. As a BR, Tasik Chini has to execute three major functional roles, comprising of development, conservation and logistic functions. Of these roles, providing a learning experience is one of the most important logistic functions. Considering that comprehending the learning approaches in Tasik Chini as a new BR in the Asian region will provide a sharing of sitespecific approaches and knowledge, this article was aimed at identifying the learning initiatives of Tasik Chini Biosphere Reserve and at understanding the learning community's responses to further improve the learning functions of the designated area.

2. BIOSPHERE RESERVE AS A LEARNING DESTINATION

2.1. The Science of Biosphere Reserve

Biosphere reserves, as defined by UNESCO, are areas of terrestrial and coastal ecosystems that promote solutions to reconcile the conservation of biodiversity with its sustainable use. Every single BR has to fulfil conservation, development, and logistic support functions; hence, a BR is commonly agreed as a "living laboratory" or "learning place" [3,4,17]. The Seville Strategy elaborates these functions. First is the conservation function which is intended to preserve landscapes, ecosystems, species, and natural resources. Second is the development function which focuses on sustainable economic and human development by considering social, cultural, and ecological issues; and third, the logistic support which is aimed at fostering education, research, and information exchange related to conservation and development of the BR [1].

With the three above aforementioned functions, each BR provides a complete ecosystem that encourages scientific ideas and tools in resolving diverse social-ecological issues. Studies of flora and fauna, ethnobotany of medicinal crops, conservation and forest, wildlife, soil biology and biochemistry of montane habitats as well as hydrology and watershed have been undertaken to comprehend and strengthen the science entity [28-30]. However, sustainable development has become an important function in the mid 1990s, and studies on environmental governance and sustainable development practices have attracted scientists to explore issues on BR [25]. Studies in the Maya forest BR, Long Point and other renowned sites explicitly positioned the science of environmental management and conservation in the biosphere reserve [13,31].

Although it has been the pure sciences that dominated the studies of the BRs, recent works, however, showed a significant integration of multi-disciplines.

2. Biosphere Reserve as a Learning Destination

[32] emphasised that research, and monitoring are fundamental components for the logistics function to provide locationspecific knowledge of ecosystems, local economies, social organizations and governance. These changing roles of logistic functions were found in four biosphere reserves - Long Point, Niagara Escarpment, Mont Saint Hilaire, and Mount Arrow Smith. The research interests have also broadened into the transition zone of biosphere reserves, communities and collaborative approaches.

On a similar vein, [31] offer an alternative for sustainability initiatives emphasising on place-based and integrated-knowledge approaches. These approaches imply a flexible combination of disciplines and types of knowledge in the context of nature-human interactions. They can be operationalised within the framework of sustainability science in three steps: 1) characterize the contextual circumstances that are most relevant for sustainability; 2) identify the disciplines and knowledge that need to be combined to appropriately address contextual circumstances; and 3) decide how these disciplines and knowledge can be effectively combined and integrated. In simple terms, the outcome of the research is relevant to scientific knowledge, and in turn, scientific knowledge makes possible for the practices within the context of local benefit. Furthermore, [3] asserts that a primary platform in learning for sustainable development is the ability to discover new ways in deepening communications with the public, private sector and civil society. All these trends ratify the needs of inventing a mechanism that allow the scientific findings to be transferred to the community and visitors.

2.2. The Learning Approaches of the Learning Sites

It is worth noting the emerging literature on diverse logistic functions of the Biosphere Reserve. The learning and logistic functions exhibit increasingly diverse interests covering topics ranging from purist ecosystem services to the applied researches concerning the visitors' acceptance as well as involvement and benefits of local community. However, recently, there has been a tendency of establishing a learning destination [5,9,14,15,19, 20], and as a result, thematic names of these sites emerged significantly. These initiatives are of various levels - area, region, target audiences and collaborations, including with the higher learning institutions, researchers and school children [24,33-37].

Looking into the initiatives shown in Table 1, it is fundamental that each learning approach has a clear and strong foundation of the site-specific knowledge even though rigourous debate on learning about the biosphere reserve is not a new agenda. Since the Seville strategy in the 70s to the Rio Summit in the 1990s, educational and learning objectives have been the focal points of logistic functions of the BRs. Being a special place that has its specific context knowledge, initiatives in documentation of diverse experiences from the newly established biosphere reserve will provide a large pool of resources on learning. To some scholars, experiential and holistic undertakings of both science and local knowledge are their preference. Similarly, educationists also stress on both tacit and explicit knowledge or deep and surface learning [38,39].

Being recognized as a learning site and at the same time, being promoted as an ecotourism destination and key area of sustainable development, most BRs will ensure a unique local knowledge as part and parcel of their logistic functions [40,41]. In this vein, the cultural heritage of the community will become the key focal experience, and authentic landscape will represent a sound product mix of the BRs. Recent trend shows that local and indigenous culture, traditional ecological knowledge and authentic experience are the buzzword in packaging these experiences.

Another crucial fact is the content of the activities introduced to the visitors. Studies on various BRs' visitors, young school children, families and youth volunteers, confirm the existence of these differences. A project entitled "breakfast: healthy-regional-sustainable" implemented by the Ryon Biosphere reserve has been recognized as a "project of the UN Decade" [11] that introduces children (from year 3) to eat healthily.

The play comprised learning about the groceries they consume, in particular the contents and nutritional values, understanding of healthy eating to climate protection and the values of harmonious family meals. This project, which uses a child-oriented method, employed a "fun" way of bringing various dimensions of healthy eating, social (family) aspects of eating and contribution to the local economy. In contrast, [6] compares the visitor centre in Sweden and Germany's BR, and found that even though the formal way of learning provides ESD contents, touristic and informal learning provide more enjoyment and remembrance.

2. Biosphere Reserve as a Learning Destination

Table 1. Approaches on learning destinations.

Scholars	Approach	Principle components/Attributes
de la Barre (2005)	Learning travel product development	Learning travel (also known as "educational travel" or "enrichment travel") is a concept. It involves a series of formal and informal learning, travel, and social activities that, when cleverly packaged, engages people in memorable "ad-ventures". Its unique selling proposition is quality-learning experiences, delivered by dynamic resource specialists. Educational travelers are willing to pay a premium for these experiences.
Khelghat-Doost et al. (2011)	Regional Centre of Expertise (RCE)	As evidenced in the Regional Centre of Expertise, (RCE) a framework of partnership that fosters capacity building and supports innovative education for sustainable development is proposed. Three key factors are leadership, partnership and networking.
Schianetz, Kavanagh, & Lockington, (2007)	Learning tourism destination	The fundamental elements of a learning tourism destination (LTD) are 1) Shared vision and goals, 2) Information system, 3) Continuous learning and co-operative research, 4) Co-operation (informal collaboration), 5) Co-ordination (formal collaboration), cultural exchange that forms the basis for mutual cooperation 6) acceptance of different worldviews and belief systems, and understanding of these differences which will enhance the dialogue between individuals and within the community, 7) Participative planning and decision making, 8) adaptive management
McCarthy et al. 2011	Social learning and sustainability	Promote social learning through sustainability workshops; ensuring an inclusive, open process and providing opportunities for increased public understanding of the purpose and role of the biosphere reserve contribute greatly to social learning.
European Commission	Tourism Learning area	A tourism learning area (TLA) is a concept aimed at improving skills in tourism. It is based on an exchange of learning experiences aimed at increasing quality, innovation and competitiveness within the industry. A TLA consists of a network of all sectors and individuals who contribute to tourism (including local authorities, entrepreneurs, learning institutions, community groups and farmers).
Ministry of Education British Columbia, 2009	CARE concept C.A.R.E. (Complexity, Aesthetics, Responsibility and Ethics)	C.A.R.E. emphasises the interdisciplinary nature of environmental concepts that leads toward deeper engagement with environmental learning in all of its forms. These principles are: Complexity: considering the complexity and interrelatedness of natural and human created systems, and how humans interact with and affect those systems; Aesthetics: developing an aesthetic appreciation for the natural world that encourages students to learn about and protect the environment; Responsibility: providing opportunities for students to take responsible action and explore the environmental impact of their decisions and actions; Ethics: providing opportunities to practice environmental ethics based on an examination of values that can give rise to new visions, possibilities and actions.

3 MATERIALS AND METHODS

3.1. The Study Site

Tasik Chini is the second largest natural freshwater lake in Malaysia. Spanning a total area of 5085 hectares, Tasik Chini was gazetted as a Reserved Area for public use, specifically for tourism activities in 1989 under the National Land Code 1964. Figure 1 shows the study area. In 2008, a larger overlapping area was designated as a UNESCO Biosphere Reserve, covering the lake catchment area and its feeders, totalling 6951.44 hectares.

The water body comprises of 12 interconnected open water bodies called "laut" by the local Orang Asli communities (Tasik Chini dossier, 2009). The indigenous Jakun tribe lives around the lakeshore and the surrounding areas [42]. The water body of the lake covers an estimated 202 hectares of open water and 700 hectares of riparian and wetland zones. Tasik Chini itself is surrounded by natural hills and lowland dipterocarp forests, disturbed forests, vegetated low hills and undulating land which constitute the watershed of the region.

Due to encroachment of catchment area, forest clearing, incompatible land use, and lack of clear guidelines for management and conservation, Tasik Chini suffers from extreme environmental degradation [43]. Considering that strong scientific research can serve as a platform to conserve and restore Tasik Chini, the Tasik Chini Research Centre (TCRC) was established in 2004. With Tasik Chini accorded the status of a UNESCO Biosphere Reserve, environmental problems need to be addressed, and restoration needs to be enhanced. In fact, it should be suitably managed, and re-established as the premier resource-based tourism destination as well as a living laboratory for research and education in Malaysia. Hence, with the situation described above, Tasik Chini has extensive resources as a Learning Destination of Biosphere Reserve.

3 Materials and Methods

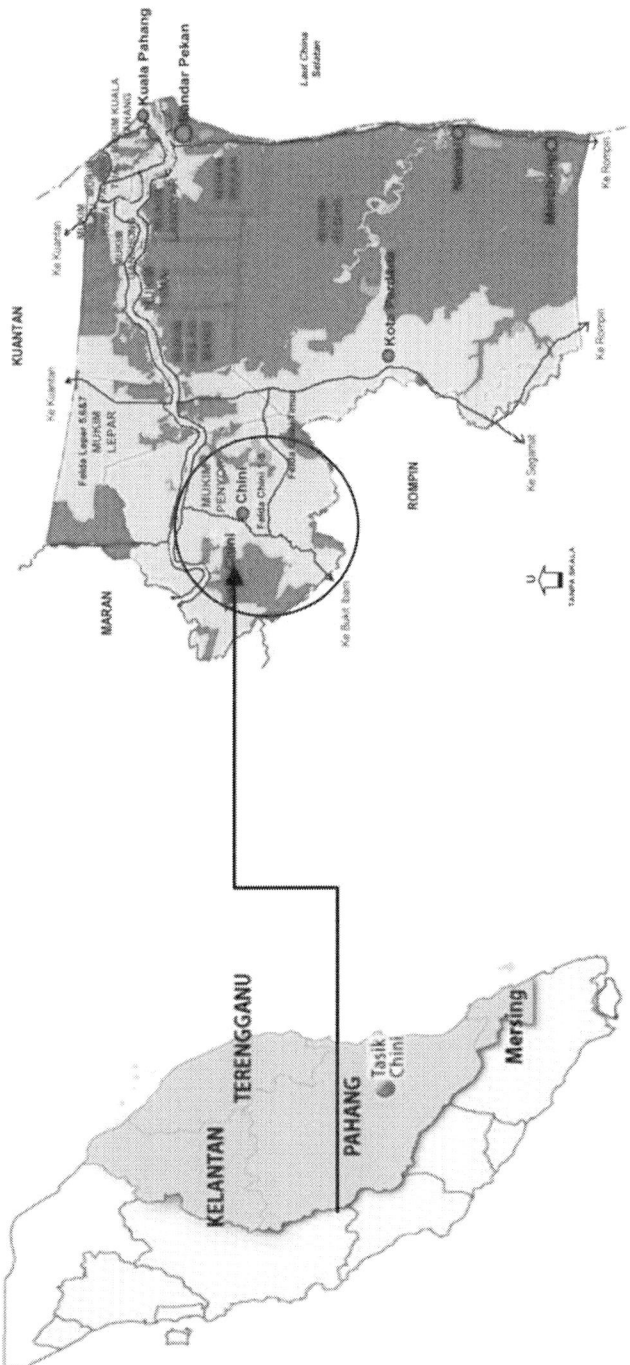

Figure 1. Location of the Study Area, Tasik Chini Biosphere Reserve.

3.2. Data Collection and Analysis

This study used a mixed method approach and the following data collection was utilized in comprehending the approaches initiated in Tasik Chini. The first was the data bank of the TCRC; which provided information regarding the series of events held from 2008 to 2013. The data provided all events or programmes held at and by the research centre, and among others, included the proposals, event reports and program book. The data has helped the authors to classify and map events into knowledge orientation, scientific and play entity, duration, location and learning objectives as well as to identify the learning community—the participants and experts involved.

The data were further divided into two stages, comprising of events during pre-dossier and events during pre-establishment of the Biosphere Reserve as well as recent events hosted by this site. The second was the qualitative data derived from various in-depth interviews with local stakeholders, participants in the focus group discussion and public consultation. The third was utilisation of the surveys conducted during each programme, especially to derive the participants' feedback, satisfaction and understanding of the overall programmes held and experienced.

4. RESULTS AND ANALYSIS

4.1. The Learning Approaches from Tasik Chini

4.1.1. Learning of the Science of Biosphere Reserve

As revealed, learning is one of the logistic functions of BRs; and thus, Tasik Chini has initiated knowledgebased programs in order to promote and provide site specific knowledge and experiences. Since its inception in 2009, a total of 14 programs have been conducted up to August 2013 during the pre-dossier preparation, preestablishment and post establishment, as depicted in Table 2.

Each programme provided a thematic understanding of the BR. To ensure visitors have satisfactory experience, each programme was tailor-made

4. Results and Analysis

according to the educational background, travel motives and duration of visit. Scientific aspects were sought from experts in the pure science and social sciences, especially in terms of knowledge, real-life experiences and hospitality of the locals. These experts provided in-depth knowledge of the major domains of the ecosystem—the lake, hydrology, forest, tourism, aborigines and culture as well as conservation initiatives, which were later moulded into holistic knowledge of this BR [44].

4.1.2. The Sustainability of Tropical Heritage

As a host, Tasik Chini is proactively developing a site specific theme of learning ecosystem. A specific theme, namely namely "The Sustainability of Tropical Heritage" is used in hosting the mobility program. A framework of learning culture in four-dimensional perspectives is utilized in this programme. Figure 2 shows the four-dimensional perspectives of the learning culture in Tasik Chini Biosphere Reserve (TCBR). Based on the details of the activities, four major themes seem to uphold the content and context of the site, namely 1) learning about and for biodiversity; 2) conservation and rehabilitation; 3) cultural, social and livelihood; and 4) local hospitality. These themes were then matched with the duration of the visit, visitors' segmentation and their knowledge background.

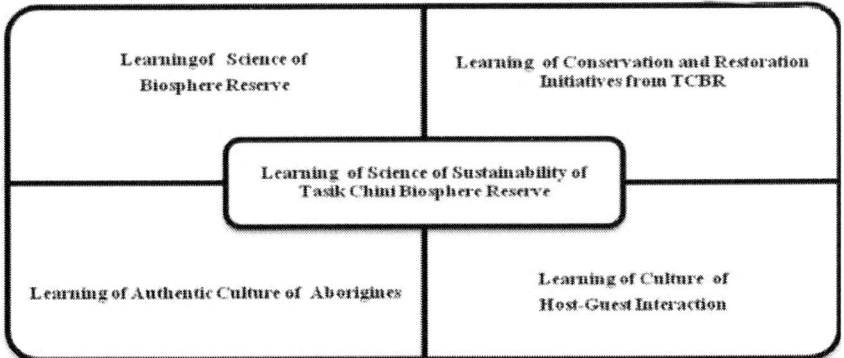

Figure 2. The Learning of Science of Sustainability of Tasik Chini Biosphere Reserve.

Table 2. Events Hosted by Tasik Chini, 2004-2013.

Year of event		Name of event	Medium of interaction	Location	Level of event	Mode of the Learning Destination
Pre-dossier preparation	2004	Tasik Chini Expedition 2004	On site experience	At the destination	State	Scientific expedition and Real experience
Pre-establishment of Tasik Chini as BR	2008	Public consultation events and World café Booth	Booth and event that stages the Tasik Chini Biosphere reserve	At the destination	National	Science of Biosphere reserve, conservation, development and education
	2009	Adopted Village of UKM	Booth and play	Johor, outside destination	State	Informed knowledge
	2010	Celebrating UKM's 40 Years	Booth and hands on activities	Bangi, Universiti Kebangsaan Malaysia	University	Disseminating knowledge to public and educational institution audience
	2010	Mobility program	On site experience, lake, forest and community living	At the destination	International	Science and play to international students
2) Post establishment and Current events	2011	UNESCO DAY	Booth and event	Kuala Lumpur, capital city	National	Disseminating knowledge to public and international audience
	2011	Mobility program	On site experience, lake, forest and community living	At the destination	International	Science and play to international students
	2011	Knowledge ecotourism event	Booth and talk	Bangi, UKM	National	Disseminating knowledge to public, tourism providers and ministry
	2012	Mobility program	On site experience, lake, forest and community living	At the destination	International	Science and play to international students
	2012	Scientific visit	On site experience, lake, forest	At the destination	Regional	Science based activities
	2012	Educational workshop	On site and living lab	At the destination	State	Science based activities
	2013	Composting awareness and knowledge transfer	On site and living lab	At the destination	Local	University-Community Engagement

4. Results and Analysis

The learning of culture of sustainability introduces the ecosystem of the Biosphere Reserve. This is because through the understanding of the ecosystem, the students would be capable of appreciating the livelihood in the area. This is particularly important as the learning provides the characteristics of the physical, biological and cultural setting. Secondly, as the site offers a different entity of an ecosystem, learning about the culture will showcase how each site approach issues of sustainability, including conservation and restoration. Thirdly, learning about the culture means a closer context with the livelihood of the community. At this stage, real examples are demonstrated to capture students' understanding, acceptance and critical mind, and most importantly, enthuseasm in the proposed programmes on conservation and sustainable Biosphere Reserve.

The final perspective deals with the hospitality of the host as well as the respect for the host-guest relationship.

This setting is aimed at fostering the learning culture of global students, comprising of the interaction between scholars-students and between students; internationals and the locals.

The students, being active participants, can experience student-centered learning and deep learning compared to the teacher-centered learning and surface learning. Between students, this environment has provided not only generic skills, but life-long learning, specifically on interactions between the locals and international societies. It is anticipated that learning about sustainability of Tropical Heritage could be achieved with the four dimensions mentioned above, especially in learning specific-site knowledge of the Biosphere Reserve (BR). Table 3 shows the details of the activities.

4.2. Learning about the Ecosystem Health of Tasik Chini

Tasik Chini's learning community comprised of various segments of visitors, ranging from young school children of the locality and from the neighbouring areas to a more educated youth segment of urban areas.

Taking into consideration that the young locals and the neighbouring school children are the "guardian of the BR", a simpler way of understanding this ecosystem was initiated. As such, activities were organised in such a way that learning about the ecosystem's health is simpler even though it was aimed at explaining the complexity of the food chain, especially on the flora and fauna found in the area. Students were engaged in conservation through activities such as planting a tree a day and art therapy. For this programme, the Kolb cycle of learning and five sensory medium of learning were utilized into hands-on activities Apparently, collages presented in scrapbooks prepared by the children, showed the high concern for TCBR's future among the young generation.

On the other hand, the scientific learning and ecotourism experience of the youth visitors were of a moderate level of complexity. Taking into consideration that they have been taught subjects such as biology, chemistry and physics at the matriculation level, programmes offered were more specialised. The programmes comprised both theoretical and practical modules, especially on the ecosystem of the lake, land ecology and diversity of local species. Practices of lake water sampling techniques, small mammal trapping and flora mapping were aligned with their matriculation learning and teaching needs.

4.3. Learning about the Culture of Tasik Chini

Envisioned as an ecotourism as well as a learning destination, Tasik Chini will definitely develop its own authentic learning experiences. In this vein, the learning processes will be iterative, tailor-made and involve the community of the BR. In the TCBR, the practice of learning with the locals varies according to the audience or visitors' needs or requests. Figure 3 shows some scenes from the learning programmes on the local culture of Tasik Chini.

4. Results and Analysis

Table 3. Activities during The Mobility Programme and The Orientation.

Activities of TCBR	Science	Local culture
Activity 1: Fishnets set-up	Cultural anthropology and fish	Gazing at the morning activities of the lake community.
Activity 2: Visit to hatchery site and acclimatization cage	Aquatic and Rehabilitation	Fish feedings and naming of fishes.
Activity 3: Water quality and hydrology	Hydrology and Geography	Boating and lake excursion
Activity 4: Macrophytes sightseeing and fish harvesting	Biology and Aquatic ecosystem	Walking under the canopy and Tropical Trails
Activity 5: Traditional Kayaking	Hydrology, Ecology	Kayaking as traditional skill, and group spirit during competition.
Activity 6: Socio-cultural activities and Kelundang dance	Cultural tourism and host's hospitality	Enjoying the local culture and interaction between participants.
Activity 7: Lotus planting and reforestation	Conservation, lotus and ecosystem	Lotus as the cultural heritage of the locals.
Activity 8: Blowpipe	Anthropological approach of aborigines	Blowpipe contest and traditional skills.

4.3.1. The Kelundang Dance

The Kelundang Dance is one of the cultural traditions inherited from the ancestors. Often staged and performed during special ceremonies among the Jakun tribe, the cultural dance fits well with members of the community. Over the years, however, this tradition has been diminishing from the community's cultural landscape and livelihood.

Their performances only seem to be staged when requested by the tourism providers. Realising that this tradition needs to be conserved and inherited, the TCRC took several initiatives in ensuring that the Kelundang Dance becomes the native tradition, including 1) stating it as one of their tradition in the dossier of Tasik Chini; in which many of their events seem to showcase this dance, and 2) forming a cultural group of young school children to ensure sustainability, with special songs/lyric to be enjoyed by the guests.

4.3.2 Weaving the Jari Lipan

The Jakun, living closely with nature, possesses untapped skills in handicrafts. Even though handicraft making is considered domestic work among females, it is through the public consultation and awareness initiated by the PPTC's events that they were informed of opportunities to produce and commercialize local crafts. Besides making them involved directly in the exhibition, the women were consulted on creative production. These opportunities ensure sustainability of small businesses within the limitation of their capabilities and know how. Feedbacks collected during thematic events held, showed that both locals and international visitors enjoyed learning the Jari Lipan. Even though they were taught by the locals who lack communication skills, especially in English, learning opportunities were noticeably appreciated.

4.3.3. The Traditional Water Kayaking

Being in the vicinity of the lake, the community is highly dependent on the lake and its ecosystem. The lake and rivers provide the cheapest mode of transportation and roaming areas, giving them the opportunity to develop their "sampan" or "kayaking" skills. These skills however, seem to have faded from the lake-scape of Tasik Chini. Taking into account the importance of conserving the local culture, this experience was re-created in the learning programmes.

4.3.4. Challenge for Survival in the Forest Trail

As the Jakun tribe usually roams the neighbouring forests for food supply, exploring what these forest trails have on offer is therefore an exciting experience. According to the village folks, the forest trails have existed for more than three decades, mostly for the use of backpackers who stayed overnight or longer. During these trails, the tourists were taught the survival skills of the natives in the forest, including cooking with bamboos, making utensils from leaves and getting water naturally from stems and roots. Of recent development, two scientific knowledge trails have been developed, namely the Tempenis and the Kempas trails. With on-going work on the flora and fauna research in Tasik Chini, these forest trails lead to a conservation of rare, threatened and endemic species.

4.4. Responses from the Learning Community

As revealed in the literature, a learning site has its own learning community; therefore, it is vital to establish who the learning communities are. To put it simply, segmentation of existing and potential learning community needs an in-depth study of their characteristics and behaviour. Table 4 shows the major segments of the learning community.

Besides recognising the learning community's characteristics, questionnaires were distributed to the participants to provide feedbacks and programme assessment. The results show that appreciation of deep learning in conservation initiatives and eco-hydrological experiments as well as cultural experience of the aboriginals achieved high mean values. Table 5 shows satisfaction of the content, staff and site scored high mean values; on the other hand, accommodation, food and cleanliness only achieved satisfactory mean values.

The results of the international students' survey indicate contradictory perspectives between the local guests - the primary school surrounding and within the BR and the matriculation students and participants of the international mobility programmes. Their comments clearly showed that the real community of the BR has yet to be established. Their expectations on various facilities demonstrated that their assessment were made in comparison to the "resort" facilities, while at this juncture, these have yet to be provided. However, their assessment on the community who hosted the cultural experiences demonstrated that they preferably favour the learning process at the Orang Asli settlement as it offered authentic and real life experience of the Orang Asli's culture. The children on the other hand, were energized with the learning of science in the area of their vicinity and had a better understanding of the Orang Asli livelihood, their closest neighbours within the BR locality. Below are some of the excerpts from their feedback:

Figure 3. The Cultural Learning at Tasik Chini.

4. Results and Analysis

Table 4. Segments of the Learning Community.

Potential Market Identified	Market Characteristics
Students, undergraduates and graduates	• Involve both local and international students who seek adventure and knowledge-added value in tour. • Work as part-time professional tour guides once completing the tour guide training module. • Apply knowledge gained in the classroom into real-life activity. • Share knowledge with tourists and local community.
Mobility programme, inbound international students	• Involve both local and international groups. • International groups act as the ambassadors of their home country by sharing their knowledge and culture with the local community. • Promote Malaysia as a tourism destination through their report writings. • Appreciate local culture and heritage through participation with the locals.
Researchers	• Local and international researchers. • Contribute to conservation through research, teaching and learning. • Assist development of the local economy through guidance, support and knowledge transfer.
Expatriates and high culture tourists	• Involve both local and international high-end k-eco-tourists. • Generally educated and demand learning process while vacationing. • Willing to spend extra money in exchange for knowledge and experience with the local heritage. • Aware of travelling impact on the environment and conscious of conservation efforts. • Choose professional tour guide services and legitimate tour agencies.

Table 5. Hospitality Rating of Thematic Programmes of TCBR.

Please rate for Hospitality of Thematic Program	Mobility program year 2010 N 37	Mobility program year 2011 N - 18	Mobility program year 2012 N 16	Kembara Pelajar Tasik Chini N 76		Matrikulasi students N-54
Year	2010	2011	2012	2010		2012
			Mean			
Services	4.03	3.61	3.75	4.60		4.46
Accommodation	3.59	2.99	2.13	Not rated, day trip only		4.57
Transportation	4.32	3.61	3.56	4.50		4.58
Food and refreshments	4.03	3.72	3.38	4.56		4.31
Activities	3.92	4.28	3.50	4.67		4.35
Social interaction	3.95	3.89	3.94	4.51		4.51
Facilitators and staffs experts	4.32	4.22	3.94	4.60		4.31
Cleanliness and safety	3.41	3.17	2.81	4.51		4.31
Information, content and approach of the programme	4.16	3.67	3.51	4.67		4.80
Site	4.00	4.00	3.40	4.58		4.07

4. Results and Analysis

Sustainable tourism-friendliness and indigenous interact with people in the city. The Fish Farm-preserves biodiversity, use of algae of research & monitoring forest.

(N 3 Hong Kong Mobility Student, 2012).

I had an excellent experience of visiting the indigenous people. I learn about their lifestyle and compare to it with Hong Kong lifestyle. What I realised - it seems that we are pursuing too much, much more than we need compared with what the aboriginals had

(N14 Hong Kong Mobility student, 2012).

I learn to be friendlier to nature and environment. Indeed I acquire new knowledge in my own area. Again, I wish to get more exposure on Biosphere reserve

(Children of Kembara Tasik Chini, 2010).

I love learning food web, culture and opportunities in this program make me understand more about Orang Asli - they are my closest neighbour indeed

(Children of Kembara Tasik Chini 2010).

Based on the above mentioned approaches, this study reveals that setting of a learning destination requires a holistic idea of the BR's key roles. Even though Tasik Chini's achievement is still at its infancy stage, several challenges need to be minimized and resolved. These include: 1) The holistic understanding of the sciences of the Biosphere Reserve as it combines multi-disciplinary approach and orientation. Pedagogical strategies in learning and teaching Biosphere reserve should be strengthened to ensure variation of knowledge and interests of learners and visitors; 2) The authenticity and creativity of the local players are required to fulfil the knowledge needs of the eco-tourists and visitors. Differences in tourists' origin may increase cultural barriers and practices; 3) While the TCBR is a rehabilitation programme, lack of facilities may cause misunderstanding and unsatisfactory experiences; 4) Collaboration between the locals and scientific community remains minimal due to limited funding. While the thrust of the locals towards collaboration is assured, ensuring participations of the locals requires a win-win negotiation and sharing of benefits in the value chain of economy and livelihood. Furthermore, as the locals possess low educational attainment, language barriers and ecotourism ethics are some of the constraints in providing quality services, highlighting the urgency of capacity building and empowerment of human capitals.

5. CONCLUSIONS

This study provides an understanding of how a Biosphere Reserve functions as a learning destination with three major principles. First, the creation and staging of experience and activities are based on site-specific knowledge of TCBR that are not only concerned with conservation and restoration practices, but in helping to build and inculcate awareness of sustainable development of the lake and wetland ecosystem. The second is the integration of players from the local communities that is equally significant in providing real life experience, tradition and involvement of the community. Third, the existence of the learning community must be ensured as this drives the sustainability of the inbound market of LTD.

The site specific knowledge, however, should not only cover scientific orientation as most of the eco-tourists or Biosphere learners also seek opportunities to experience the cultural flavour of the locals. Staging of the local cultural tradition in the villages is more appreciated as their surroundings are real and authentic. Additionally, taking ecotourism into conservation and rehabilitation programme is also appreciated but should not be overemphasized because of the short duration of visits and leisure time to freely experience the natural surroundings. Longer duration will be recommended only when these learners need to engage in scientific experiments.

Overall, even though this initiative is still at its very infancy, undertakings the knowledge eco-tourists experiences could be the platform to segment the learners to ensure better function of the Learning Destination. It is recommended that the principles of the scientific and local knowledge become the thrust of LTD of Tasik Chini.

6. ACKNOWLEDGEMENTS

The authors are grateful to all stakeholders in Tasik Chini who kindly devoted their precious time for the interview process and data collection. The work was funded by the Top Down-FRGS Grant, entitled Lake Ecosystem Assessment of Tasik Chini, and Long Term Research Grant: Managing the Science of Biosphere Reserve, from the Ministry of Education, Malaysia as well as the Faculty of Social Sciences and Humanities Research Development Grant DPP-2013-169, 2013.

REFERENCES

1. UNESCO, "The Man and the Biosphere (MAB) Programme," 2012. http://www.unesco.org/new/en/natural-sciences/ environment/ ecologicalsciences/man-andbiosphere-programme/

2. UNESCO, "UNESCO Biosphere Reserves: Learning Laboratories for Sustainable Development," 2007. http://unesdoc. unesco.org/ images/ 0015/001516/151607e.pdf

3. N. Ishwaran, "Science in Intergovernmental Environmental Relations: 40 Years of UNESCO's Man and the Biosphere (MAB) Programme and Its Future," Environmental Development, Vol. 1, No. 1, 2012, pp. 91-101.

4. N. Ishwaran, A. Persic and N. H. Tri, "Concept and Practice: The Case of UNESCO Biosphere Reserves," International Journal of Environment and Sustainable Development, Vol. 7, No. 2, 2008, pp. 118-131.

5. D. Kušová, J. Těšitel, K. Matějka and M. Bartoš, "Biosphere Reserves—An Attempt to Form Sustainable Landscapes: A Case Study of Three

Biosphere Reserves in the Czech Republic," Landscape and Urban Planning, Vol. 84, No 1, 2008, pp. 38-51.

6. J. Kriesel, "Education for Sustainable Development in the Biosphere Reserves Schaalsee, Germany, and Kristianstads Vattenrike, Sweden," Diploma Thesis, University of Greifswald, 2011. http://www.mnf.unigreifswald.de/fileadmin/Geowissenschaften/geographie/angew_geo/Diplomarbeiten/Janin_Kriesel _Diplomar beit_ BNE.pdf

7. Y. Luo and J. Deng, "The New Environmental Paradigm and Nature-Based Tourism Motivation," Journal of Travel Research, Vol. 46, No. 4, 2008, pp. 392-402.

8. M. Batisse, "Biosphere Reserves, a Challenge for Biodiversity Conservation and Regional Development," Environment, Vol. 39, No. 5, 1997, pp. 7-33.

9. J. J. Kay, H. Regier, M. Boyle and G. R. Francis, "An Ecosystem Approach for Sustainability: Addressing the Challenge of Complexity," Futures, Vol. 31, No. 7, 1999, pp. 721-742.

10. C. Canning, "Conservation and Local Communities: Exploring the Upper Bay of Fundy Biosphere Reserve Initiative in Nova Scotia," 2005. http://www.bofep.org/PDFfiles/Caroline_Canning_Final_ 20Thesis.pdf

11. L. Kruse-Graumann, "Education for Sustainable Development in German Biosphere Reserves," 2007. http://www.unesco. de/fileadmin/medien/Dokumente/unesco-heute/uh2-07-p22-26.pdf

12. J. N. Pretty, I. Guijt, J. Thompson and I. Scoones, "Participatory Learning and Action: A Trainer's Guide," IIED, London, 1995.

13. M. F. Price, "The Periodic Review of Biosphere Reserves: A Mechanism to Foster Sites of Excellence for Conservation and Sustainable Development," Environmental Science & Policy, Vol. 5, No. 1, 2002, pp. 13-18.

14. D. McCarthy, G. Whitelaw, P. Jongerden and B. Craig, "Sustainability, Social Learning and the Long Point World Biosphere Reserve," Environments Journal, Vol. 34, No. 2, 2006, pp. 1-15.

15. J. Tippett, B. Searle, C. Pahl-Wostl and Y. Rees, "Social Learning in Public Participation in River Basin Management-Early Findings from Harmony COP European Case Studies," Environmental Science and Policy, Vol. 8, No. 3, 2005, pp. 287-299.

16. T. Van Mai and O. J. H. Bosch, "Systems Thinking Approach as a Unique Tool for Sustainable Tourism Development: A Case Study in the Cat Ba Biosphere Reserve of Vietnam," 2010. www.systemdynamics.org/conferences/2010/proceed/.../P1312.pdf

17. N. C. Nguyen, O. J. H Bosch and K. E. Maani, "The Importance of Systems Thinking and Practice for Creating Biosphere Reserves as Learning Laboratories for Sustainable Development," 2009. http://journals.isss.org/index.php/proceedings53rd/article/view/1161/398

18. K. Schianetz, L. Kavanagh and D. Lockington, "The Learning Tourism Destination: The Potential of a Learning Organisation Approach for Improving the Sustainability of Tourism Destinations," Tourism Management, Vol. 28, No. 6, 2007, pp. 1485-1496.

19. K Schianetz, J. Tod, L. Kavanagh, P. A. Walker, D. Lockington and D. Wood, "The Practicalities of a Learning Tourism Destination: A Case Study of the Ningaloo Coast," International Journal of Tourism Research, Vol. 11, No. 6, 2009, pp. 567-581.

20. H. Gibson, "The Educational Tourist," Journal of Physical Education, Recreation and Dance, Vol. 69, No. 4, 1998, pp. 32-34.

21. A. Holdnak and S. Holland, "Edutourism: Vacationing to Learn," Parks and Recreation, Vol. 31, No. 9, 1996, pp. 72-75.

22. Ministry of British Colombia, "The Environmental Learning and Experience. Curriculum Map Environment and Sustainability across

Bc's K-12 Curric," 2009. http://www. bced.gov. bc.ca/ environment_ed/ele_maps.pdf

23. S. De la Barre, "Learning Travel Product Development Workbook: A Step-By-Step Guide for Yukon and Northern Entrepreneurs, North to Knowledge (N2K), Whitehorse, Yukon," 2005. http://www.tc.gov.yk.ca/pdf/LearningTravelProductDevelopmentWorkbook.pdf

24. M. Flitner, U. Matthes, G. Oesten and A. Roeder, "The Ecosystem Approach in Forest Biosphere Reserves: Results from Three Case Studies," Albert-Ludwigs-Universität Freiburg, Freiburg, 2006. http://www.bfn.de/fileadmin/MDB/documents/skript168.pdf

25. H. L. Ballard, M. E. Fernandez-Gimenez and V. E. Sturtevant, "Integration of Local Ecological Knowledge and Conventional Science: A Study of Seven CommunityBased Forestry Organizations in The USA," Ecology and Society, Vol. 13, No. 2, 2008, p. 37. http://www.ecologyandsociety.org/vol13/iss2/art37/

26. G. E. Yates, T. V. Stein and M. S. Wyman, "Factors for Collaboration in Florida's Tourism Resources: Shifting Gears from Participatory Planning to Community-Based Management," Landscape and Urban Planning, Vol. 97, 2010, pp. 213-220.

27. C. Lashley and P. Barron, "The Learning Style Preferences of Hospitality and Tourism Students: Observations from an International and Cross-Cultural Study," International Journal of Hospitality Management, Vol. 25, No. 4, 2006, pp. 552-569.

28. J. Purkayastha, S. C. Nath and M. Islam, "Ethnobotany of Medicinal Plants from Dibru-Saikhowa Biosphere Reserve of Northeast India," Fitoterapia, Vol. 76, No. 1, 2005, pp. 121-127.

29. K. S. Rao, R. K. Maikhuri, S. Nautiyal and K. G. Saxena, "Crop Damage and Livestock Depredation by Wildlife: A Case Study from Nanda Devi Biosphere Reserve, India," Journal of Environmental Management, Vol. 66, No. 1, 2002, pp. 317-327.

REFERENCES

30. S. K., Singh, J. P. N. Rai and A. Singh, "Influence of Prevailing Disturbances on Soil Biology And Biochemistry of Montane Habitats at Nanda Devi Biosphere Reserve (NDBR), India During Wet and Dry Seasons," Geoderma, Vol. 162, No. 3-4, 2011, pp. 296-302.

31. D. Manuel-Navarrete, S. Slocombe and B. Mitchell, "Science for Place-Based Socioecological Management: Lessons from the Maya Forest (Chiapas and Petén)," Ecology and Society, Vol. 11, No. 1, 2006, p. 8. http://www.ecologyandsociety.org/ vol11/ iss1/ art8/

32. G. Francis and G. Whitelaw, "Biosphere Reserves in Canada: Exploring Ideals and Experience," Environment, Vol. 32, No. 3, 2004, pp. 61-78.

33. V. Christidou, "Interest, Attitudes and Images Related to Science: Combining Students' Voices with the Voices of School Science, Teachers, and Popular Science," International Journal of Environmental & Science Education, Vol. 6, No. 2, 2011, pp. 141-159.

34. A. Watson, L. Alessa and B. Glaspell, "The Relationship between Traditional Ecological Knowledge, Evolving Cultures, and Wilderness Protection in the Circumpolar North," Conservation Ecology, Vol. 8, No. 1, 2003, p. 2. http://www. consecol.org/ vol8/iss1/ art2

35. G. Boucher, C. Conway and E. V. Der Meer, "Tiers of Engagement by Universities in their Region's Development," Regional Studies, Vol. 37, No. 9, 2003, pp. 887- 889.

36. D. Buss, "Secret Destinations. Creativity or Conformity? Building Cultures of Creativity in Higher Education," A Conference Organised by the University of Wales Institute, Cardiff in Collaboration with the Higher Education Academy, Cardiff, 8-10 January 2007. http:www.creativityconference07.org/presented_papers/Buss_Secret.doc

37. O. Zbyranyk, "Collaboration between Researchers and Biosphere Reserve Practitioners: A Case Study of Redberry Lake Biosphere Reserve, Canada," Thesis, University of Saskatchewan Saskatoon, 2012. http://ecommons.usask.ca/bitstream/handle/10388/ETD-2012-09-654/ZBYRANYK-THESIS.pdf?sequence=4

38. L. Schultz and C. Lundholm, "Learning for Resilience? Exploring Learning Opportunities in Biosphere Reserves," Environmental Education Research, Vol. 16, No. 5, 2010, pp. 645-663.

39. P. K. Ankomah and R. T. Larson, "Education Tourism: A Strategy to Strategy to Sustainable Tourism Development in Sub-Saharan Africa," 2000.
http://www.unpan1.un.org/intradoc/groups/public/documents/.../UNPAN002585

40. J. Coria and E. Calfucura, "Ecotourism and the Development of Indigenous Communities: The Good, the Bad, and the Ugly," Ecological Economics, Vol. 73, 2012, pp. 47-55.

41. M. Galliford, "Touring 'Country', Sharing 'Home': Aboriginal Tourism, Australian Tourists and The Possibilities for Cultural Transversality," Tourist Studies, Vol. 10, No. 3, 2010, pp. 227-244.

42. A. Habibah, J. Hamzah, I. Mushrifah, A. Buang, M. E. Toriman and K. Jusoff, "The Success Factors of Public Consultation in the Establishment of a Biosphere Reserve -Evidence from Tasik Chini," World Applied Science Journal, Vol. 13, 2011, pp. 74-81.

43. A. Habibah, I. Mushrifah, J. Hamzah, M. E. Toriman, A. Buang, K. Jusoff, M. J. Mohd Fuad, A. C. Er and A. M. Azima, "Assessing Natural Capital for Sustainable Ecotourism in Tasik Chini Biosphere Reserve," Advances in Natural and Applied Sciences, Vol. 6, No. 1, 2012, pp. 1-9.

44. A. Habibah, R. Mohamed, I. Mushrifah, J. Hamzah, M. N. Aimi Syairah and A. Buang, "Positioning University as Knowledge Ecotourism Destination: Key Success Factors," International Business Management, Vol. 6, No. 1, 2012, pp. 32-40.

CHAPTER 10

Tourism Waste Management in the European Union: Lessons Learned from Four Popular EU Tourist Destinations

Chukwunonye Ezeah[1], Jak Fazakerley[2], Timothy Byrne[3]

[1]*Faculty of Sciences and Engineering, University of Wolverhampton, City Campus-South, Wolverhampton, UK*
[2]*Royal Haskoning DHV UK Ltd., Honey Comb, Liverpool, UK*
[3]*School of Sciences and Technology, The University of Northampton, Northampton, UK*

ABSTRACT

From a sustainability perspective, achieving greater efficiencies in environmental waste management is at the heart of current academic discussion on climate change science. Over the last few decades the tourism industry has developed exponentially and is now considered one of the most dynamic economic activities worldwide. Solid waste is a commonly identified and ever increasing aspect of tourism; the improper management of which can lead to substantial and irreversible direct and indirect environmental, economic and social impacts. However, the management of solid waste in tourism dominated island communities is particularly problematic due to climatic conditions, topography, financial restraints, planning issues, changing consumption patterns, transient population, and seasonal variations in solid waste quantity and composition. In addition, there is often a lack of momentum to implement new initiatives and programs as stakeholders involved in the design, construction and operation of tourist resorts have conflicts of interest. Using information gathered from key informant interviews, participation observations and literature reviews, this article appraises current waste management practices in four European tourist destinations, namely: Mallorca, Tenerife, Kefalonia and Rhodes. Findings indicate that, although there are signs of compliance with global best practice, a variety of locally-based measures need to be implemented to enhance sustainability.

KEYWORDS

Waste Management, Tourist Destination, Island Communities, European Union, Greenhouse Gas Emission

1. INTRODUCTION

From a sustainability perspective, the improvement of waste management practices of the European Union (EU) hospitality industry has become very necessary. This is partly because of reported empirical evidence linking improper waste management with greenhouse gas (GHG) emissions. In this regard, management of solid waste in tourism dominated small island communities presents particular challenges due to their peculiar climatic conditions, topography, transient population, and seasonal variations in the quantity and composition of waste materials. A critical ingredient towards achieving this objective is Directive 2008/98/EC of the European Parliament 2008, also known as the Waste Framework Directive (WFD). Based on the requirements of the WFD, all EU member states must have systems in place for sustainable management of municipal solid waste (MSW), which should incorporate the following 1) waste minimization 2) reuse 3) recycling 4) energy recovery 5) landfill deposal. It must be said however, that how much waste is actually generated varies significantly between countries within the

1. Introduction

EU; richer countries tend to produce more waste per person. For instance, municipal waste generated per person varies from 294 kg in the Czech Republic to 801 kg in Denmark [1]. Out of all the municipal waste generated in the EU, 42% is landfilled, 38% is recovered and 20% is incinerated. Poorer countries tend to rely mostly on landfills for waste disposal while richer countries are the biggest users of incineration. The highest amount of waste is landfilled in Bulgaria, Romania, Lithuania, Malta and Poland (90% or more); Germany, Belgium, the Netherlands and Austria recycle or compost the most (59% or more); while Denmark, Luxembourg, and Sweden incinerate the largest proportion (47% or more). From a pan European position, though landfills are generally regarded as the least desirable of options for waste disposal, closing landfills may not be feasible until all member states and regions have developed capacities for alternative disposal options. This is especially so for new member states and regions with challenging socio-economic and environmental conditions. Touristic Islands exemplify EU regions with characteristic limitations which often affect how waste is collected, transported, treated and disposed. Such locations have common waste management problems which may include:

- Reduced number of facilities for waste treatment or disposal.
- Significant variations in waste arising based on tourism season.
- High population density.
- Limited land mass to locate landfills and other waste treatment infrastructure.
- Difficulties in achieving economies of scale [2].

Over the last few decades EU tourism industry has developed considerably and is now considered one of the most dynamic globally. Despite the phenomenal growth recorded by the sector, research has lagged behind with respect to waste arising from its operations; this situation is compounded by the near absence of reliable data on waste management in the sector.

Notwithstanding, solid waste remains an important aspect of tourism both with the EU and globally. It has been estimated that the industry produces around 35 million tonnes of solid waste annually [3] -[5]. Mateu-Sbert et al. [6] estimated that on the average, 1.41 kg/day of solid waste was generated per resident in EU-27 countries during 1998-2010. Solid waste management (SWM) is therefore a critical, complex and multi-dimensional challenge facing the EU hospitality sector in particular and the world in general [7] [8] It is therefore important that the tourism industry continues to improve and adapt its operation towards waste minimization; following that, waste should be collected, transported and disposed of in an environmentally sound and cost-effective manner, so as to help achieve overall EU climate change objectives [6] . Improper management of waste can lead to substantial and irreversible environmental, economic and social impacts, such as increases in greenhouse gas emission, land degradation, resource deprivation, surface and groundwater water pollution, loss of biodiversity and the loss of aesthetic value of tourism locations [9] [10] .

The management of solid waste in tourism dominated island communities is particularly problematic due to climatic conditions, land mass and topography, financial restraints, poor planning, changing consumption patterns, transient population (i.e. tourism flows), and seasonal variations in solid waste quantity and composition [3] . The isolated geographies of most islands communities also mean that materials are most times imported with little or no thought as to how to manage waste that arise after those materials might have been used. This situation often exacerbates the pressure on carrying capacity of the islands waste management infrastructure and further challenges the inefficient and unsustainable management practices of waste collection and disposal [10] .

In addition, there is also sometimes a noticeable lack of momentum to implement new initiatives and programs designed to ameliorate the situation. Many stakeholders are involved in the design, construction and operation of tourist resorts, including municipal managers and planners, project developers and constructors, real estate brokers, hotel managers, tour operators and local residents [10] ; all of whom operate in a highly complex environment and have different priorities [11] . For example, the tourists themselves do not feel responsible for the waste problem as they stay for only short periods of time and do not develop social ties with places they visit [6] . Furthermore, as Von Bertrab et al. [11] argue, waste management is still

1. Introduction

largely perceived as an additional cost instead of as an income generating resource. Evidently, giving the above situation, finding ways to involve all stakeholders and promoting sustainable solid waste management (SSWM) in most tourism establishments has continued to remain a challenging and complex issue.

As a solution, it has been suggested that public-private partnerships (PPP) could play a key role in this direction. Using information gained from key informant interviews, participant observations and literature reviews, this article attempts an appraisal of waste management practices of four Mediterranean tourism destinations: Mallorca, Tenerife, Kefalonia and Rhodes, focusing particularly on collection, transportation, disposal as well as stakeholder collaboration. The paper concludes with a number of key recommendations that could be utilised to inform future decisions on sustainable waste management in the regions. Whilst island regions are not usually subject to the same level of legislative control as larger mainland states (for example, with regard to key EU Directives on waste management), it is usual for Island local authorities to seek to adopt strategies and to introduce facilities and technologies that at least go some way towards reflecting the high standards set out in European or other similar legislation. Within EU legislation, including that relating to waste based on the principles of the EU Waste Framework Directive and Landfill Directive. Having said this, the challenge of seeking to apply these strict standards to small islands must be recognized and in such cases exemptions or derogations could apply.

1.1. Engaging the Public-Private Partnership Model for Sustainable Waste Management in Small Islands

Public-private partnerships (PPP) have been identified as effective tools for improving service provision in the waste management sector [10] . This is more so because, collaborative approaches to governance and decision making help to address problems too complex to be effectively resolved by independent action [7] . Willmott and Graci [3] argue that traditional top-down, regulatory and end of pipe approaches to waste management often leads to a variety of challenges particularly in Island communities, including technical concerns, financial restraints, a lack of capacity, limited education and awareness, corruption, stakeholder influence and poor planning. However, an inclusive working relationship between, for example, government agencies, local authorities, technical consultants, hotels, waste companies, social enterprises, banks, residents and tourists can allow for access to resources, networks, financial support, training, technical assistance, knowledge exchange and experience [11] . These facets can then best be utilised to identify problems, set direction and implement initiatives to improve solid waste management under a common strategy [7] . McDevitt [12] believes that when a system is reflective of stakeholder concerns, individuals and organisations are more likely to provide support. Engaged and facilitated dialogue can therefore lead to a better regulatory and institutional framework, based upon complementarity, subsidiarity and neutrality, which can ultimately improve transparency and accountability within the waste sector [10] . However, for a high-quality waste and resource management service to develop and expand, PPPs need to remain dynamic; incorporating a number of environmental, economic, social and technical criteria which are underpinned by relevant legislation [6] .

In the case of Small island locations for instance, key stakeholders such as tour operators could play a vital role in facilitating Public-private partnerships engagement for efficient delivery of waste management services. It has been noted however that in some instances, external factors may affect the sustainability of services [13] . For example, a destination's wastewater treatment system, waste management scheme, policies on the protection of cultural and natural heritage or social conditions may all influence the sustainability of performance. Where such external factors limit improvements in sustainability performance, it may be necessary to work in partnership with local and national government authorities [14] . It is important to recognize that the circumstances and priorities of some destinations may be very different, so it is necessary to avoid approaches that would impose undue burdens on local authorities [15] .

1. Introduction

1.2. Models for Sustainable Waste Management in the Tourism Sector

1) **Sustainable waste management in the tourism sector of the Mexican Caribbean**

In October 2008, GIZ, the Swiss travel agent Kuoni Travel Holding, the non-governmental organisation Amigos de Sian Ka'an (with financing from Travel Foundation Netherlands) and the State of Quintana Roo's Ministries for the Environment and Tourism agreed on measures to promote integrated waste management in the region as part of a development partnership with the private sector [16].

Specifically, the project partners made the following contributions:

- GIZ: Advisory services on formulating pilot plans for integrated waste management in hotels, project management, workshop moderation, appointing consultants to provide a market analysis of local waste collection and the recycling industry;
- Kuoni Travel Holding: Financing workshops to raise awareness among the tourism industry;
- Amigos de Sian Ka'an (Travel Foundation Netherlands): Financing and conducting campaigns to raise awareness among hoteliers and tourists, coordinating activities on the ground; and
- Quintana Roo Ministry for the Environment and Quintana Roo Ministry for Tourism: Creating the legislative and regulatory conditions to introduce waste management plans in the tourism sector, taking care of experts on the ground.

The project created a legislative framework for standard waste management plans in hotels in the Mexican Caribbean, allowing the plans to be introduced across the board. The project also raised awareness among tourists, hoteliers and their employees with a focus on environmental issues on the ground and providing practical tips on preventing waste [16].

2) Bay of Bengal Green Model for Eco-Tourism

The Welcomegroup Bay Island (WGBI) hotel, owned and operated by ITC-Welcomegroup, is located in Port Blair, the capital city of the Andaman and Nicobar Islands, an archipelago in the Bay of Bengal. As a result of its geographic location, the islands rely mostly almost completely on the Indian mainland for all food supplies, medicines and items of daily use. Initially there was a lack of systems, procedures and policies surrounding waste management incentives and disincentives at the hotel. As such, the hotel created a green strategy, based on the Four-R principle, to produce an eco-tourism model for the island. The hotel addressed the issues by adopting both internal and external strategies [17].

Internal Strategy—Initially, tourists went for sightseeing trips to other islands in Port Blair with lunches packed in cardboard boxes. These boxes were invariably left behind on the islands thereby creating a waste problem. The hotel replaced cardboard boxes with steel lunch boxes. Users had to return the boxes or pay a fine. Through discussion and policy change, the system eliminated the recurring cost of buying cardboard boxes and reduced the garbage levels on the other islands.

External Strategy—The Andaman & Nicobar Tourism Guild (ANTG) convinced the local government to position the island for high-end tourism in order to seed the local economy and reduce the pressure on the already scarce resources by restricting the number of tourist through cost measures. The chain has since become the first eco-responsible chain in India. Initial costs of the hotel's activities were shared with its competitors, the local administration and hotel guests, leading to a cascading effect in terms of echoing eco-tourism practices. In recognition for their work they received The British Airways Tourism Award for Environment, subsequently leading to image and operational advantages [17].

2. METHODOLOGY

Figure 1 is an outline of the multi-stage methodological approach adopted for this investigation, covering desktop study, fieldwork and data

2. Methodology

analysis/prescription stages. The preliminary stage of the investigation involved the identification and review of relevant literature on tourism waste management in the European Union [18] [19] as well as a key informant interview with 8 stakeholders in waste management such as tourists, residents, hotel managers, local authority staff etc. Following this, four key EU tourist destinations, Mallorca, Tenerife, Kefalonia and Kalithea-Rhodes were selected as case studies using the purposive sampling technique. Apart from being geographically representative, the four Islands together account for a significant percentage of EU bound tourism and as such provide ideal case studies for this investigation. Adopting participant observation methodology, primary data on waste management were collected from the four case study locations between 2008 and 2012. A focus group discussion comprising 10 experts drawn mostly from the tourism and hospitality sector in Kalithea—Rhodes was also held to aggregate and validate some of the findings from the study. Data collected were mostly qualitative in nature; analysis of quantitative data was carried out using Microsoft Excel for Windows.

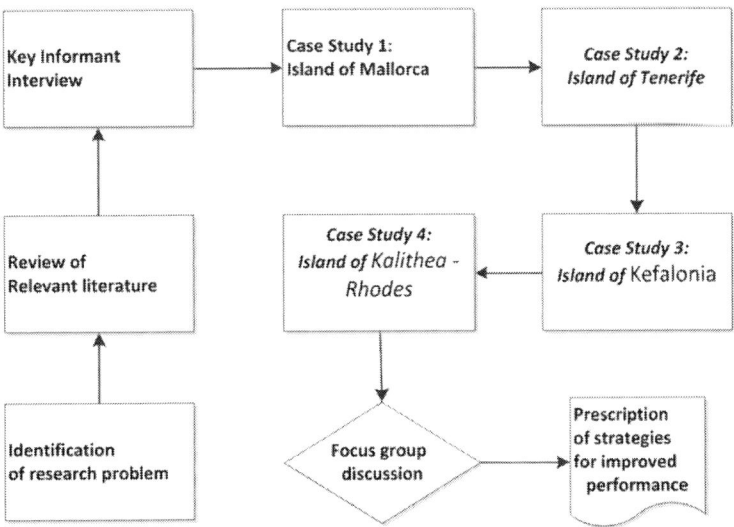

Figure 1. Outline of the research method.

2.1. Case Study 1: Tourism Waste Management on the Island of Mallorca

Mallorca is a Spanish island located in the Mediterranean Sea. It is the largest island in the Balearic Islands archipelago (Figure 2). It has a population of approximately 870,000 (2010 census) and covers a total land area of 3640 km^2. The climate is largely Mediterranean, with mild to cool winters and hot, bright summers [6] . This facet, along with picturesque beaches, rugged coastlines and high-quality amenities, makes the island a popular tourist destination—with over 6 million holiday makers visiting the island in 2010 [20] . It is estimated that as much as half of the population now work in the tourist sector, which accounts for approximately 80% of Majorca's GDP. In 2014, three additional beaches in Mallorca, Sant Elm, Camp de Mar and the Sailing Club in Port Andratx gained Blue Flag status, a recognition of the very high water quality, environmental management, safety and services [21].

The Island of Mallorca has a modern, integrated waste management system. In 1992, the Government of Mallorca signed a major contract for the operation and management of five waste transfer stations for the local municipalities. Municipal and commercial waste is delivered to the Son Reus energy from waste plant near Palma [21] . In 2010, the energy from waste plant was expanded with the commissioning of two grate-fired thermal treatment lines each having a throughput of 30 tonnes per hour. A new tipping hall was also constructed for the additional two lines and the plant expansion can now deal with municipal and commercial waste, biomass, sewage sludge and small quantities of waste tyres [22] . The overall throughput of the energy from waste facility is now 800,000 tonnes per annum and since the plants expansion, has ended the use of landfill for municipal and commercial waste disposal on the island of Mallorca [23] .

As with most areas in Spain, waste collection on the Island is at night with collection vehicles starting their route around 2300 hours. However, the energy from waste facility and transfer stations are operational on a 24 hour basis [22] . As part of a separate contract, the transportation of waste materials from the five transfer stations around Mallorca to the energy from

2. Methodology

waste facility in Son Reus was signed with ALCUDIA SA [23]. This company utilises a semi-trailer fitted with Marrel hook loader equipment which is then pulled by a tractor unit fitted with the necessary hydraulics to operate the hook loader system fitted to the semitrailer. It was observed that staffs who work at these transfer stations comply with all European health and safety guidelines [23].

However, Zero Waste Europe [24] believe that Mallorca can be seen as "the best and the worst" of waste management in southern Europe. On one hand some, zero waste municipalities have implemented ambitious door-to-door source separation schemes which would allow them to recycle more than 75% of the waste (i.e. Esporles and Puigpunyent). On the other hand, the island currently only produces around 540,000 tonnes of waste per annum meaning that vast quantities are imported from many km away to keep up with the operational capacity of the incinerator. This contradicts the proximity principle, the waste hierarchy and European Resource Efficiency Roadmap. As such, it is unlikely that the targets set by the EU will be accomplished unless the economic and legislative drivers are changed to prioritise recycling.

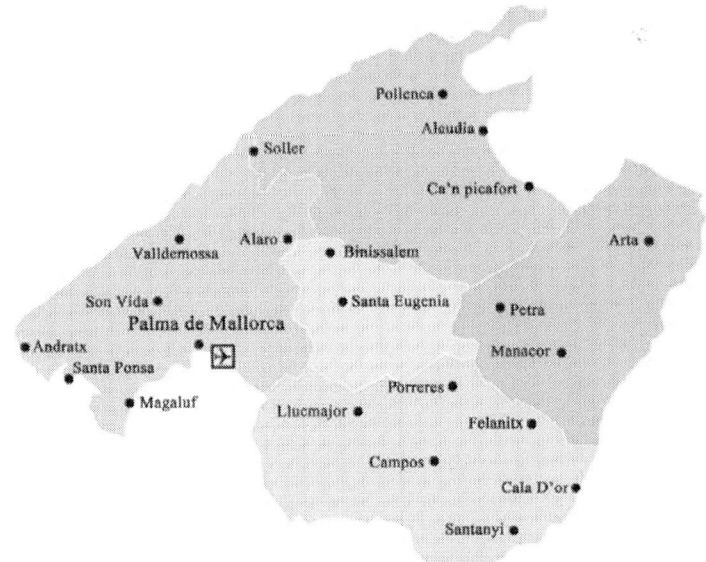

Figure 2. Map of the Island of Mallorca, Spain.

2.2. Case Study 2: Tourism Waste Management on the Island of Tenerife

Tenerife, the largest of the Canary Islands, is one of Europe's most popular tourist destinations. The island's warm climate attracts five million holidaymakers every year, so managing the collection and treatment of municipal solid waste is a very important task [25] . On the south west coast of the island is the district of Santiago Del Teide, located close to the island's active volcano Mount Teide, the third largest in the world. The district is made up of several towns, villages and coastal resorts, i.e. El Molledo, El Retamar, Tamaimo, Los Gigantes (which also has the highest volcanic cliffs in Europe), Puerto de Santiago, and Playa de la Arena (Figure 3) [26] .

1) Waste Collection

Waste is collected on a district basis, mostly by small family owned businesses [21] . Waste is collected in 660 litre containers, since there is limited space to place 1100 litre containers in the district (Figure 4). The containers are located in communal collection points along the street, where up to six containers could be placed. Although these containers are smaller than the 1100 litre option, the number of 660 litre containers stored in each communal point has adequate capacity for the waste produced [27] . Tourists, residents, retail outlets and bars place their residual waste into nearby communal containers. Usually, there is also an igloo in each compound for glass, bottles, paper and card recycling. The recyclables are collected by a third party on behalf of the Island's council. There is also an underground waste collection system in operation in the three coastal resorts. Waste is placed into loading apertures at street level, and then falls into a roll on/off container stored below street level. This helps to improve the area's appearance and keep odour and vermin to a minimum in the warm weather [26] .

In Santiago Del Teide waste is collected at night from 22.00 hours until 05.00 hours. The collection service is operated seven days a week, throughout the year [25] . Waste collection vehicles are manned in the spring season by a driver and one loader, the driver helping the loader place the containers at

2. Methodology

the rear of the refuse collection vehicle to be emptied. In the summer season, when tourist numbers increase, each vehicle is manned with a driver and two loaders, to handle the extra waste produced in the district.

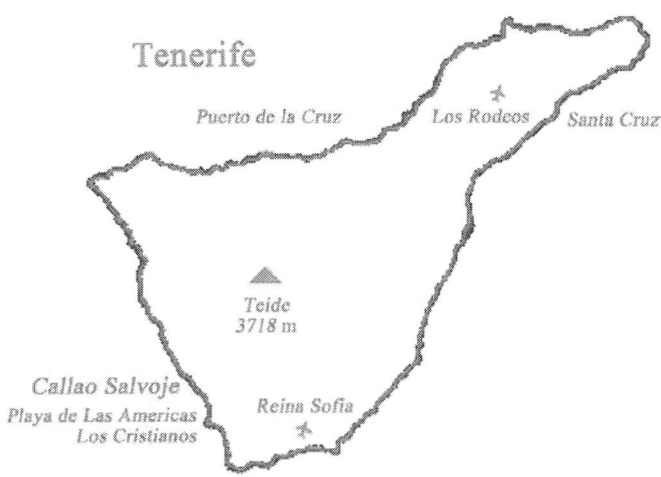

Figure 3. Map of Tenerife, the Canary Islands.

Figure 4. A private contractor emptying 660 litre waste containers in a hotel in playa de la arena— Tenerife (authors photograph).

There is also a roll on/off vehicle fitted with Multilift HL5 hook lift mounted on a two axle Iveco Eurocargo chassis which collects the full roll on/off containers from the underground waste collection systems. The vehicle travels to the various underground waste collection systems with an empty roll on/off container so reloading of fresh waste can continue throughout the night.

Each refuse collection vehicle collects one to two loads a night depending on the number of tourists in the resort and the amount of waste being generated [25] . Once all of the vehicles have finished their collection rounds they travel together to the waste transfer station.

2) Bin and street washing

The 660 litre containers are steam cleaned once they have been emptied. Vehicles carrying out this task have the capacity to transport as much as 7000 litres of water [27] . The water is heated so that when the containers are washed they are disinfected to eradicate any odours or bacteria left from the waste collection process. This service is carried out in all districts of Santiago Del Teide. The bin washer also has the ability to wash the streets after the waste collection service has taken place [21] [26] . Street washing is carried out through the two water jets fitted in the front bumper of the vehicle (Figure 5).

3) Waste Transfer Station Buzanada-El Malpaso-Arona

All waste produced from Santiago Del Teide, and from the neighbouring districts of Guia de Isora, Costa Adeje, Playa de Las Americas-ARONA-Los Cristianos, San Miguel de Abona and Granadilla de Abona is delivered to the waste transfer station at Buzanada-El Malpaso-Arona [27] . The facility is one of four operated on the island by UTE—a company formed by Vertresa and the Cabildo Insular de Tenerife. Vetresa is owned by Urbaser S.A., Spain's second largest waste management contractor [21] . The facilities are located at: Buzanada- El Malpaso-Arona, La Guancha, La Orotava-Puerto de la Cruz and El Rosario-Santa Cruz de Tenerife [25] . On arrival, all refuse collection vehicles are weighed on the computerised weighbridge [21] . Once weighed, the refuse collection vehicles proceed to one of two static compactor loading apertures [25]. The number of compacting apertures varies at each waste transfer station depending on the throughput of waste. The largest waste transfer station on the island is at El Rosario-Santa Cruz de Tenerife [27]. The collection vehicles reverse up to the unloading aperture and the driver discharges his load into the compactor. The waste is compacted into hermetically sealed roll on/off compaction containers. Once the containers are full they are collected from the compactor by a Marrel roll on/off vehicle. A fresh container is unloaded and fixed to the compactors' loading apertures so that the reloading of waste can continue [28].

4) Dirty MRF and Sanitary landfill site of Arico

The waste from all four waste transfer stations is delivered to the dirty MRF at the site of the sanitary landfill at Arico, located in the south east of the island [27]. The dry recyclables and food waste is extracted and segregated from the remaining municipal waste to help the island meet its EU Landfill Directive (1999/31/EC) targets as well as targets set by the Revised Waste Framework Directve (2008/98/EC). The dry recyclables are sorted through a series of trommels inside the MRF while the organic waste fraction is fermented in composting tunnels. The rejects are land filled under sanitary conditions [21]. The facility incorporates a methane extraction system, with the methane produced from the waste decomposition extracted from the landfill body and harnessed through a turbine [25]. The electricity produced from this process is exported to the national grid. Leachate is treated on site in the leachate lagoons at the desalination plant. Once it has been purified, the leachate is discharged into the ground water sewerage system. The facility also incorporates storage facilities for recyclable materials collected from across the island before they are transported for reprocessing.

2.3. Case Study 3: Tourism Waste Management on the Island of Kefalonia, Greece

Kefalonia is the largest of the Ionian Islands. Between the months of May to October, it is home to many thousands of tourists. Dealing with waste management on the island is a very important task because of the large volumes of waste produced by tourists combined with Mediterranean temperatures reaching 40 degrees in the months of July and August [29] [30] . All waste and recyclables are collected and managed by the island of Kefalonia council at a purpose built depot in Argostoli, the capital of the island (Figure 6) [21] . This houses eleven waste collection vehicles, eight of which are deployed on scheduled collection rounds while the other three are for additional rounds deployed in the summer months to manage the increase of waste generated by tourists. The depot has a fully equipped workshop and can carry out all minor and major repairs to the fleet. A washing bay has also been constructed outside so that all of the waste collection vehicles can be washed at the end of their days work to eradicate any germs or flies. The council also has two waste collection vehicles which work in the town and the outskirts of Lixouri. The smaller of the two collection vehicles also collects waste in Zola, Assos and the Port of Fiskardo.

Figure 5. Bin washing in the municipality of Santiago del Teide—Tenerife (authors photograph).

There are four principal waste systems for the island: plastic bin liner (city of Argostoli's narrow back streets), 240 and 360 litre containers for the collection of dry recyclables, 660 litre roll top steel and plastic flat top containers and 1100 litre roll top steel and plastic flat top containers. The 660 and 1100 litre containers are used for the collection of both recyclable

2. Methodology

and non-recyclable waste, distinguished in colour; blue represents the collection of dry recyclables while green is for the collection of non-recyclable waste. Containers are placed at communal collection points along the street where domestic dwellings, apartments, shops and taverns can place their waste and dry mixed recyclables for collection. Hotels are mostly provided with 1100 litre roll top containers but these are stored on their own grounds. The collection service begins at 3-5 am every day to avoid traffic and disrupting tourists. On a Sunday, the waste collection service covers the whole island in the summer months and in the winter months the service is only carried out in Argostoli. Each collection crew is manned by a driver and two personnel who position the containers to the rear of the collection vehicles for emptying by the vehicle's lifting equipment.

There are two waste collection rounds which collect waste from the City of Argostoli and the tourist resort of Lassi. One concentrates on collecting the waste in the long streets while the other vehicle concentrates on collecting waste in the areas of Argostoli and Lassi with restricted access. A small satellite 7.5 tonne waste collection vehicle also collects waste in the tight access areas and provides a second collection service for the larger inner city streets mid morning with the round finishing between 2-3 pm. There are also separate collection vehicles which collect waste from Sami, Antisamos and Agia Effimia, Poros, Skala and Katelios, Lourdas, Pesada and Gerasimos, Minia, Metaxata, Lakithra, Spartia and the International Airport.

From 1996, plan was put in place to update existing fleet of waste collection vehicles. The island purchased a new Mercedes SK 2024 18 tonne two axle chassis with Faun Variopress waste collection bodywork and trunnion lifting equipment mounted to it. This vehicle was assigned to collect waste in the town of Sami, Antisamos and Agia Effimia. The second purchase was in 1999, a Mercedes Atego 1523 two axle chassis of 15 tonnes with 12 cubic metres Kaoussis CRV 1600 rotating drum compaction equipment. This equipment featured a trunnion and bar lift for the emptying of containers of 80 - 1100 litre capacity DIN 30,700/DIN 30,740. This vehicle was purchased for accessing the narrows streets across the island, for example, in Lixouri,

Zola, Assos and the Port of Fiskardo as well as some small villages near to Argostoli.

Figure 6. Fleet of waste collection vehicles used to collect municipal and commercial waste at their purpose built depot in Argostoli (authors photograph).

In 2004 the island council also purchased a Mercedes Atego 8157.5 tonne two axle chassis with Mazzocchia satellite waste collection equipment mounted to it from Greek waste collection vehicle manufacturer Kaoussis. This collection vehicle also featured a bar lift which could empty containers from 80 -1100 litres capacity DIN 30,700 and DIN 30,740 types. This collection vehicle was purchased to collect waste in the narrow streets of Argostoli and once full, it would off load its waste into the Mercedes Actros 2640 three axle chassis with Kaoussis Norba RL300 22 cubic metre bodywork for onward transport to the sanitary landfill site of the island [29] [31] .

In 2008, four more larger waste collection vehicles were purchased; three Iveco Eurocargo 190EL28 two axle 18 tonne chassis with Kaoussis Norba RL300 16 cubic metre bodywork and an Iveco Trakker 450 three axle 26 tonne chassis with Kaoussis Norba RL300 22 cubic metre bodywork. The purpose for buying these four new collection vehicles was to replace the last of the remaining Mercedes SK Faun/Kuka Rotopress rotating drum waste collection vehicles initially bought from Germany second hand as well as also replacing one of the Mercedes Actros 1831 two axle Kaoussis Norba RL300 16 cubic metre waste collection vehicles purchased new in 2002.

2.4. Case Study 4: Tourism Waste Management in the Municipality of Kalithea-Rhodes

To the south of the city of Rhodes lies the Bay of Faliraki in the district of Kalithea. This is a busy tourist resort for holidaymakers worldwide (CIWM, 2011). Built along the bay is the busy resort of Faliraki, which generates fifty tonnes of waste a day, mostly produced from the hotels, apartments, bars and taverns (Figure 7). Typical composition of waste samples is: household waste (37%), Paper (10%), cardboard (5%), plastics (10%), glass (7%), metal (4%), textiles (2%) wood (0.5%) organic fraction (24.5%) (Table 1). The collection of this waste is carried out by the municipality of Kalithea. The municipality operate three waste collection vehicles, two of which service the hotels and the town of Faliraki, while the latter services the villages which lie inland.

Along the public highways are communal collection points with 120, 240, 660 and 1100 litre containers. Tourists from apartments, bars and tavernas deposit their waste into these containers to await collection. The hotels are supplied with 660 and 1100 litre containers into which they place their waste. Due to the warm temperatures, each hotel has a cold room where they can store several of their 660 litre containers. Food waste is deposited into these containers by employees at the hotels. The cold room is fitted with a sealed door and the waste is chilled by a thermostat to eradicate the possibility of the waste decaying in the warm temperatures while it awaits collection, preventing a health threat to tourists and hotel employees.

As a result of the recent economic recession, the municipality doubled shift two of its waste collection vehicles. The first shift starts at 5.00 am and finishes at 10.00 am, servicing central Faliraki and its villages, while the second shift starts at 11.00 am finishing at 4.00 pm. The second shift services all of the hotels located along Faliraki Bay. The waste collection crews consist of a driver and two loaders. The loaders place the containers to the rear of the refuse collection vehicle for emptying, and also load any excess waste into the refuse collection vehicle that has built up from tourism. The wearing of high visibility clothing by operatives is not enforced in Greece, so none of the operatives were wearing protective clothing, except gloves.

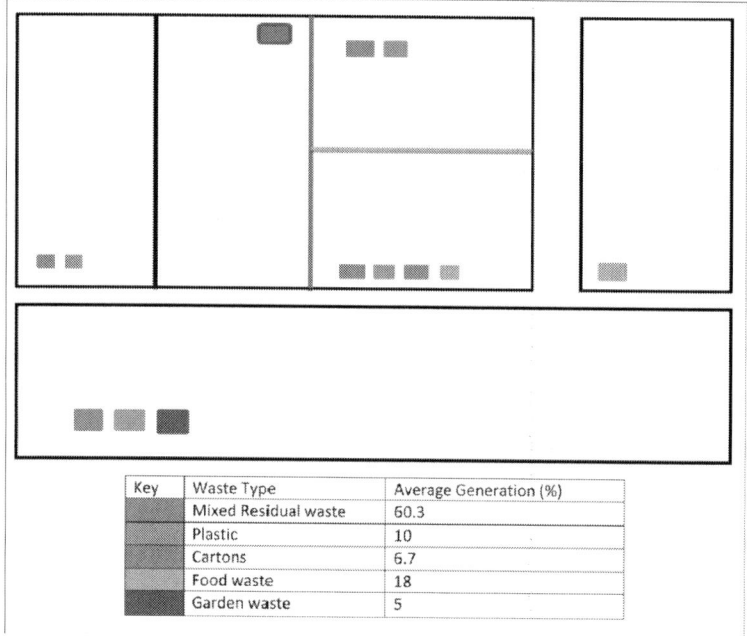

Figure 7. A waste map for a typical hotel in Kalithea-Rhodes based on Pirani, et al. [4].

Once the vehicle is full, the vehicle and crew travel to the sanitary landfill site in the centre of the Island. Upon arrival, the waste collection vehicles enter the weighbridge and the vehicles' gross weight and payload are accurately recorded. Once weighed, the refuse collection vehicle proceeds up the steep incline to the tip face of the landfill where the driver discharges its load. Plant machinery then covers the waste with soil, to prevent it blowing off site. The vehicle then returns to the weighbridge area and the driver collects his waste transfer note from the weighbridge operator. The landfill site has been operational since 2005, and is now nearing its end of life. At the

2. Methodology

present, neither a new landfill site has been provided nor has any alternative waste treatment infrastructure been built. The existing landfill is built to a sanitary specification and incorporates leachate treatment facilities, as well as methane being flared off site.

Table 1. Types of non-hazardous waste in the hotel industry (based on Zein et al. [32]).

Non-hazardous waste type	Components	Source
Household wastes	Food/kitchen waste, used or dirty paper and wrapping, plastic wrapping or bags, composite wrappers	Hotel's different departments
Cardboard	Packaging	Hotel's different departments
Paper	Printed documents, brochures, menus, maps, magazines, newspaper	Administration, reception, guest rooms, restaurants
Plastic	Bags, bottles (that did not contain hazardous material), household goods, individual portion wrappers for various products	Kitchen, restaurants, bars, guest rooms, administration
Metal	Tin cans, jar lids, soda cans, food containers, mayonnaise, mustard and tomato puree tubes, aluminum packaging	Kitchen, restaurants, bars, guest rooms
Glass	Bottles, jars, flasks	Kitchen, restaurants, bars, guest rooms
Cloth	Tablecloths, bed-linen, napkins, clothes, rags	Kitchen, restaurants, bars, bathrooms, guest rooms
Wood	Wooden packaging, pallets	Purchasing department
Organic waste	Fruit and vegetable peelings, flowers and plants, branches, leaves, grass	Kitchen, restaurants, bars, guest rooms, gardens

Near to the entrance to the site, and located near the weighbridge, is an area where waste electronic and electrical equipment is stored. This waste is collected by authorised transporters and delivered to authorised treatment facilities on the mainland, or shipped by authorised exporters for further processing. Wood waste is also stored here for reprocessing whilst wood waste which is contaminated is shredded on site and pre-treated before it is sent for disposal.

Recycling infrastructure is also increasing on the Island. Recent recycling initiatives on the Island have been driven mostly by the private sector. Perme [33], a mainland contractor which has been in existence since 1999, has rolled out a glass recycling and cardboard waste collection services to hotels, bars, tavernas and shops. This is to help Greece meet its statutory EU recycling and landfill diversion targets. Each hotel is supplied with a bottle bank which is emptied twice a week by one of Perme's collection vehicles. This vehicle consists of a tipper with crane to lift the bottle banks and discharge their contents into the collection vehicle. Once the collection vehicle has finished its round, it delivers its load to Perme's waste transfer station in Kalithea. The glass is collected by private hauliers and shipped to the mainland for reprocessing. A rear loading refuse collection vehicle also collects the paper and cardboard waste and delivers the collected material to the waste transfer station where it is baled, awaiting transport for reprocessing.

Following recent Greek economic recession, additional powers have devolved to Local Authorities. Although each municipality has its own waste collection vehicles, each municipality on the Island now has to seek guidance from Rhodes town before they can make any local decision. Rhodes town has decided that privatisation is the way forward for waste collection services. So far, there are two private waste collection contractors collecting municipal solid waste on behalf of municipalities—Helesi S.A. (Hellenic Environmental Systems Industry) in Ialysos and Afantou and Perme (Environmental Transport Ltd) in Archangelos and Lindos [28] . Helesi S.A. has a proven track record in collecting municipal waste on behalf of Greek municipalities and was involved in a joint venture in 2004 with Spanish waste collection contractor Urbaser S.A. in collecting all of the waste generated by the Olympic games in Athens [34] .

2.5. Strategies for Minimizing Waste in the Hospitality/Tourism Industry

Waste minimisation, through redesigning products, changing societal patterns (education and training) and purchasing with eco-intelligence ("green" purchasing), is widely considered the most environmentally resourceful, economically efficient, and cost effective long term approach to managing waste during the life cycle of a product [35] .

Very often, those involved with the tourism industry are faced with barriers to reducing waste due to a lack of information and guideline, time constraints, space and finance [36] . Furthermore, waste management practices are not uniform and so waste prevention can often require altering the way business is conducted. Nevertheless, there are a number of simple, yet effective, tools to preventing facility waste at source. These can be

2. Methodology

characterised into accommodation, food and beverage, open space, and administrative functions [37].

Accommodation

1) Introducing sorting and recycling plastic bottles, tins and glass from guest rooms;
2) Providing bulk dispensers and eliminate the use of individual bottles;
3) Returning laundered clothes to guests in reusable cloth bags/baskets thereby eliminating plastic bags;
4) Instructing housekeeping not to replace half-used rolls of toilet paper/tissue boxes and leave replacements for guests to use when required;
5) Using partially used items from guestrooms in employee restrooms or donating to charities; and
6) Extending the lifespan of equipment by having it serviced regularly.

Food and Beverage

1) Establishing purchasing guidelines to encourage the use of durable equipment and high-quality, reusable products such as linens and tableware;
2) Serving yoghurt, butter and jam in serving dishes rather than individual jars;
3) Using refillable containers for such items as sugar, salt, pepper, flour, soda, syrup and cream;
4) Replacing plastic/foam cups, utensils and plates with washable cups, dishes and utensils;
5) Delivering juices in bulk containers and made available from dispensers;
6) Using cloth rolls towels or hand dryers instead of paper products;
7) Donating unused food to local food banks or other charitable organizations; and
8) Collecting unusable food scraps and giving or selling them to local pig farmers for animal feed.

Open Spaces

1) Phasing out the use of hazardous materials where possible; and
2) Using organic gardening techniques and products.

Administrative Functions

1) Using bulletin boards for memos, pamphlets and brochures instead of circulating copies to all staff;
2) Using e-mail;
3) Where printing is required, print on both sides;
4) Reuse envelopes for internal staff documents; and
5) Purchasing refillable pens and toner cartridges.

3. CONCLUSIONS

It is evident from this investigation that the environmental impacts of a linear buy-use-dispose economy seriously undermine long-term sustainability and climate change objectives. A shift towards a much more integrated approach to address the interdependent social, economic and environmental systems is required so as to address this persistent and systemic challenge. This includes decoupling, in absolute terms, the lock-ins between resource efficiency and other environmental pressures associated with such destinations, including transport, agriculture and industry. However, neither environmental policies nor economic and technology-driven efficiency gains alone are likely to be sufficient to achieve the vision; a fundamental transition in the character of institutions, practices, technologies, policies, lifestyles and perceptions are needed, underpinned by well designed, long term physical and socio-economic planning.

Waste reduction provides a number of long-term benefits to tourism facilities and their host communities such as cost savings, greater operational efficiency, environmental protection, improved image and customer satisfaction. Based on the case studies that have been described, it is clear that waste management presents a great test for island tourist destinations.

3. Conclusions

Barriers constraining the efficient collection, transfer and disposal of waste include, climatic conditions, land mass and topography, financial restraints, planning issues, limited infrastructure, changing consumption patterns, dynamic population and seasonal variations in waste quantity and composition. To ensure sustainable waste management practices in these settings, dynamic, location based strategies must be adopted; no single system is ideal for all four regions studied. Planning for such systems should take a staged approach, incorporating a range of human, physical, technical, fiscal and legal instruments, with decision support tools to hand. That way, it will be possible to:

- Generate accurate waste management data;
- Model future scenarios;
- Undertake robust feasibility analysis to assess costs, optimize service delivery, capacities and efficiencies.

In the destinations examined, much of the waste produced is disposed of in communal open skips. Such mechanisms quickly become filled with recyclable material, compostable material and general waste in an uncontrolled way, thus being an inefficient economic and environmental disposal route. In areas where space is restricted, balers and compactors should be utilised to produce clean, segregated, recyclable materials which will attract a higher value, as opposed to low value unprocessed waste. In addition, it would also allow for a thorough understanding of the character and composition of waste produced from tourism related services as well as establishes the sector's overall GHG emission contribution.

Recommendations

- A detailed quantitative assessment of waste generation from each of the Islands is required to be able to more accurately predict quantities and composition of the waste streams. In this regard, adopting a time series analysis approach, rather than year-to-year, would allow for the implementation and verification of long term robust management strategies.

- A reliable and well-planned (local) monitoring programme concerning the number of tourists, their socio- economic situation and waste generation habits is required so as to be able to predict more accurately which part of the Island's waste arising is related to tourism and which part relates to other activities.
- Explore waste management strategies that go beyond conventional recycling, to include Extended Producer Responsibility (EPR), green procurement and eco-taxes, so as to challenge producers, re-conceptualise product designs and ultimately move towards a zero waste target.
- Encourage greater public education and community engagement initiatives, i.e. public meetings, waste specific training for businesses and island clean ups assisted by local school, shops, businesses, Island residents and tourists, so as to raise awareness of waste related issues, as well as overcome the divide that commonly exists between local residents and the tourism industry.
- Adopt a more proactive approach towards the development and application of the public-private partnerships (PPP) model for waste management on touristic Islands. In this regard, collaborative approaches to governance and decision making will help to address some of the ongoing issues faced in the case studies, i.e. financial constraints, a lack of capacity and limited education, thus allowing current initiatives to be assessed and future opportunities identified.

REFERENCES

1. Hall, D. (2010) Waste Management in Europe: Framework, Trends and Issues. Public Services International Research Unit (PSIRU). http://brecht.tttp.eu/IMG/pdf/it_10_2010-02_Waste_trends-3.pdf

2. Santamarta, J.C., Rodríguez-Martín, J., Paz Arraiza, M.P. and López, J.V. (2014) Waste Problem and Management in Insular and Isolated Systems. Case Study in the Canary Islands (Spain). IERI Procedia, 9, 162-167.

References

3. Willmott, L. and Graci, S.R. (2012) Solid Waste Management in Small Island Destinations: A Case Study of Gili Trawangan, Indonesia. TEOROS, Special Issue, 71-76.

4. Denafas, G., Ruzgas, T., Martuzevicus, D., Shmarin, S., Hoffmann, M., Mykhaylenko, V., Ogorodnik, S., Romanov, M., Neguliaeva, E., Chusov, A., Turkadze, T., Bochoidze, I. and Ludwig, C. (2014) Seasonal Variation of Municipal Solid Waste Generation and Composition in Four East European Cities. Resources, Conservation and Recycling, 89, 22-30.

5. Pirani, S.I. and Arafat, H.A. (2014) Solid Waste Management in the Hospitality Industry: A Review. Journal of Environmental Management, 146, 320-336.

6. Mateu-Sbert, J., Ricci-Cabello, I., Villalonga-Olives, E. and Cabeza-Irigoyen, E. (2013) The Impact of Tourism on Municipal Solid Waste Generation: The Case of Menorca Island (Spain). Waste Management, 33, 2589-2593.

7. Skordilis, A. (2004) Modelling of Integrated Solid Waste Management Systems in an Island. Resources, Conservation and Recycling, 41, 243-254.

8. Denafas, G., Ruzgas, T., Martuzevicus, D., Shmarin, S., Hoffmann, M., Mykhaylenko, V., Ogorodnik, S., Romanov, M., Neguliaeva, E., Chusov, A., Turkadze, T., Bochoidze, I. and Ludwig, C. (2014) Seasonal Variation of Municipal Solid Waste Generation and Composition in Four East European Cities. Resources, Conservation and Recycling, 89, 22-30.

9. Kariminia, S., Ahmad, S.S. and Hashim, R. (2012) Assessment of Antarctic Tourism Waste Disposal and Management Strategies towards a Sustainable Ecosystem. Procedia—Social and Behavioural Sciences, 68, 723-734.

10. .Shamshiry, E., Nadi, B., Bin Mokhtar, M., Komoo, I., Saadiah Hashim, H. and Yahaya, N. (2011) Integrated Models for Solid Waste Management in Tourism Regions: Langkawi Island, Malaysia. Journal of Environmental and Public Health, 2011, Article ID: 709549.

11. Von Bertrab, A., Hernandez, J.D., Macht, A. and Rodriguez, M. (2009) Public-Private Partnerships as a Means to Consolidate Integrated Solid Waste Management Initiatives in Tourism Destinations: The Case of the Mexican Caribbean. http://www.iswa.org/uploads/ tx_ iswaknowledgebase/3-340paper_long.pdf

12. McDevitt, C. (2008) Sustainable Waste Management within the British Virgin Islands. http://www.globalislands.net/userfiles/bvi_2.pdf

13. Xin, T.K. and Chan, J.K. (2014) Tour Operator Perspectives on Responsible Tourism Indicators of Kinabalu National Park, Sabah. Procedia—Social and Behavioural Sciences, 144, 25-34.

14. De Cantis, S., Parroco, A.M., Ferrante, M. and Vaccina, F. (2014) Unobserved Tourism. Annals of Tourism Research, 50, 1-18.

15. Chen, M.C., Ruijs, A. and Wesseler, J. (2005) Solid Waste Management on Small Islands: The Case of Green Island, Taiwan. Resources, Conservation and Recycling, 45, 31-47.

16. German Federal Ministry for Economic Cooperation and Development (BMZ) (2010) Sustainable Waste Management in the Tourism Sector of the Mexican Caribbean. http://www.giz.de/en/worldwide/24475.html

17. .World Business Council for Sustainable Development (2005) ITC-Welcomegroup Hotels—A Green Model for Eco-Tourism. http://www.wbcsd.org/web/publications/case/itc_gsc_full_case_final_web.pdf

18. Papachristou, E. and Chatziaggelou, H. (1991) Qualitative and Quantitative Analysis of Municipal Solid Waste of Rhodes, AUTH.

19. Gidarakos, E., Havas, G. and Ntzamilis, P. (2005) Municipal Solid Waste Composition Determination Supporting the Integrated Solid Waste Management System in the Island of Crete. Waste Management, 26, 668-679.

20. Hitachi Zosen Inova (2011) Mallorca/Spain: Energy from Waste Plant. http://www.hz-inova.com/cms/images/stories/pictures/download/hzi_ref_mallorca_en.pdf

References

21. Byrne, T. (Ed.) (2012) Waste Collection and Transfer in Touristic Locations from a Practical Perspective. D-Waste, Athens. http://www.d-waste.com/reports/waste-collection-and-transfer-in-touristic-locations-from-a-practical-perspective2012-09-05-13-36-45_-detail.html#.UyxGb4XPvbo

22. TIRME (2014) Overview of Processes at TIRME WTE Facility. http://www.yokogawa.com/success/newenergy/suc-TIRME.htm

23. ONWASTE (2013) Changing the Face of Waste on Mallorca. http://www.onwaste.com/images/stories/residuos/recogida_transporte/waste_mallorca.pdf

24. Zero Waste Europe (2012) Mallorca; Sun & Waste! http://www.zerowasteeurope.eu/2012/11/mallorca-sun-waste-the-sunny-and-shady-sides-of-zero-waste/

25. INTEREMPRESAS (2013) Marcos Gorrin Linares Manages the Collected of Waste in Tenerife. http://www.interempresas.net/Recycling/News/108120-Marcos-Gorrin-Linares-manages-collected-waste-in-Tenerife.html

26. ONWASTE (2013) Tenerife's Trash, Trucks and Transfer Stations. http://www.onwaste.com/images/stories/residuos/recogida_transporte/RWW_757_Tenerife.pdf

27. Mark Allen Group (2011) Tenerife's Trash, Trucks and Transfer Stations. Recycling and Waste World Magazine, June, 4.

28. ONWASTE (2013) Marcos Gorrin Linares Manages Waste Collection on the Island of Tenerife. http://www.onwaste.com/images/stories/residuos/recogida_transporte/marcos_gorrin_linares.pdf

29. Penwell (2013) Case Study—Waste Collection on the Island of Kefalonia. http://www.waste-management-world.com/articles/2013/12/case-study-waste-collection-on-the-island-of-kefalonia.html

30. Faversham House Group (2014) From One Island to Another. Local Authority Waste and Recycling Magazine, February, 12-13.

31. INTEREMPRESAS (2013) A Examen el Sistema de Recogida de Residuos en la isla de Cefalonia (Grecia). http://www.interempresas.net/Equipamiento_Municipal/Articulos/115468-A-examen-el-sistema-de-recogida-de-residuos-en-la-isla-de-Cefalonia-%28Grecia%29.html

32. Zein, K., Wazner, M.S. and Meylan, G. (2008) Best Environmental Practices for the Hotel Industry [WWW Document]. Sustainable Business Associates. http://www.sbaint.ch/spec/sba/ download/ BGH/ SBABGEHOTELLERIEENG2008.pdf

33. Perme (Environmental Transport Ltd). Accessed on 23 March 2014. http://www.perme.gr/index.php?cPath=45&sess=da784687d655dd065d0a573cd0041740

34. Hellenic Environmental Systems Industry (2007) Services. http://www.helesi.com/english/services_gb.html

35. Remolador, M.A. (2011) Guide to Greening Hotels through Waste Management & Green Purchasing. http://nerc.org/documents/green_hotels_guide.pdf

36. United Nations Environment Programme (2003) A Manual for Water and Waste Management: What the Tourism Industry Can Do to Improve Its Performance. http://www.unep.fr/shared/ publications/ pdf/WEBx0015xPA-WaterWaste.pdf

37. Radwan, H.R., Jones, E. and Minoli, D. (2010) Managing Solid Waste in Small Hotels. Journal of Sustainable Tourism, 18, 175-190.

INDEX

B

Bin and street washing, 376

Biosphere Reserve, 336

C

Catering Analysis, 310

climate change, 3, 148, 157, 209, 248, 261, 275, 276, 277, 278, 280, 281, 282, 283, 284, 285, 286, 287, 288, 289, 290, 291, 292, 293, 294, 295, 296, 297, 298, 363, 366, 386

D

Deep Learning, 336

destination appeal, 276

Destination management, 87, 88, 93, 96, 101, 105, 115, 118, 123, 124, 125, 126

Destination marketing, 88, 92, 122, 124, 256

DMS, 69, 70, 81, 82, 83, 319, 320, 321, 322, 323, 324, 325, 330, 332, 333

E

Economics, 66, 69, 70, 102, 113, 125, 155, 217, 219, 220, 223, 256, 319, 331, 362

Economy, 77, 157, 173, 224, 317, 320, 331, 332, 334

Ecotourism, 34, 234, 336, 362

entropy change, 1, 2, 12, 14, 15, 18, 21, 23, 24, 29

European Union, 363, 364, 371

Event tourism, 127, 132, 133, 137, 138, 148, 196, 210, 216, 227, 234

G

Greenhouse Gas Emission, 364

Guangzhou, 42, 299, 300, 304, 305, 306, 307, 308, 310, 311, 312, 313, 314, 315, 316, 317

I

Island Communities, 364

K

Knowledge creation, 195

L

leadership, 75, 81, 103, 276, 278, 288, 290, 291, 294, 322

Local government, 88, 95, 101, 108, 125

M

Madagascar, 69, 70, 71, 72, 73, 74, 76, 77, 78, 79, 80, 81, 82, 83, 84, 85, 319, 320, 321, 323, 324, 325, 326, 328, 329, 330, 331

Marketing, 45, 46, 48, 59, 60, 62, 63, 66, 67, 85, 86, 106, 107, 122, 123, 124, 125, 156, 215, 225, 226, 231, 232, 236, 241, 242, 243, 246, 251, 252, 256, 261, 263, 264, 268, 269, 270, 271, 272, 275, 276, 277, 279, 298, 317, 323, 331, 332, 333

O

Online Travels, 299, 304, 305

Ontology, 127

R

Regional tourism organizations, 88

S

sustainable tourism, 2, 3, 11, 32, 33, 34, 35, 36, 37, 38, 39, 43,

96, 120, 125, 219, 294, 296, 298, 330

SWOT Technique, 46

T

Tasik Chini Biosphere Reserve, 335, 337, 343, 345, 362

tourism destination ecosystem, 1, 2, 11, 13, 14, 15, 16, 17, 18, 21, 22, 25, 26, 27, 28, 29

Tourists Behavioral Patterns, 46

W

Waste Management, 363, 364, 367, 369, 372, 374, 377, 381, 388, 389, 390, 392